Lecture Notes in Statistics 116

Springer
New York
Berlin
Heidelberg
Barcelona
Budapest
Hong Kong
London
Milan
Paris
Santa Clara
Singapore
Tokyo

Genshiro Kitagawa
Will Gersch

Smoothness Priors Analysis of Time Series

Springer

Genshiro Kitagawa
The Institute of Statistical Mathematics
4-6-7 Minami-Azabu, Minato-ku
Tokyo 106
Japan

Will Gersch
University of Hawaii
Department of Information and Computer Science
Honolulu, HI 96822

CIP data available.
Printed on acid-free paper.

Camera ready copy provided by the author.
Printed and bound by Braun-Brumfield, Ann Arbor, MI.
Printed in the United States of America.

9 8 7 6 5 4 3 2 1

ISBN 0-387-94819-8 Springer-Verlag New York Berlin Heidelberg SPIN 10523945

Preface

The work described here is an attempt to integrate into a cohesive form, the papers that we have written and some previously unpublished work on the subject of "smoothness priors" and in particular, on the state space smoothness priors modeling of time series.

Originally smoothness priors as done in Shiller (1973) and Akaike (1980), was a linear model Gaussian stochastic regression least squares computational treatment of scalar and bivariate stationary and scalar nonstationary mean time series. Beginning in 1981, our own work independently and together, continued on the Shiller-Akaike path and also introduced linear Gaussian state space smoothness priors modeling of time series that were not considered by Shiller or Akaike.

The smoothness priors method is essentially quasi-Bayesian. A prior distribution on the model coefficients is parameterized by hyperparameters which in turn have a crucial role in the analysis. The maximization of the likelihood of a small number of hyperparameters permits the robust modeling of a time series with relatively complex structure and a very large number of implicitly inferred parameters. The critical statistical ideas in smoothness priors are the likelihood of the Bayesian model and the use of likelihood as a measure of the goodness of fit of the model. The Bayesianess provides a framework for doing statistical inference.

The state space modeling of not necessarily linear, not necessarily Gaussian time series is due to Genshiro Kitagawa, beginning with Kitagawa (1987). That work evolved in several stages. Kitagawa (1987) achieved general state space modeling in which the recursive conditional distributions for prediction, filtering and smoothing were successfully realized using numerical integration. That methodology was satisfactory for the modeling of non stationary mean data with abrupt discontinuities, nonstationary variance, nonstationary covariance data, nonhomogeneous discrete time series data, and non-linear time series data. Each of those models was realized with a relatively small number of states. The modeling of linear not necessarily Gaussian state space problems, with a larger number of states motivated a Gaussian mixture distribution-two filter smoothing formula to approximate the conditional densities involved. The most recent development is a general state space, (hence suitable for nonlinear modeling), Monte Carlo "particle-path tracing" method in which the distributions are approximated by many realizations. The Monte Carlo method is applicable to non Gaussian, nonstationary, nonlinear state space modeling for both small and large numbers of states.

Akaike (1968) and Pagano (1978) introduced an alternative instantáneous response orthogonal innovations representation of the multivariate stationary autoregressive time series model. We exploited that representation in a "one channel at-a-time" paradigm to realize the modeling of multivariate nonstationary covariance time series.

In addition to presenting the various methods of state space modeling, a substantial number of applications of the modeling methodologies including, seasonal time series, discrete time processes, quasi-periodic processes, nonlinear state estimation and smoothing, the modeling of a large data set with missing data and outliers and a hidden Markov state classification procedure are shown. We expect that some of the data and programs will be distributed through S-news.

We are indebted to Professor Hirotugu Akaike, former Director General of the Institute of Statistical Mathematics, Tokyo, Japan first for the intellectual stimulation which motivated our own work in smoothness priors, and for his continuing encouragement in our efforts and also for encouraging us to work together in Tokyo to complete this manuscript. We are also indebted to Dr. David Findley, who suggested that we work together in Washington D.C. at the Division of Statistical Research of the Census Bureau of the U.S. Department of Commerce in 1981-1982. We were supported in that work by an American Statistical Association grant.

We also appreciate and thank the coauthors of several of our papers. Takanami, Jiang and Matsumoto contributed to some of the Kitagawa papers and Stone contributed to work with Gersch. We are both very appreciative of the tireless efforts of Mrs. Ono for her help in preparing the manuscript. Also we are especially grateful to and indebted to Professor David Brillinger, University of California, Berkeley, whose careful editing of our manuscript and his very constructive suggestions have helped improve the manuscript. Finally, Will Gersch is grateful to the Ministry of Education, Science, Culture and Sports of Japan for providing financial support to work in Japan.

Tokyo, Japan and Honolulu, Hawaii
March 1996

Contents

1

Introduction

Some of the problems of modeling scalar and multivariate, linear and nonlinear, Gaussian and non-Gaussian stationary and nonstationary time series are addressed here primarily from a smoothness priors state space point of view. Smoothness priors is a stochastic regression treatment of stationary and nonstationary time series. Least squares and standard (Kalman filter) state space methods are used in linear Gaussian time series modeling. A general state space approach is introduced and exploited to model not necessarily Gaussian-not necessarily linear time series with abrupt discontinuities and outliers.

First we discuss some of the background and personal history of our approach to this work. Then somewhat in the context of distinguishing between stationary and nonstationary time series some of the problems of time series analysis that are addressed in this book are illustrated.

1.1 Background

The paper Akaike (1980a) motivated our interest in a quasi-Bayesian linear Gaussian modeling of time series. Akaike's paper was a penalized likelihood constrained least squares computational approach to that modeling. Kitagawa (1981) and Brotherton and Gersch (1981) introduced Kalman filter type-state space approaches to that modeling. Gersch and Kitagawa (1983a), Kitagawa and Gersch (1984, 1985a, 1985b) adopted and retained the term "smoothness priors" (from Shiller 1973). Subsequently we attempted to redo conventional time series analysis from the linear Gaussian smoothness priors approach as well as extend that approach to the analysis of previously unaddressed time series modeling problems including the modeling of multivariate nonstationary covariance time series.

Kitagawa (1987) initiated the extension of smoothness priors to include the state space modeling of non-Gaussian and nonlinear time series. That development continued (for example with Kitagawa 1988, 1989, 1991, 1993 and 1996) with emphasis on the development of computational methods and on new applications. Initially, in Kitagawa (1987), conditional distributions were computed by numerical integration methods. That approach was adequate for systems with a small number of states. For the analysis of seasonal time series with as many as 13 states, Kitagawa (1989) used a Gaussian sum approximation to state estimation. Kitagawa (1994), a two-filter formula for Gaussian sum smoothing, completes the Gaussian sum approximation modeling. The most recent development, (Kitagawa 1993, 1996), is a Monte Carlo approach to state estimation. This book is an attempt to coherently integrate our work and present the smoothness priors state space approach to time series analysis. It is important to note that the AIC, Akaike's information theoretic criterion (Akaike 1973, 1974a), for evaluating the parametric models fitted to data has a crucial role in our approach to time series modeling.

1.2 What is in the Book

The book is in two parts. The first part, labeled A: Introduction and Background, introduces the subject and history of smoothness priors modeling. Chapter 2 presents essentially all of the technical background and prerequisites for the book. This includes a discussion of the Kullback Leibler measure of dissimilarity between distributions, a development of Akaike's AIC criterion for selecting the best of alternative parametric models, the Householder transformation method of least squares modeling, a brief introduction to state space modeling, maximum likelihood estimation and brief treatments of quasi-Newton optimization and information square root filtering and smoothing. The concept of smoothness priors and the least squares computational approach to smoothness priors in Akaike (1980a) are shown separately in Chapter 3. Chapter 4 includes a smoothness priors least squares modeling of nonstationary mean time series, power spectrum estimation and transfer function estimation. In the first of those problems, the prior distribution on the model parameters is expressed in terms of a time domain difference equation constraint in which the unknown constraint equation variance is a "hyperparameter", (Lindley and Smith 1972). The estimation of a hyperparameter is in effect function estimation. The concept of frequency domain priors is introduced and exploited in the smoothness priors "long" autoregressive modeling for spectral estimation and in smoothness priors transfer function estimation and permits the analysis to remain in the domain of linear least squares modeling.

The theory of state space modeling, also in Part A, is described in two chapters. In Chapter 5, linear Gaussian state space analysis and the now classical Kalman filter are reviewed . That material is relevant for a variety of state space smoothness priors modeling including the linear Gaussian modeling of time series with irregularly spaced observations and nonstationary mean and nonstationary covariance time series. Applications of linear Gaussian state space modeling are shown in Chapter 7. General state space modeling is treated separately in Chapter 6. There the non-Gaussian and nonlinear state space modeling that is necessary to model time series with abrupt discontinuities and outliers is treated. A unique derivation of the Kalman filter is shown starting from the general state space model.

Part B: Applications of State Space Modeling, is the largest part of the book. In the last 10 chapters in this part of the book, (Chapters 7 through Chapter 16), applications of linear-Gaussian, not necessarily linear not necessarily Gaussian modeling are applied to a considerable variety of problems. In Chapter 8 the modeling of nonstationary mean time series by linear Gaussian and by linear non-Gaussian and nonlinear non-Gaussian methods are contrasted. Chapter 9 extends the the modeling of nonstationary mean time series to time series with seasonal and other components that are relevant in seasonal adjustment. The modeling of time series with non-constant variance is in Chapter 10. In Chapter 11, scalar nonstationary covariance are modeled by state space linear Gaussian and non-Gaussian methods. Chapter 12 treats the modeling of multivariate nonstationary covariance time series. The smoothness priors state space modeling of nonhomogeneous discrete time series appears in Chapter 13. The modeling of quasi-periodic time series is in Chapter 14. Chapter 15 is a state space treatment of nonlinear state estimation and smoothing. Finally Chapter 16, "Other Applications", includes an analysis of a large (one half million observation), data set with missing data and outliers in a component decomposition application, a (hidden) Markov state time series classification problem and an extension of the smoothness priors long AR modeling method for scalar spectrum estimation to multivariate spectrum estimation.

1.3 Time Series Examples

This book addresses the modeling of linear and nonlinear, stationary and nonstationary, Gaussian and non-Gaussian, scalar and multivariate time series from a state space smoothness priors approach. A stationary time series is one whose statistical description is invariant with respect to time origin or time shifts. Brockwell and Davis (1991) and Caines (1988) include formal definitions and properties of stationary processes. Nonstationary mean time series refer to time series whose average or mean value is not constant. Time series with trends or seasonalities are nonstationary mean time series. Nonstationary covariance time series are time series whose correlation or covariance structure changes with time. The data sets shown below illustrate the various time series categories. These are some of the data sets that are analyzed in the book.

Figure 1A is the $n = 114$ Canadian lynx data. It is used first in Chapter 4 as a stationary time series. In Chapter 4 a smoothness priors long autoregressive model is introduced to estimate the power spectrum of the lynx data. This data is also analyzed in Chapter 7 in an illustration of stationary time series state space modeling. Finally, the lynx data is also analyzed in a nonlinear state space model in Chapter 14 where we think of the lynx data as a quasi-periodic process.

Figure 1B is the $n = 296$ Box-Jenkins Series J bivariate stationary time series input-output data from a real physical chemical plant. It is analyzed by two different smoothness priors methods in Chapter 4. First the impulse response or distributed lag model response of the system is analyzed. Secondly, the transfer function between input and output is analyzed

Figure 1C is an example of a seismic event record that consists of a superposition of background noise, a P-wave and an S-wave as well as observation noise. It is analyzed in Chapter 7 as the first illustration of the state space analysis decomposition of a time series into component parts. The P wave and S wave events are represented as stationary autoregressive processes. A similar seismic event time series is analyzed in Chapter 10 in the analysis of a time series with changing variance. It is also analyzed in Chapter 11 to illustrate the analysis of a time series with changing covariance structure.

Figure 1D is the illustration of a synthetic nonstationary mean or time series with trend. It is analyzed in Chapter 8 by state space methods.

Figure 1E is scatter diagram of 21 years of the daily maximum temperatures recorded in Tokyo 1971-1991. The multiple time series is analyzed in Chapter 8 as a common annual trend model with individual autoregressive components for each year of data.

Figure 1F is the illustration of a synthetic data set with abruptly changing mean value. It is analyzed in Chapter 8, by three different methods, as the first example of non-Gaussian state space analysis.

Figure 1G is the monthly wholesale hardware data 1967–1990 BHDWWS obtained from the U.S. Census Bureau. It is analyzed in Chapter 9 as an example of a seasonally adjusted time series.

Figure 1H is the quarterly business inventory data in Japan 1965–1983. It is analyzed in Chapter 9 as an example of a seasonal time series with abruptly changing trend and seasonal structure.

Figure 1I illustrates the simultaneously observed two horizontal and vertical components of a seismic event. It is analyzed in Chapter 12 as an example of the analysis of multivariate nonstationary covariance data.

Figure 1J is an illustration of the Wolfer sunspot series. It is analyzed in a nonlinear state space model in Chapter 13 as an example of a quasi-periodic time series.

Figure 1K is simulated nonhomogeneous mean Poisson process data. It is analyzed in Chapter 14 as an example of the analysis of nonhomogeneous discrete data.

Figure 1L is an example of a synthetic nonlinear time series. It is analyzed in Chapter 15 as an example of nonlinear time series smoothing.

Figure 1M is an example of underground water level data that is analyzed in Chapter 16 as an example of the decomposition into components analysis of a very large data set with missing data and outliers.

Figure 1N is an example of a synthetic data set that is generated by a switching autoregressive model with four different component models. It is analyzed in Chapter 16 as an example of Markov state classification of time series.

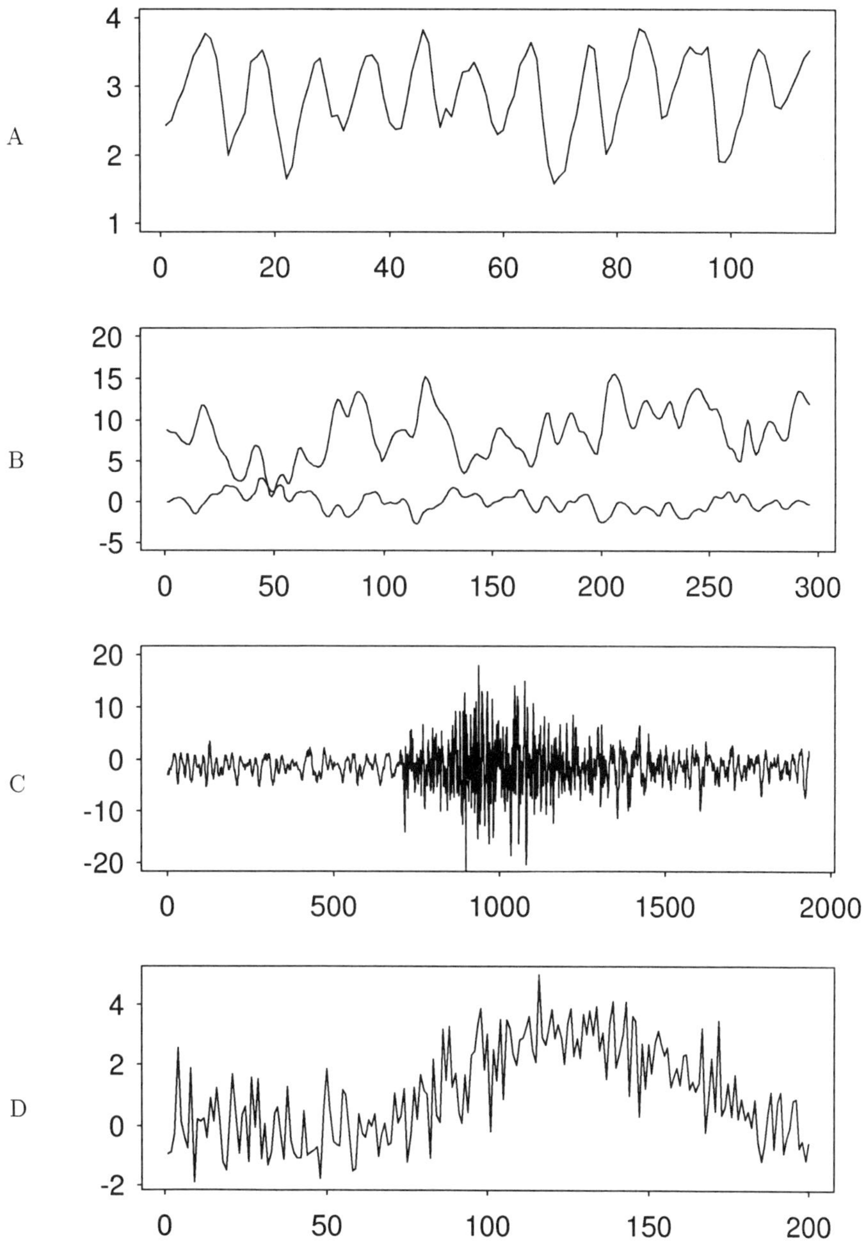

Figure 1. Some examples of time series.

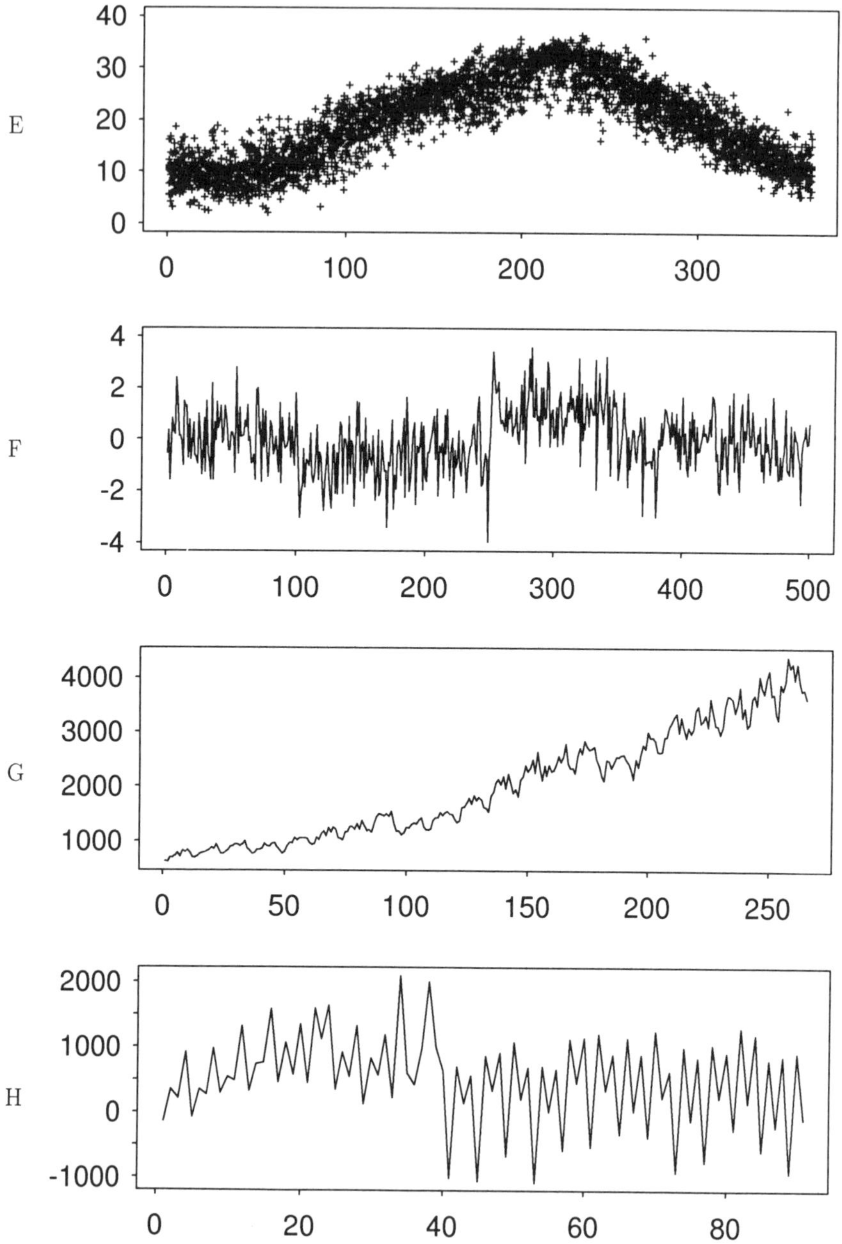

Figure 1. Some examples of time series (continued).

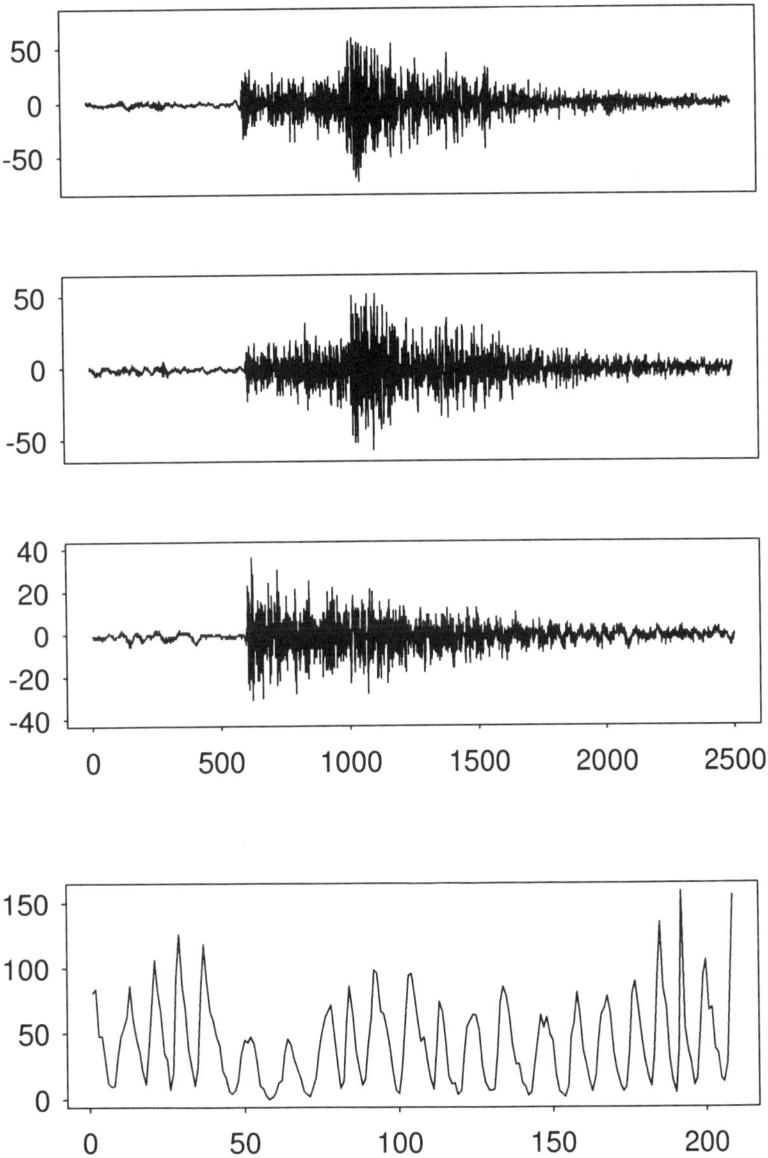

Figure 1. Some examples of time series (continued).

K

L

M

N

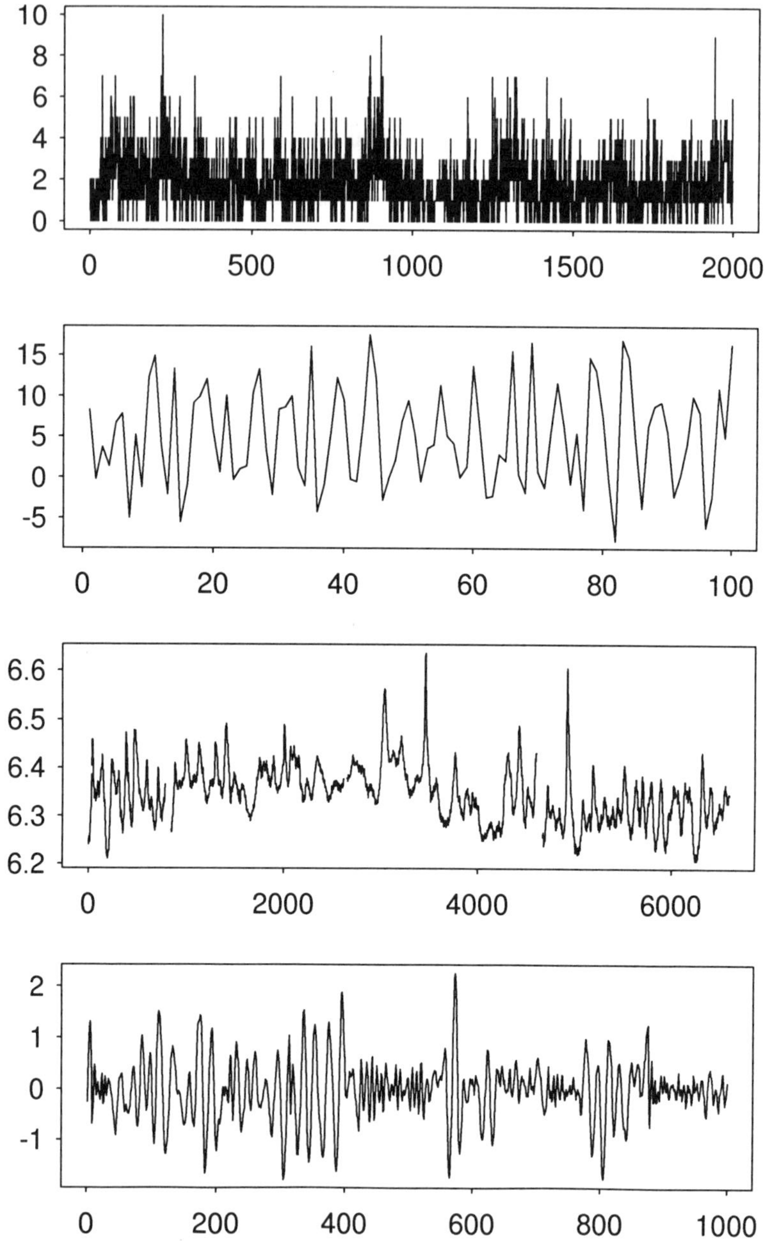

Figure 1. Some examples of time series (continued).

2

Modeling Concepts and Methods

In this chapter the ideas that we identify as the basis of our approach to time series analysis are outlined. Several topics in parameter estimation and model selection are treated. Akaike's AIC for parametric model selection is treated first. That treatment includes a discussion of the Kullback-Leibler information, and a theoretical development of the AIC. Also included are treatments of the Householder transformation based least squares estimation, the maximum likelihood method of parameter estimation, (including a method of minimizing a function of several variables), a fairly general discussion of state space modeling including Kalman filter for standard linear Gaussian state space modeling, and general state space modeling.

2.1 Akaike's AIC: Evaluating Parametric Models

Akaike's AIC (Akaike 1973, 1974a) is an information theoretic criterion for the selection of the best of alternative parametric models based on observed data. The AIC has proven to be extensively applicable in statistical data analysis. It is an approximately unbiased estimate of the expected log-likelihood which is the essential part of the Kullback-Leibler information. The Kullback-Leibler information or the K-L number is an information theoretic measure of the dissimilarity between two distributions (Kullback and Leibler 1951; Kullback 1958). The larger the measure, the greater the difference between the two distributions.

Based on a survey of leading statisticians, the AIC was identified in "*Breakthroughs in Statistics*" Volume 1, (Kotz and Johnson 1992), as one of the nineteen breakthroughs of the twentieth century in the foundations and basic theory of statistics. Some of the recent applications of the AIC in the selection of models, in addition to time series analysis are, factor analysis, regression analysis, multinomial logistic regression, cluster analysis, discriminant analysis and covariance structure models. Respectively these are treated in Akaike (1987a), Terasvirta (1987), Findley (1988), Fornell and Rust (1989), Shibata (1981), Takane *et al.* (1987), Bozdogan and Sclove (1987), Sambamoorthi (1989) and Jöreskog (1992). Linhart and Zucchini (1986) emphasizes social science applications of the AIC. Bozdogan (1994) is a collection of articles that involve the AIC and variations of it.

The AIC is relevant in this book in linear and nonlinear, Gaussian and non-Gaussian scalar and multivariate time series modeling. Akaike (1969) was Akaike's first paper on scalar time series AR model order selection. That problem had been under consideration for a long time and distinguished scientists, including John Von Neumann, (Neumann 1941), had worked on that problem. Akaike (1971) is an extension of AR model order determination to multivariate AR models. Akaike (1973) was the first formal paper on the AIC. Other important contributions by Akaike on the AIC include Akaike (1974a, 1977a, 1981a, 1985).

The AIC has provoked the intellectual curiosity of and competition from many statisticians. Akaike (1978) is a Bayesian interpretation of the AIC for scalar AR model order determination. Other contributions related to AR model order selection, subset selection in regression and more general model selection issues include, Mallows (1973), Gersch and Sharpe (1973), Jones (1976), Parzen (1978), Bhansali and Downham (1977), Schwartz (1978), Rissanen (1978, 1989), Sugiura (1978), Hannan and Quinn (1979), Quinn (1980), Shibata (1980, 1981, 1986), Chow (1981), Findley (1985b), Lutkepohl (1985), Bozdogan (1987), Sclove (1987), and Hurvich and Tsai (1989, 1992, 1993). Choi (1992) is a survey of some of the alternative model order selection schemes for AR and ARMA models that have been stimulated by the appearance of the AIC.

In Section 2.1.1 the Kullback-Leibler measure of information, its properties and relation to the AIC are discussed. Section 2.1.2 shows applications of the AIC to scalar and multivariate AR model order selection. Section 2.1.3 is a theoretical development of the AIC. Following that, in view of the extensive literature provoked by the AIC, in Section 2.1.4, is some additional discussion which includes some of the claimed disadvantages of the AIC.

2.1.1 The Kullback-Leibler Measure and the Akaike AIC

Consider a probability density function $g(y)$ of a continuous distribution function g, and a probability density function $f(y)$, an approximation to g. The Kullback-Leibler information, $I(g;f)$, between g and f is defined as the following:

$$\begin{aligned} I(g;f) &= E_Y\left\{\log\frac{g(Y)}{f(Y)}\right\} \\ &= \int_{-\infty}^{\infty} g(y)\log g(y)dy - \int_{-\infty}^{\infty} g(y)\log f(y)dy. \end{aligned} \tag{2.1}$$

For a specific probability density function $g(y)$, $\int_{-\infty}^{\infty} g(y)\log g(y)dy$ is a constant in (2.1). Therefore the crucial term in (2.1), in selecting a minimum Kullback-Leibler information model, is $E_Y \log f(Y) = \int_{-\infty}^{\infty} g(y)\log f(y)dy$, the expected log-likelihood of the model's probability density function $f(y)$ with respect to $g(y)$. In selecting a minimum Kullback-Leibler number model, one seeks to maximize $E_Y \log f(Y)$.

Two important properties of the K-L number are:

1. $I(g;f) \geq 0$ and

2. $I(g;f) = 0$ if and only if $g(y) = f(y)$ almost everywhere.

Let f_a and f_b be two contending models of the true distribution g. Then, we say that f_a is closer to g than is f_b if $I(g; f_a) < I(g; f_b)$. Thus in choosing a K-L number best model of g from a family of candidate models $f_m, m = 1, \ldots, M$, we are required to determine m^*, $1 \leq m^* \leq M$ such that $I(g; f_{m^*}) \leq I(g; f_m)$, $m = 1, \ldots, M$. Equivalently, the best K-L number model satisfies

$$E_Y \log f_{m^*}(Y) \geq E_Y \log f_m(Y), \quad m = 1, \ldots, M. \tag{2.2}$$

Consider n observations of independent and identically distributed random variables, $\{y_1, \ldots, y_n\}$ from an unknown probability density function $g(y)$. Assume a set of alternative or candidate models $f(\,\cdot\,|\theta_m)$, where θ_m is the parameter vector of the density f_m. We wish to estimate $I(g; f_m(\,\cdot\,|\theta_m))$, $m = 1, \ldots, M$, the K-L number for the dissimilarity between each of the alternative models f_m and the true density g.

From (2.2), the best K-L number model is the one with the greatest expected log likelihood, $E_Y \log f(Y|\,\theta_m)$. Assume that the parameter vectors θ_m's are known. Then from the law of large numbers, the expected log likelihood of the model can be estimated by

$$E_Y \log f_m(Y|\,\theta_m) \;=\; \int_{-\infty}^{\infty} g(y) \log f_m(y|\,\theta_m) dy \;\doteq\; \frac{1}{N} \sum_{n=1}^{N} \log f_m(y_n|\theta_m). \tag{2.3}$$

Therefore the use of the log-likelihood

$$\ell(\theta_m) = \sum_{n=1}^{N} \log f_m(y_n|\theta_m) \tag{2.4}$$

is motivated. Then taking exponential, we obtain the likelihood of the model as

$$L(\theta_m) = \prod_{n=1}^{N} f_m(y_n|\theta_m) \tag{2.5}$$

We emphasize here that in (2.3), the expected log-likelihood is estimated by the log-likelihood. The parameter vectors, θ_m, for the model distributions $f_m(\,\cdot\mid\theta_m), m = 1, \ldots, M$ are unlikely to be known in practice. They must be estimated from the data and the assumed model $f_m(\,\cdot\mid\theta_m)$. It is natural to use the maximum likelihood estimate $\hat{\theta}_m$ that maximizes the right hand side of (2.3). (Maximum likelihood estimation is treated in Section 2.3.) In this case, the law of large numbers equality in (2.3) does not hold when the parameter vectors θ_m are replaced by their maximum likelihood estimates $\hat{\theta}_m$, (Akaike 1973, Sakamoto *et al.* 1986). That is,

$$E_x E_y \log f_m(y|\,\hat{\theta}_m) \neq \frac{1}{N} E_x \sum_{n=1}^{N} \log f_m(y_n|\,\hat{\theta}_m). \tag{2.6}$$

This is due to the fact that the same data set was used twice for the estimation of the parameters and for the estimation of the expected log-likelihood.

Thus, from (2.6), the maximized log likelihood is a biased estimator of the average expected log likelihood. The bias is approximately equal to the number of parameters estimated in the model, (Akaike 1973). An approximate correction of the bias is reflected in the definition of the AIC given below.

$$\begin{aligned}
\mathrm{AIC}(m) &= -2(\text{maximized log likelihood of the model}) \\
&\quad +2(\text{number of estimated parameters in the model}) \qquad (2.7) \\
&= -2\sum_{n=1}^{N} \log f_m(y_n \mid \hat{\theta}_m) \;+\; 2|\,\hat{\theta}_m| \;.
\end{aligned}$$

In (2.7), $|\,\hat{\theta}_m|$ denotes the dimension of the vector $\hat{\theta}_m$.

2.1.2 Some Applications of the AIC

In this section applications of the AIC to scalar and multivariate autoregressive model order selection are illustrated. The use of the AIC in regression is discussed in Section 2.2, in the context of Householder transformation least squares.

Scalar Autoregressive Model Order Selection

With regard to autoregressive model order selection, in practice, both the true model parameter and the true model order of the data generating system are generally unknown. In fact, the true model is very likely not to be a finite order autoregressive model. Akaike (1974a), Parzen (1978) and Shibata (1980) address the oftentimes realistic situation in which the true model is infinite-dimensional. Correspondingly, we adopt the policy that fitting an AR model to data is only an approximation to the unknown infinite-dimensional model and recognize that model order determination procedures allow us to select a finite-dimensional approximating model. When using finite-dimensional autoregressive models to estimate the assumed infinite-dimensional truth, the AIC provides an asymptotically efficient solution under a quadratic loss function, (Shibata 1983), to selecting the best fit to the data.

Let the family of candidate autoregressive models be defined by

$$y_n = \sum_{j=1}^{p} \alpha_j y_{n-j} + \varepsilon_n \qquad (2.8)$$

where ε_n is independently identically distributed and $\varepsilon_n \sim N(0,\sigma^2)$. The joint probability density function for this model is

$$\begin{aligned}
f(y_1,\ldots,y_N) &= \prod_{n=1}^{N} f(y_n|\alpha_1,\cdots,\alpha_p,\sigma^2) \\
&= \prod_{n=1}^{N} \left(\frac{1}{2\pi\sigma^2}\right)^{\frac{1}{2}} \exp\left\{-\frac{1}{2\sigma^2}\left(y_n - \sum_{j=1}^{p}\alpha_j y_{n-j}\right)^2\right\}. \qquad (2.9)
\end{aligned}$$

The corresponding log-likelihood function is

$$\sum_{n=1}^{N} \log f(y_n|\alpha_1,\ldots,\alpha_p,\sigma^2) = -\frac{N}{2}\log 2\pi\sigma^2 - \frac{1}{2\sigma^2}\sum_{n=1}^{N}\left(y_n - \sum_{j=1}^{p}\alpha_j y_{n-j}\right)^2. \qquad (2.10)$$

Maximizing (2.10) with respect to the unknown parameters, yields the maximum likelihood estimates of the AR coefficients and of σ^2. The estimate $\hat{\sigma}^2$ of σ^2 is given by,

$$\hat{\sigma}^2 = \frac{\text{SSE}}{N}, \quad \text{SSE} = \sum_{n=1}^{N}\left(y_n - \sum_{j=1}^{p}\hat{\alpha}_j y_{n-j}\right)^2. \tag{2.11}$$

In (2.11), the initial conditions, that is the data $y_{1-p}, \ldots, y_0$, are assumed to be available.

Substituting the quantities in (2.11) into (2.10), and taking into account the definition of the AIC, yields the AIC for the scalar AR model of order p

$$\text{AIC}(p) = N\left\{\log(2\pi) + \log\hat{\sigma}^2 + 1\right\} + 2(p+1). \tag{2.12}$$

Multivariate AR Model Order Selection

Let y_n be a d-dimensional stationary time series. Consider the multivariate AR model of order P of that process

$$y_n = \sum_{m=1}^{P} A_m y_{n-m} + \varepsilon_n \tag{2.13}$$

with $\varepsilon_n \sim N(0, V)$, i.i.d.. The A_m's are $d \times d$ AR coefficient matrices and V is the $d \times d$ innovations covariance matrix. Assuming independence of the ε_n's yields the joint probability density function

$$\begin{aligned} f(y_1, \ldots, y_N) &= \prod_{i=1}^{N} f(y_i|A_1, \cdots, A_P, V) \\ &= (2\pi)^{-\frac{dN}{2}}|V|^{-\frac{N}{2}}\exp\left\{-\frac{1}{2}\sum_{i=1}^{N}\varepsilon_i^T V^{-1}\varepsilon_i\right\}. \end{aligned} \tag{2.14}$$

In (2.14) and subsequently the notation A^T denotes the transpose of the matrix A. The log-likelihood function corresponding to the distribution in (2.14), (Sakamoto, Ishiguro and Kitagawa 1986), is

$$\sum_{i=1}^{N}\log f(y_i|A_1, \cdots, A_P, V) = -\frac{N}{2}\left\{d\log(2\pi) + \log|V|\right\} - \frac{1}{2}\sum_{i=1}^{N} e_i^T V^{-1} e_i. \tag{2.15}$$

where e_i is the i-th vector residual (e_i is approximately an estimate of ε_i). As a result, the AIC for the P-th order d dimensional multivariate AR model is

$$\text{AIC}(P) = N\left\{d\log(2\pi) + \log|\hat{V}|\right\} + \sum_{i=1}^{N} e_i^T\hat{V}^{-1}e_i + 2\left(Pd^2 + \frac{d(d+1)}{2}\right) \tag{2.16}$$

where $\hat{V}$ is the maximum likelihood estimate of the innovations covariance matrix V.

In determining the AIC best multivariate AR model order, a sequence of AR models are fitted to the observed data for example, recursively by the Whittle (1963) algorithm or by a direct Yule-Walker solution method, (Lutkepohl 1993). The model which gives the smallest AIC is defined to be the AIC best model.

2.1.3 A Theoretical Development of the AIC

In this section, Akaike's AIC criterion is developed starting from the definition of the Kullback-Leibler information. From Section 2.2.1 the Kullback-Leibler measure of information of f with respect to g is

$$I(g; f) = E_Y \log g(Y) \; - \; E_Y \log f(Y). \tag{2.17}$$

In (2.17) the right hand side term, $E_Y \log f(Y)$, the expected log-likelihood is given by $E_y \log f(Y) = \int_{-\infty}^{\infty} g(y) \log f(y) dy$. The quantity $\ell = \sum_{n=1}^{N} \log f(y_n)$ is referred to as the log-likelihood. Then, by the law of large numbers, assuming the mean to exist, the average log-likelihood converges to the expected log-likelihood, $E_y \log f(Y)$.

Similarly, if the distribution is parameterized by the parameter vector θ, by the law of large numbers, the average of the parametrized log-likelihood converges to the expected value of the parameterized log-likelihood. That is, $\frac{1}{N}\ell(\theta) = \frac{1}{N}\sum_{n=1}^{N} \log f(x_n|\theta)$ $\rightarrow E_Y \log f(Y|\theta)$. However, if the parameter vector of the distribution is unknown and is estimated by the maximum likelihood method, it is not true that the average of the maximized log-likelihood converges to the expected value of the parameterized log-likelihood. In that case, $\frac{1}{N}\ell(\hat{\theta}) = \frac{1}{N}\sum_{n=1}^{N} \log f(x_n|\hat{\theta}) \not\rightarrow E_Y \log f(Y|\hat{\theta})$. The AIC is an approximate correction for this bias. Next the bias is evaluated.

Let B denote the bias between the expected maximized log-likelihood and the average of the maximized log likelihood over a conceptually infinite number of realizations. Then B is given by

$$B \; = \; E_X \left\{ E_Y \log f(Y|\hat{\theta}) \; - \; \frac{1}{N} \sum_{n=1}^{N} \log f(x_n|\hat{\theta}) \right\}. \tag{2.18}$$

In what follows we demonstrate that B the bias satisfies $B \approx -\frac{k}{N}$ where k is the number of independent parameters estimated for the model. Then using substitution we have the result that

$$\text{AIC} \; = \; -2 \sum_{n=1}^{N} \log f(x_n|\hat{\theta}) \; + \; 2k. \tag{2.19}$$

The key to this development is the decomposition of B into three components and an evaluation of the contributions of each those components. That is let $B \; = \; B_1 + B_2$ $+ \; B_3$. The expansion of B is

$$\begin{aligned} B \; &= \; E_X \left\{ E_Y \log f(Y|\hat{\theta}) \; - \; E_Y \log f(Y|\theta_0) \right\} \\ &+ \; E_X \left\{ E_Y \log f(Y|\theta_0) \; - \; \frac{1}{N} \sum_{n=1}^{N} \log f(x_n|\theta_0) \right\} \\ &+ \; E_X \left\{ \frac{1}{N} \sum_{n=1}^{N} \log f(x_n|\theta_0) - \frac{1}{N} \sum_{n=1}^{N} \log f(x_n|\hat{\theta}) \right\} \end{aligned} \tag{2.20}$$

with the components B_1, B_2, B_3 identified sequentially in (2.20).

The relationship between the bias components is shown graphically in Figure 2.1. The B_1, B_2, B_3 components are evaluated in what follows next. Here θ_0 is the value of θ that maximizes $E_y \log f(Y|\theta)$ and thus satisfies $\frac{\partial}{\partial \theta} E_y \log f(Y|\theta_0) = 0$.

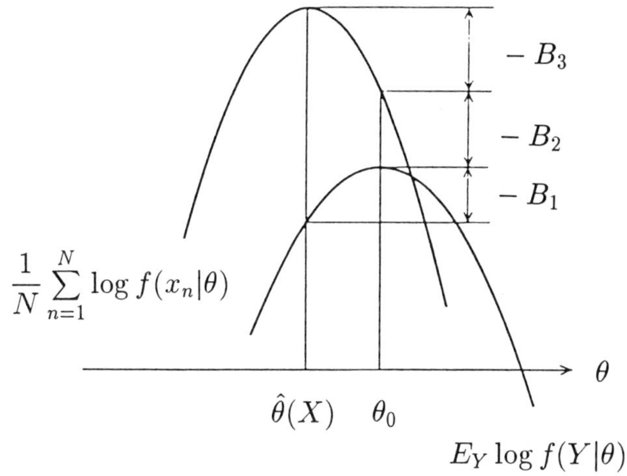

Figure 2.1 Decomposition of the bias.

Evaluating B_1

Here in order to evaluate B_1, consider the Taylor series expansion of $E_Y \log f(y|\hat{\theta})$ around $E_Y \log f(y|\theta_0)$ up to second order term. That is,

$$\begin{aligned} E_Y \log f(Y|\hat{\theta}) &\approx E_Y \log f(Y|\theta_0) + \left\{\frac{\partial}{\partial\theta} E_Y \log f(Y|\theta_0)\right\}^T (\hat{\theta}-\theta_0) \\ &\quad + \frac{1}{2}(\hat{\theta}-\theta_0)^T \left\{\frac{\partial^2}{\partial\theta\partial\theta'} E_Y \log f(Y|\theta_0)\right\}(\hat{\theta}-\theta_0), \\ &= E_Y \log\ f(Y|\theta_0) - \frac{1}{2}(\hat{\theta}-\theta_0)^T J(\hat{\theta}-\theta_0), \end{aligned} \tag{2.21}$$

where J is defined as the negative of the expected Hessian matrix,

$$J \equiv -E_Y \left\{\frac{\partial^2}{\partial\theta\partial\theta'} \log f(Y|\theta_0)\right\}. \tag{2.22}$$

Also, if the family of models contain the true distribution from the theory of the maximum likelihood, asymptotically with increasing N, $\sqrt{N}(\hat{\theta} - \theta_0)$, the scaled difference between the estimate $\hat{\theta}$ of θ_0 and θ_0, converges in distribution to a normally distributed zero mean vector with covariance matrix, $I(\theta_0)^{-1}$, (Dzhapridze 1986), where

$$I(\theta_0) = E_Y \left[\left\{\frac{\partial}{\partial\theta}\log f(Y|\theta_0)\right\}\left\{\frac{\partial}{\partial\theta}\log f(Y|\theta_0)\right\}^T\right], \tag{2.23}$$

is the Fisher information matrix.

Huber (1965) proved consistency and asymptotic normality of maximum likelihood estimators under the relatively weak conditions that i) it was not assumed that the true distribution underlying the observations belongs to the parametric family defining the maximum likelihood estimator and ii) the regularity conditions do not involve the 2nd derivatives of the likelihood function. Then from Huber (1965), using the asymptotic normality under nonstandard conditions, $\sqrt{N}(\hat{\theta}(X) - \theta_0) \sim N(0, J^{-1}IJ^{-1})$, and we have that

$$E_X\left\{(\hat{\theta} - \theta_0)^T J(\hat{\theta} - \theta_0)\right\} = \frac{1}{2N}\text{tr}(IJ^{-1}). \tag{2.24}$$

Therefore we have that the contribution of the first term B_1 to B is

$$B_1 \equiv E_X\left\{E_Y \log\ f(y|\hat{\theta})\right\} \approx -\frac{1}{N}\text{tr}(IJ^{-1}). \tag{2.25}$$

Evaluation of B_2

Taking into account that

$$E_X\left\{\frac{1}{N}\sum_{n=1}^{N}\log f(X_n|\theta_0)\right\} = \frac{1}{N}\sum_{n=1}^{N}E_X\left\{\log f(X_n|\theta_0)\right\} = E_X \log f(X|\theta_0) \tag{2.26}$$

we observe that the contribution of B_2, the second term in the expansion to B is zero. That is,

$$B_2 = E_X\left\{E_Y \log f(Y|\theta_0) - \frac{1}{N}\sum_{n=1}^{N}\log f(x_n|\theta_0)\right\} = 0. \tag{2.27}$$

Evaluation of B_3

In order to evaluate the third component in the decomposition of the expected bias, consider the Taylor series expansion of $\frac{1}{N}\log f(x_n|\theta_0)$ around $\hat{\theta}$. That yields

$$\begin{aligned}\frac{1}{N}\sum_{n=1}^{N}\log f(x_n|\theta_0) &\approx \frac{1}{N}\sum_{n=1}^{N}\log f(x_n|\hat{\theta}) + \left\{\frac{1}{N}\sum_{n=1}^{N}\frac{\partial}{\partial\theta}\log f(x_n|\hat{\theta})\right\}^T(\theta_0 - \hat{\theta}) \\ &+ \frac{1}{2}(\theta_0 - \hat{\theta})^T\left\{\frac{1}{N}\sum_{n=1}^{N}\frac{\partial^2}{\partial\theta\partial\theta'}\log f(x_n|\hat{\theta})\right\}(\theta_0 - \hat{\theta}). \end{aligned} \tag{2.28}$$

By the law of large numbers, in the limit as N increases

$$\frac{1}{N}\sum_{n=1}^{N}\frac{\partial^2}{\partial\theta\partial\theta'}\log f(x_n|\hat{\theta}) \longrightarrow E_Y\left\{\frac{\partial^2}{\partial\theta\partial\theta'}\log f(Y|\theta_0)\right\} = -J. \tag{2.29}$$

Taking that asymptotic equality into account, we have that

$$\frac{1}{N}\sum_{n=1}^{N}\log f(x_n|\theta_0) \approx \frac{1}{N}\sum_{n=1}^{N}\log f(x_n|\hat{\theta}) - \frac{1}{2}(\theta_0 - \hat{\theta})^T J(\theta_0 - \hat{\theta}).$$

Again taking into account the average over a conceptually infinite number of possible realizations yields

$$B_3 = E_X\left\{\frac{1}{N}\sum_{n=1}^{N}\log f(x_n|\theta_0) - \frac{1}{N}\sum_{n=1}^{N}\log f(x_n|\hat{\theta})\right\} \approx -\frac{1}{2N}\mathrm{tr}(IJ^{-1}). \quad (2.30)$$

Finally adding the contributions of B_1, B_2 and B_3 together yields the result that the expected bias is given by

$$B \equiv E_X\left\{E_Y \log f(Y|\hat{\theta}) - \frac{1}{N}\sum_{n=1}^{N}\log f(x_n|\hat{\theta})\right\} \approx -\frac{1}{N}\mathrm{tr}(\mathrm{IJ}^{-1}). \quad (2.31)$$

In the special situation where the true distribution belongs to the family of models, it is known that $J = I$ and B can be simply expressed by

$$B = \mathrm{tr}(IJ^{-1}) = \mathrm{tr}(I_k) = k, \quad (2.32)$$

where I_k is $k \times k$ identity matrix.

That is then

$$\frac{1}{N}\ell(\hat{\theta}) + B \approx \frac{1}{N}(\ell(\hat{\theta}) - k) \quad (2.33)$$

which directly yields that

$$\mathrm{AIC} = -2\ell(\hat{\theta}) + 2k. \quad (2.34)$$

Expressed in words that is,

$$\begin{aligned}\mathrm{AIC} = & -2(\text{maximized log-likelihood}) \\ & +2(\text{number of fitted parameters}).\end{aligned} \quad (2.35)$$

2.1.4 Further Discussion of the AIC

Substantial attention in the literature has been given to the "magic number" 2, the crucial structure dependent term in the AIC, (note (2.7) and (2.35)). Rissanen (1978) suggests it is coincidental, Bhansali and Downham (1977) did simulation studies considering the magic number to be between 1 and 4. The issue of the propriety of the number 2 is convincingly addressed in Akaike (1979, 1985) and Atkinson (1980).

Another objection that has been raised is that minimizing the AIC does not produce an asymptotically consistent estimate of model order, (Shibata 1976, Schwarz 1978, Woodroofe 1982 and numerous others). Any consideration of consistency of the order presupposes the existence of a "true order" of a model. That of course may not be the case when the model itself is an approximation. So, in the case of real data true order is not known and the concept may be quite suspect. In fact the AIC was not designed to be consistent, (Akaike 1981b), nor is its inconsistency necessarily a defect (Hannan 1986). The objective of the AIC was to minimize the negentropy, (expected log likelihood). Shibata (1983), in an extensive study of asymptotic behavior of the AIC and its variants in terms of its asymptotic efficiency under a quadratic loss function concludes that the AIC fulfills Akaike's motivating goal.

2.2 Least Squares Regression by Householder Transformation

Because of its flexibility and widespread applicability we make extensive use of the Householder transformation, an orthogonal decomposition method for the solution of linear equations, (Lawson and Hanson 1974). A Householder transformation algorithm is first described here in the context of fitting a scalar AR model to data via an ordinary linear least squares regression computation. In Chapter 12, we use the Householder transformation in the fitting of multivariate AR models by successive scalar AR model fits.

Consider the least squares regression model fitting of an AR model to the data y_n, $n = 1, \ldots, N$. Assume that the mean of the data has been removed. Then the AR model of order m is

$$y_n = \sum_{i=1}^{m} \alpha_i y_{n-i} + \varepsilon_n \tag{2.36}$$

where the ε_i, $i = 1, \ldots, N$ are independently identically distributed zero mean random variables with unknown variance σ^2. Then define the $(N-m) \times m$ matrix Z, and respectively the $N \times 1$ and $p \times 1$ vectors y and α by

$$Z = \begin{bmatrix} y_m & y_{m-1} & \cdots & y_1 \\ y_{m+1} & y_m & \cdots & y_2 \\ \vdots & \vdots & \ddots & \vdots \\ y_{N-1} & y_{N-2} & \cdots & y_{N-m} \end{bmatrix}, \quad y = \begin{bmatrix} y_{m+1} \\ y_{m+2} \\ \vdots \\ y_N \end{bmatrix}, \quad \alpha = \begin{bmatrix} \alpha_1 \\ \vdots \\ \alpha_m \end{bmatrix}. \tag{2.37}$$

The least squares estimate of the vector α is obtained by minimizing $||y - Z\alpha||^2$ where $|| \cdot ||$ denotes the Euclidean norm. That estimate is achieved here using a Householder transformation following the indicated procedure:

We construct an $(N-m) \times (m+1)$ matrix by augmenting the matrix Z with the vector y on the right

$$X = [\, Z \mid y \,] = \left[\begin{array}{cccc|c} y_m & y_{m-1} & \cdots & y_1 & y_{m+1} \\ y_{m+1} & y_m & \cdots & y_2 & y_{m+2} \\ \vdots & \vdots & \ddots & \vdots & \vdots \\ y_{N-1} & y_{N-2} & \cdots & y_{N-m} & y_N \end{array}\right]. \tag{2.38}$$

A Householder transformation is realized by applying an orthogonal matrix U to the left side of (2.35). This operation can be understood to reduce X to an upper triangular form

$$X \longrightarrow UX = [\, UZ \mid Uy \,] = S = \begin{bmatrix} s_{11} & s_{12} & \cdots & s_{1m} & s_{1,m+1} \\ & s_{22} & \cdots & s_{2m} & s_{2,m+1} \\ & & \ddots & \vdots & \vdots \\ & & & s_{mm} & s_{m,m+1} \\ & & & & s_{m+1,m+1} \\ & & 0 & & \end{bmatrix} \tag{2.39}$$

The structure and computation of UX are described in Golub (1965). From the orthogonality of U we have that $\| y - Z\alpha \|^2 = \| Uy - UZ\alpha \|^2$ which in turn can be written as

$$\| Uy - UZ\alpha \|^2 = \| \begin{bmatrix} s_{11} & \cdots & s_{1m} \\ & \ddots & \vdots \\ & & s_{mm} \\ & & \\ & 0 & \\ & & \end{bmatrix} \begin{bmatrix} \alpha_1 \\ \vdots \\ \alpha_m \end{bmatrix} - \begin{bmatrix} s_{1,m+1} \\ \vdots \\ s_{m,m+1} \\ s_{m+1,m+1} \\ 0 \\ \vdots \\ 0 \end{bmatrix} \|^2$$

$$= \| \begin{bmatrix} s_{11} & \cdots & s_{1m} \\ & \ddots & \vdots \\ & & s_{mm} \end{bmatrix} \begin{bmatrix} \alpha_1 \\ \vdots \\ \alpha_m \end{bmatrix} - \begin{bmatrix} s_{1,m+1} \\ \vdots \\ s_{m,m+1} \end{bmatrix} \|^2 + s_{m+1,m+1}^2. \tag{2.40}$$

It is clear that the minimum of this quantity is $s_{m+1,m+1}^2$ and that the least squares estimate $\hat{\alpha} = a$, with $\hat{\alpha}_i = a_i$, $i = 1, \ldots, m$, is given by the solution of the linear equation

$$\begin{bmatrix} s_{11} & \cdots & s_{1m} \\ & \ddots & \vdots \\ & & s_{mm} \end{bmatrix} \begin{bmatrix} a_1 \\ \vdots \\ a_m \end{bmatrix} = \begin{bmatrix} s_{1,m+1} \\ \vdots \\ s_{m,m+1} \end{bmatrix}. \tag{2.41}$$

These equations can be solved recursively for one parameter at-a-time using scalar operations solving the last equation first and continuing "from the bottom up". That is,

$$\hat{\alpha}_m = a_m = \frac{s_{m,m+1}}{s_{mm}} \tag{2.42}$$

$$\hat{\alpha}_i = a_i = \frac{(s_{i,m+1} - s_{i,i+1}a_{i+1} - \cdots - s_{i,m}a_m)}{s_{ii}}, \quad i = m-1, \ldots, 1 \tag{2.43}$$

Then because the first m observations are used as initial conditions in the least squares algorithm, the estimate σ^2 of the m-th order AR model, denoted $\hat{\sigma}_m^2$ is given by

$$\hat{\sigma}_m^2 = \frac{s_{m+1,m+1}^2}{N - m}. \tag{2.44}$$

AR Model Order Selection by AIC

From Section 2.1.2 we have that the maximized log-likelihood of the AR model is

$$\ell(\theta) = -\frac{N-m}{2} \log 2\pi\hat{\sigma}_m^2 - \frac{N-m}{2}. \tag{2.45}$$

Note that the initial m observations $y_1, \cdots, y_m$ are used for initial condition and the actual number of observations is $N - m$. Using this quantity yields the result that the value of the AIC for fitting an AR model of order m using the Householder transformation method is,

$$\begin{aligned} \mathrm{AIC}_m &= -2\ell(\hat{\theta}) + 2(\text{number of fitted parameters}) \\ &= (N-m)(\log 2\pi\hat{\sigma}_m^2 + 1) + 2(m+1). \end{aligned} \tag{2.46}$$

Fitting Models With Successively Increasing Orders

Once the triangular matrix S is obtained, we can fit the autoregressive model of order $k \le m$ by solving

$$\begin{bmatrix} s_{11} & \cdots & s_{1m} \\ & \ddots & \vdots \\ & & s_{km} \end{bmatrix} \begin{bmatrix} a_1 \\ \vdots \\ a_k \end{bmatrix} = \begin{bmatrix} s_{1,m+1} \\ \vdots \\ s_{k,m+1} \end{bmatrix}. \tag{2.47}$$

The estimates of the innovation variance and the value of the AIC are then given by

$$\hat{\sigma}_k^2 = \frac{1}{N-m} \sum_{i=k+1}^{m+1} s_{i,m+1}^2, \tag{2.48}$$

$$\mathrm{AIC}_k = (N-m)(\log 2\pi\hat{\sigma}_k^2 + 1) + 2(k+1). \tag{2.49}$$

The triangular matrix in (2.39) contains all of the information necessary for the fitting of AR models with orders up to m. It should be noted that AIC_k, (2.49), can be computed without solving the linear equation (2.47).

Successive Matrix Reduction

Computer storage efficiency can be achieved with the Householder transformation algorithm using successive matrix reductions. Consider the partition of the data matrix X given by

$$X = \begin{bmatrix} X_1 \\ X_2 \\ \vdots \\ X_L \end{bmatrix}. \tag{2.50}$$

First the matrix X_1 is reduced to an upper triangular form S_1 by applying a Householder transformation U_1. Then the matrix S_1 is augmented with the matrix X_2 below it and the resulting matrix is reduced to an upper matrix S_2 by a second Householder transformation U_2. By repeating the process, we can achieve the Householder reduction of the matrix X regardless of the number of observations, storing only $2(m+1)$ rows. For example, consider the matrix $X_{1,2}$ corresponding to $N+M$ observations $y_1, \ldots, y_{N+M}$,

$$X_{1,2} = \left[\begin{array}{ccc|c} y_m & \cdots & y_1 & y_{m+1} \\ y_{m+1} & \cdots & y_2 & y_{m+2} \\ \vdots & \ddots & \vdots & \vdots \\ y_{N+M-1} & \cdots & y_{N+M-m} & y_{N+M} \end{array}\right], \tag{2.51}$$

After X_1 is reduced to the matrix S_1 and S_1 is augmented by the matrix X_2 we have

$$\left[\frac{U_1 X_1}{X_2}\right] = \left[\frac{S_1}{X_2}\right] = \left[\begin{array}{cccc} s_{11} & \cdots & s_{1m} & s_{1,m+1} \\ & \ddots & \vdots & \vdots \\ & & s_{mm} & s_{m,m+1} \\ & & & s_{m+1,m+1} \\ \hline y_N & \cdots & y_{N-m+1} & y_{N+1} \\ \vdots & & \vdots & \vdots \\ y_{N+M-1} & \cdots & y_{N+M-m} & y_{N+M} \end{array}\right]. \tag{2.52}$$

The Householder reduction of $X_{1,2}$ obtained by the direct computation $U_{1,2}X_{1,2}$ is identical to the Householder reduction of (2.52).

2.3 Maximum Likelihood Estimation and an Optimization Algorithm

The method of maximum likelihood estimation was motivated in Section 2.1.1 by information theoretic principles. That is in contrast with what has been an essentially ad-hoc motivation for this otherwise richly developed subject. Cramér (1946) includes the original development of the theory and the asymptotic properties of maximum likelihood estimation for independent, identically distributed observations of a random variable under certain smoothness conditions. Cox and Hinkley (1974) is a more recent treatment and LeCam (1990) is an interesting review of maximum likelihood estimation that contains a few surprises. Some of the relatively recent works that treat maximum likelihood estimation with particular emphasis on its applications to time series analysis are Dzhapridze (1986), Caines (1988) and Brockwell and Davis (1991). In view of the extensive treatment in the literature, in this section we treat the subject very briefly and merely summarize some of the properties of maximum likelihood estimation. In addition we describe some details of our approach to the quasi-Newton numerical optimization procedure that we employ for maximization of the likelihood. Fletcher (1987), is a very adequate reference for this material.

Maximum Likelihood Estimation

The method of maximum likelihood parameter estimation uses all of the information in a model by using the complete distribution of the observations. This is in contrast with the method of least squares estimation in which only the first two moments of a distribution are used. Of course, it is well known that if the observations are assumed to be normally distributed, the least squares parameter estimators are in fact maximum likelihood estimators. The decomposition of the likelihood function for the state space model by the Kalman filter and a general non-Gaussian filter in terms of its one-step-ahead prediction errors is described in Chapters 5 and 6. That fact is extensively exploited in Chapters 7–16. That property of state space models helps motivate their use in time series modeling.

The objective of maximum likelihood parameter estimation is to determine the values of the parameters which are "most likely", given the observations. Let $y_1, \ldots, y_N$

denote a set of observations distributed in accordance with an assumed joint parametrically expressed density function, say $f(y_1, \ldots, y_N \mid \theta)$, where θ is parameter vector of the distribution. The principle of maximum likelihood is that the maximum likelihood estimate of θ is given by $\hat{\theta}^*$ where $\hat{\theta}^*$ is a value such that

$$L(\hat{\theta}^* | \, y) \geq L(\hat{\theta} \mid y \,) \tag{2.53}$$

for all other possible estimates $\hat{\theta}$.

The classical theory of maximum likelihood estimation is based on the situation in which the observations are independent and identically distributed. In that case, the likelihood function is

$$L(\theta|y) \;=\; \prod_{i=1}^{N} f(y_i|\theta). \tag{2.54}$$

Because the log function is monotonic, the parameters which maximize the likelihood function also maximize the log of the likelihood function and the log-likelihood is often easier to work with. The m vector-valued first derivative of the log-likelihood, $\ell(\theta) \equiv \partial L(\theta|y)/\partial\theta$ is referred as the gradient vector. The $m \times m$ matrix of second derivatives, $\partial^2 L(\theta|y)/\partial\theta\partial\theta'$ is called the Hessian. The maximum likelihood estimator is obtained as a solution to the likelihood equations

$$\frac{\partial \ell(\theta)}{\partial \theta} \;=\; 0. \tag{2.55}$$

That solution might not be unique, but verifying that the Hessian is negative definite will at least confirm that the solution corresponds to a maximum of the likelihood function in the neighborhood. It does not guarantee that the maximum is global rather than local.

In general the maximum likelihood estimate, (MLE), must be computed by an iterative numerical procedure. In the following section following we briefly outline the quasi-Newton algorithm that we have found useful. It is included in the programs distributed in TIMSAC-84, (Akaike *et al.* 1985). Before turning to that section, we briefly mention of the asymptotic properties of maximum likelihood estimators.

From Dzhapridze (1986), the maximum likelihood estimators satisfy the Cramér-Rao inequality $\text{Var}(\hat{\theta}) \geq I(\theta)^{-1}$, where $\hat{\theta}$ is the MLE, θ is some other parameter estimator and $I(\hat{\theta})$ is the Fisher information evaluated at the MLE $\hat{\theta}$. That is, the MLE's are statistically efficient, in that their variance-covariance matrix is less than or equal to that of any other estimate.

On the assumption that the average information matrix converges to a positive definite matrix as the number of observations increases, and subject to certain regularity conditions, $\sqrt{N}(\hat{\theta} - \theta_0)$ has a limiting multivariate normal distribution with mean vector zero and covariance matrix $I(\theta_0)^{-1}$. The information matrix for several varieties of times series models are worked in Dzhapridze (1986) and Brockwell and Davis (1991).

An Optimization Procedure

In general we are interested in minimizing $-\ell(\theta|y)$, the negative of the log-likelihood with respect to the parameter vector θ. To simplify the notation in this section, we refer to that quantity as $f(\theta)$. Here we report a Newton-Raphson procedure, (simply referred to as a Newton procedure), which employs the first and second derivatives of $f(\theta)$.

Consider the Taylor series expansion, up to second order terms of $f(\theta)$ around the (unknown) argument of the minimum value of $f(\theta)$, θ^*,

$$f(\theta) \;\approx\; f(\theta^*) \;+\; (\theta-\theta^*)^T g(\theta^*) + \frac{1}{2}(\theta-\theta^*)^T(\theta-\theta^*), \tag{2.56}$$

where in (2.56), θ is the m-vector, $\theta = (\theta_1, \ldots, \theta_m)^T$ and $g(\theta)$ and H respectively refer to the gradient, the derivative of $f(\theta)$ with respect to θ, and the Hessian, the second derivative with respect to θ, of f,

$$g(\theta) \;=\; \begin{bmatrix} \dfrac{\partial f(\theta)}{\partial \theta_1} \\ \vdots \\ \dfrac{\partial f(\theta)}{\partial \theta_m} \end{bmatrix}, \quad H \;=\; \begin{bmatrix} \dfrac{\partial^2 f}{\partial\theta_1\partial\theta_1} & \cdots & \dfrac{\partial^2 f}{\partial\theta_1\partial\theta_m} \\ \vdots & \ddots & \vdots \\ \dfrac{\partial^2 f}{\partial\theta_m\partial\theta_1} & \cdots & \dfrac{\partial^2 f}{\partial\theta_m\partial\theta_m} \end{bmatrix}. \tag{2.57}$$

Differentiating (2.56) with respect to θ yields

$$g(\theta) \;\approx\; g(\theta^*) \;+\; H(\theta-\theta^*) \tag{2.58}$$

Since θ^* is the argument of the minimum of $f(\theta)$, $g(\theta^*) = 0$. Therefore, (2.58) yields

$$g(\theta) \;\approx\; H(\theta-\theta^*). \tag{2.59}$$

If $f(\theta)$ were quadratic, (2.59) would be exact. Since in time series models, the log-likelihood is in general a fairly complex function of the parameters and not quadratic, (2.59) suggests the iterative computational scheme

$$\theta^* \;\approx \tilde{\theta} \;-\; H^{-1}g(\tilde{\theta}). \tag{2.60}$$

In (2.60), $\tilde{\theta}$ is the current estimate of θ^* and the Hessian is evaluated at $\theta \;=\; \tilde{\theta}$. Equation (2.60) then denotes a revised estimate which is (hopefully) closer to the minimum of $f(\theta)$ than $f(\tilde{\theta})$. A recursive expression of (2.60), starting from an initial estimate θ_0 and expressing the dependence of H^{-1} on the current value of θ as H_{k-1}^{-1} is,

$$\theta_k \;=\; \theta_{k-1} \;-\; H_{k-1}^{-1}g(\theta_{k-1}). \tag{2.61}$$

As a result of the complexity of the log-likelihood, an analytic evaluation of H is in general quite difficult. This fact motivated a variety of modifications referred to as quasi-Newton methods. The Davidon-Fletcher-Powell and Broyden-Fletcher-Goldfarb-Shanno formulas, (Fletcher 1987, Dennis and Schnabel 1983), respectively in (2.62) and (2.63) are variations that we have used for the direct computation of H_k^{-1} without requiring computation of H_k or its inverse. The sequence of steps in these procedures are as follows:

1. Initialize the procedure and let $k = 0$ with a suitable starting value θ_0 and with an initial estimate of H^{-1}, say $H_0^{-1} = I$, (the $m \times m$ identity matrix).

2. Numerically evaluate the gradient $g(\theta_{k-1})$.

3. Identify a "direction" vector h_k by the expression $h_k = -H_{k-1}^{-1} g(\theta_{k-1})$. (The geometry of this situation may be seen for example in Fletcher 1987.)

4. Using a linear search procedure, estimate a scalar quantity λ_k such that $f(\theta_k)$ is minimal, (and $f(\theta_k) \le f(\theta_{k-1})$ because h_k is in a descent direction of $f(\theta)$ at $\theta = \theta_{k-1}$). That is let $\theta_k = \theta_{k-1} - \lambda_k h_k$.

5. Estimate H_{k-1}^{-1} by the procedures either in (2.62) or (2.63).

$$H_k^{-1} = H_{k-1}^{-1} + \frac{\Delta\theta_k \Delta\theta_k^T}{\Delta\theta_k^T \Delta g_k} + \frac{H_{k-1}^{-1} \Delta g_k \Delta g_k^T H_{k-1}^{-1}}{\Delta g_k^T H_{k-1}^{-1} \Delta g_k} \qquad (2.62)$$

$$H_k^{-1} = \left\{ I - \frac{\Delta\theta_k \Delta g_k^T}{\Delta\theta_k^T \Delta g_k} \right\} H_{k-1}^{-1} \left\{ I - \frac{\Delta\theta_k \Delta g_k^T}{\Delta\theta_k^T \Delta g_k} \right\} + \frac{\Delta\theta_k \Delta\theta_k^T}{\Delta\theta_k^T \Delta g_k}, \qquad (2.63)$$

where $\Delta\theta_k = \theta_k - \theta_{k-1}$ and $\Delta g_k = g(\theta_k) - g(\theta_{k-1})$.

6. Finally update k as in, $k = k + 1$ and repeatedly return to step 2 until $\Delta\theta_k$ and Δg_k are sufficiently small.

2.4 State Space Methods

The concept of *state* is one of the fundamental concepts of system theory. Informally, the state of a system is a summary of the past behavior of the system. The state taken together with the future system inputs determines all future states and system outputs. Furthermore, the output is a function of the current state and current input values only. In effect, the state of a process of a system is a perpetually renewing sequence of initial conditions for that system. The utility of state space models for time series analysis is a consequence of the Markov process property of the state which facilitates computation of the likelihood of a state space model of observed data. Caines (1988), contains a formal treatment of the concept of state defined on a probability space. In Chapters 5, 6 and 15 we deal with linear Gaussian, general and nonlinear state space models respectively.

A fairly general, (nonlinear), state space model is defined by a measurement or observation equation, which relates the observed series $Y_N = \{y_1, \ldots, y_N\}$ to a k-dimensional vector of unobservable states, x_n, and an evolution or state equation which defines the Markov transition density. That is,

$$x_n = f(x_{n-1}) + w_n \qquad (2.64)$$

$$y_n = h(x_n) + \varepsilon_n, \qquad (2.65)$$

where (2.65) is the state equation and (2.64) is the observation equation and where h and f denote arbitrary single-valued functions. An initial state distribution $p(x_0)$ is assumed. The general problem of state space modeling is is to evaluate $p(y_n|Y_m)$, the distribution of y_n given the observations $Y_m = (y_1, \ldots, y_m)$, $n > m$, and the initial distribution.

In Chapter 5 the discussion is specialized to treat the well known linear state space model.

$$\begin{aligned} x_n &= F_n x_{n-1} + G_n w_n \\ y_n &= H_n x_n + \varepsilon_n, \end{aligned} \tag{2.66}$$

where the input $w_n \sim p_w$ and $\varepsilon_n \sim p_\varepsilon$ and $E(w_n, \varepsilon_m^T) = 0$, for all n and m. The state at time n is x_n, the state space is the set of all possible values of x_n, $n = 1, \ldots, N$, F_n is the state transition matrix at time n. The input process w_n, $n = 1, \ldots, N$ and the observation error process ε_n, $n = 1, \ldots, N$ are orthogonal and orthogonal to each other, their covariances may vary instantaneously, and the input process is orthogonal to the initial state x_0, where x_0 is an initial state distributed in accordance with an arbitrary distribution $p(x_0)$, These latter properties account for the Markovian property of the state process, (Caines 1988, Theorem 4.2).

In Chapter 6 we consider a system described by a general state space model

$$\begin{aligned} x_n &\sim q(x_n|x_{n-1}) \\ y_n &\sim r(y_n|x_n), \end{aligned} \tag{2.67}$$

where y_n is the time series and x_n is the unknown state vector and the symbol $\sim$ denotes distribution. q and r are conditional distributions of x_n given x_{n-1} and of y_n given x_n, respectively. The initial state vector x_0 is distributed according to the distribution $p(x_0|Y_0)$. Linear, nonlinear and discrete process state space models are special cases of the general state space model and applications of each of those state space model types appear in this book.

Kalman (1960) and Kalman and Bucy (1961) essentially introduced linear Gaussian state space methods into the system theory community. The popularity of the state space modeling of time series is a more recent development. Hannan (1970) and Anderson (1971) were among the first books to include state space time series methods. Currently, state space representations of time series appears in almost every book on the subject. Kailath (1980), Aoki (1990), Brockwell and Davis (1991), Caines (1988), Harvey (1989) among many others, and Chapter 5 in this book, illustrate state space representations for a variety of time series models. The crucial points are that essentially every type of stationary and nonstationary linear and nonlinear time series model can be cast in state space form, and because of the recursive computational properties associated with the Markovianness of the state process, the likelihood of a time series model can be computed in $O(N)$ time, (independent of the consideration of optimization procedures which are external to state space computations). That is in contrast with ordinary least squares computations which are $O(N^3)$.

The limitations of the Kalman-Bucy linear Gaussian (Kalman filter), framework for some engineering applications motivated activity in nonlinear systems modeling. Starting in 1970, with $n = 1$, that topic was emphasized in a sequence of publications of the *n-th Symposium on Nonlinear Estimation.* Sorenson (1985) is an edited collection of articles on linear and nonlinear Kalman filtering.

Our own interest in state space modeling of time series was propelled by the realization of state space smoothness priors models of nonstationary mean time series, Brotherton and Gersch (1981), Kitagawa (1981). Those initial linear Gaussian state space modeling papers were motivated by and were extensions of Akaike (1980a). Kitagawa (1987) and in subsequent papers demonstrated state space modeling for not necessarily linear not necessarily Gaussian time series. This topic of general state space methods is treated in detail in Chapter 6.

The concern of most contemporary text's treatment of state space modeling is on the Gaussian disturbances linear state space model. Such modeling exploits the Kalman filter algorithm. That algorithm realizes the mean and covariance matrices of the normally distributed state evolution algebraically. The Kalman filter algorithm is shown in Chapter 5, it is derived from the general state space setting in Chapter 6. In Chapters 7, 8 and 9 the Kalman filter is applied to smoothness priors Gaussian disturbance linear state space modeling of nonstationary mean and nonstationary covariance time series. The general state space model is also used in those chapters and the results obtained by the Gaussian disturbance linear state space modeling and the general state space modeling are contrasted.

3

The Smoothness Priors Concept

The "history" of smoothness priors essentially starts with a problem addressed in Whittaker (1923). It was followed by Shiller (1973), and Akaike (1980a) in which the framework initiated by Shiller was continued. Akaike (1980a) was a quasi-Bayesian Gaussian disturbances linear regression, least squares computations, model framework. Stochastic difference equation constraints were placed on the prior distributions of the model parameters. The critical computation was that of the likelihood of hyperparameters of those distributions. Our own work and a considerable amount of other work was motivated by Akaike (1980a). Here we identify some of that work and some relationship of that work to other research as well as developments and extensions. The least squares computational framework of smoothness priors is also presented here.

3.1 Introduction

The problem addressed by Whittaker (1923) in the estimation of a smooth trend, (the mean of a nonstationary mean time series), embedded in white noise was the first work in the subject. The term "smoothness priors" is very likely due to Shiller (1973). Shiller did not appear to be aware of Whittaker's work. He modeled the distributed lag (impulse response) relationship between the input and output of economic time series under difference equation "smoothness" constraints on the distributed lags. A tradeoff of the goodness-of-fit of the solution to the data and the goodness-of-fit of the solution to a smoothness constraint was determined by a single smoothness tradeoff parameter. Shiller did not offer an objective method of choosing the smoothness tradeoff parameter. Akaike, (1980a), continued the analysis initiated by Shiller. Akaike developed and exploited the concept of the likelihood of the Bayesian model and used a maximization of the likelihood procedure for determining the smoothness tradeoff parameter. (In Bayesian terminology, the smoothness tradeoff parameter is a "hyperparameter", Lindley and Smith, 1972.) The smoothing problem context is understood to be common to a large variety of other statistical data analysis problems including density estimation and image analysis, (Titterington 1985).

Akaike (1980a) motivated our own efforts in this area of time series. Following Akaike and initially for us, (Brotherton and Gersch 1981, Kitagawa 1981), smoothness priors was primarily a normal distribution theory, linear model, stochastic regression treatment of stationary and nonstationary time series. Subsequently we appropriated the term "smoothness priors" for use in several papers, (Gersch and Kitagawa 1983a, 1988, Kitagawa and Gersch 1984, 1985a, 1985b, Gersch 1992). In a very significant extension, Kitagawa (1987) showed a smoothness priors state space modeling of nonstationary time series in which neither the system noise or the observation noise are necessarily Gaussian distributed. Among other papers, Kitagawa (1988, 1989, 1991, 1993, 1994, 1996), are further developments, extensions and applications of the not necessarily linear - not necessarily Gaussian state space modeling of time series.

The smoothness priors method is Bayesian. The Bayesianness provides a framework for doing statistical inference. A prior distribution on the model parameters is expressed in the form of a stochastic difference equation and parameterized by hyperparameters which in turn have a crucial role in the analysis. The maximization of the likelihood of a small number of hyperparameters permits the modeling of a time series with relatively complex structure and a very large number of implicitly inferred parameters. The critical statistical ideas in smoothness priors are the likelihood of the Bayesian model and the use of likelihood as a measure of the goodness-of-fit of the model.

3.2 Background, History and Related Work

A conceptual predecessor of smoothness priors can be seen in a smoothing problem posed by Whittaker (1923). In that problem the observations $y_n, \quad n = 1, \ldots, N$ are given. They are assumed to consist of the sum of a "smooth" function f and observation noise or,

$$y_n = f_n + \varepsilon_n, \tag{3.1}$$

where $\varepsilon_n \sim N(0, \sigma^2)$. The problem is to estimate the unknown $f_n, \quad n = 1, \ldots, N$. In a time series interpretation of this problem, $f_n, \quad n = 1, \ldots, N$ is the trend of a nonstationary mean time series. A typical approach to this problem is to approximate f by a class of parametric polynomial regression models. The quality of the analysis is dependent upon the appropriateness of the assumed model class. A flexible model is desirable. In this context, Whittaker suggested that the solution balance a tradeoff of goodness-of-fit to the data and goodness-of-fit to a smoothness criterion. This idea was expressed by minimizing

$$\sum_{n=1}^{N} (y_n - f_n)^2 + \mu^2 \sum_{n=k+1}^{N} (\nabla f_n^k)^2 \tag{3.2}$$

for an appropriately chosen smoothness tradeoff parameter μ^2. In (3.2), ∇f_n^k expresses a k-th order difference constraint on the solution f, with $\nabla f_n = f_n - f_{n-1}$, $\nabla^2 f_n = \nabla(\nabla f_n)$, etc. (Whittaker's original solution was not expressed in a Bayesian context.

Whittaker and Robinson (1924) is a Bayesian interpretation of this problem. Greville (1957) showed that there is a unique solution to (3.2).

The properties of the solution to the problem in (3.1)–(3.2) are apparent. If $\mu^2 = 0$, $f_n = y_n$ and the solution is a replica of the observations. As μ^2 becomes increasingly large, the smoothness constraint dominates the solution and the solution satisfies a k-th order constraint. For large μ^2 and $k = 1$, the solution is a constant, for $k = 2$, it is a straight line etc.. Whittaker left the choice of μ^2 to the investigator.

In a closely related direction, Schoenberg (1964) suggested an adoption of Whittaker's smoothing method to the fitting of a continuous function to the observed data points, with the data not necessarily evenly spaced. In that case the data model is,

$$y_i = f(x_i) + \varepsilon_i, \tag{3.3}$$

where the ε_i are as in (3.1) and f is assumed to be "smooth" on the interval $[\, a, b \,]$ and the observations are at the n points $x_1, \ldots, x_n$. An estimate of f is assumed to be the minimizer of

$$\frac{1}{n}\sum_{i=1}^{n}[y_i - f(x_i)]^2 + \lambda \int_a^b [f^{(m)}(x)]^2 dx, \tag{3.4}$$

with $f \in C^{m-1}$ and $\lambda > 0$. As in the regularly spaced data-discrete function solution problem, again we have a tradeoff between infidelity of the solution to the data, and the "roughness" of the solution as represented by the square integral of the m-th derivative. The nonnegative parameter λ controls the tradeoff. From this nonparametric function estimation interpretation, here too, the parameter of interest is a function.

Parzen (1961, 1963) in a development unrelated to smoothness priors, introduced a reproducing kernel Hilbert space, (RKHS), approach to time series. DeBoor and Lynch (1966) was a RKHS treatment of spline approximation. Kimmeldorf and Wahba (1970a, 1970b, 1971) exploited both developments and treated the general spline smoothing problem from an RKHS-stochastic equivalence perspective. A key result of Kimmeldorf and Wahba, is that minimum norm interpolation and smoothing problems with quadratic constraints imply an equivalent Gaussian stochastic process. Their solutions are Bayesian estimates. Following from the minimum norm interpretation of the smoothing problem, an RKHS is the natural mathematical framework for smoothness priors. Weinert (1972) and Weinert *et al.* (1980) exploited the equivalence to express spline smoothing algorithms in a computationally efficient state space-recursive computational framework. Subsequently Wecker and Ansley (1983), and Ansley and Wecker (1983), and Kohn and Ansley (1988), generalized and realized state-space recursive computational algorithms and applied them to practical data analysis problems. Gersch (1992) includes a review of the RKHS state space approach to smoothness priors.

The Kimmeldorf-Wahba minimum norm-stochastic equivalence implies that the extensively studied signal extraction problem and the smoothing problem are equivalent problem statements. The significance of that result is that the smoothing problem context, and hence smoothness priors, is common to a large variety of other statistical data analysis problems, (i.e. smoothing problems), including density estimation, image restoration, X-ray computed tomographic image reconstruction, and NMR imaging, (for example see Titterington 1985 and Mohamad-Djafari 1994a,b).

Smoothness priors also relates to the ill-posed problems or inverse problems and problems described as statistical regularization, Tikhonov (1963). In that context, scalar inverse problems are discretized and a quadratic regularization criterion is imposed in which, as in the Whittaker problem, the solution balances a tradeoff between infidelity of the solution to the data and infidelity of the solution to the regularization criterion. (Nashed and Wahba 1974 treat statistical regularization in the context of RKHS's.) Penalized likelihood methods, introduced in Good and Gaskins (1980) in the context of density estimation, have been used for example in regression for both Gaussian and non-Gausian data (Wahba 1990, Silverman 1985, O'Sullivan *et al.* 1986, Gu 1990, Hastie and Tibshirani 1993, Green and Silverman 1994), density estimation (Good and Gaskins 1971, Leonard 1978, Silverman 1982, Gu and Qiu 1993, Gu 1993a), hazard rate function estimation (Anderson and Senthiselvan 1980, Gu 1993b) estimation for the intensity function of a Poisson process (Gu and Qiu 1993), and time varying coefficient modeling (Hastie and Tibshirani 1993), which relates closely to our own work in time series.

Further, the Bayesian framework easily provides estimates of the precision of the estimate. In regression for Gaussian data, the commonly used quadratic roughness penalty was shown by Wahba (1978) to be equivalent to a partially improper Gaussian prior, in the sense that the penalized likelihood estimator is identical to the mean of the corresponding Gaussian posterior. Gu (1992) showed that when the sampling likelihood is non-Gaussian, that under Wahba's prior, with appropriate approximations, the penalized likelihood estimator is the mean of the approximate posterior. Bayesian interpretations for the penalized likelihood models, including Bayesian confidence intervals, have for example also been given by Good and Gaskins (1980), Kohn and Ansley (1987), Leonard (1978), Nychka (1988), Silverman (1985), Wahba (1983) and Hastie and Tibshirani (1993). We also note that in a methodology that is relevant in high-dimensional regression problems, the relationship between penalized least squares and estimation in linear additive models, (Buja *et al.* 1989), is the key tool in establishing many of the results.

Vigorous work, initially at the Institute of Statistical Mathematics, Tokyo, and more recently elsewhere in Japan, has resulted in the application of linear Gaussian smoothness priors methods to a variety of applications, other than the ones we discuss here. These applications include the seasonal adjustment of time series, (Akaike 1980b, Kitagawa 1989, 1994), tidal analysis (Ishiguro *et al.* 1981), binary regression (Ishiguro and Sakamoto 1983), distributed lag estimation (Akaike 1986), cohort analysis (Nakamura 1986), density estimation (Tanabe *et al.* 1988), decomposition of spacecraft optical observations, (Higuchi *et al.* 1988), detection of temporal changes in seismic P-waves, (Hurukawa and Imoto 1989), geodetic data inversion Yabuki and Matsu'ura (1992), evaluation of gravitational anomalies, (Murata 1993), and spatial smoothing (Kashiwagi 1993).

Finally, we note that the papers on smoothing serial Poisson data counts by Kashiwagi and Yanagimoto (1992), and Markov state switching in an application to endocrinology by Komaki (1993), have been directly influenced by the not necessarily linear-not necessarily Gaussian work in Kitagawa (1987).

3.3 Smoothness Priors Bayesian Modeling

The theoretical and computational approach in the seminal paper, Akaike (1980a) is described here. Consider the Gaussian disturbances stochastic linear regression model

$$y = X\theta + \varepsilon. \tag{3.5}$$

The dimensions of the matrices in (3.5) are $y: n \times 1$; $X: n \times p$; $\varepsilon: n \times 1$; θ: $p \times 1$. ε is normally distributed with mean 0 and covariance matrix $\sigma^2 I_n$, and θ is a normally distributed prior parameter vector independent of ε with mean 0 and covariance matrix $\lambda^{-2}D^{-1}D^{-T}$, D nonsingular. y is the vector of observed data, X and D are assumed known, ε is the observation noise vector, and σ^2 and λ^2, (where λ is referred to as a hyperparameter, Lindley and Smith (1972)), are unknown. That is,

$$\begin{bmatrix} y \\ \theta \end{bmatrix} \sim N\left(\begin{bmatrix} X\theta \\ 0 \end{bmatrix}, \begin{bmatrix} \sigma^2 I & 0 \\ 0 & \lambda^{-2}D^{-1}D^{-T} \end{bmatrix} \right). \tag{3.6}$$

The dimensions of the matrices in (3.6) are y: $n \times 1$; X: $n \times p$; θ: $p \times 1$. σ^2 and λ^2 are unknown. y is the vector of observed data, X and D are assumed known. θ is the normally distributed prior parameter vector. The observation noise variance is σ^2. In this conjugate family Bayesian situation (Zellner 1971, Berger 1985), the mean of the posterior normal distribution of the parameter vector θ minimizes

$$||y - X\theta||^2 + \lambda^2 ||D\theta||^2. \tag{3.7}$$

If λ^2 were known, the computational problem in (3.7) could be solved by an ordinary least squares computation. The solution for $\hat{\theta}$, the posterior mean, is the minimizer of

$$\left\| \begin{bmatrix} y \\ 0 \end{bmatrix} - \begin{bmatrix} X \\ \lambda D \end{bmatrix} \theta \right\|^2. \tag{3.8}$$

That solution is

$$\hat{\theta} = \left(X^T X + \lambda^2 D^T D \right)^{-1} X^T y \tag{3.9}$$

with the residual sum of squares,

$$\mathrm{SSE}(\hat{\theta}, \lambda^2) = y^T y - \hat{\theta}^T \left(X^T X + \lambda^2 D^T D \right) \hat{\theta}. \tag{3.10}$$

For a Bayesian smoothness priors interpretation of the problem, multiply (3.7) by $-1/2\sigma^2$ and exponentiate. Then, the θ that minimizes (3.7) also maximizes

$$\exp\left\{ -\frac{1}{2\sigma^2} ||y - X\theta||^2 \right\} \exp\left\{ -\frac{\lambda^2}{2\sigma^2} ||D\theta||^2 \right\}. \tag{3.11}$$

From (3.11), the posterior distribution interpretation of the parameter vector θ is that it is proportional to the product of the conditional data distribution (likelihood), $p(y|X,\theta,\sigma^2)$, and a prior distribution, $\pi(\theta|\lambda^2,\sigma^2)$ on θ,

$$\pi(\theta|y,\lambda^2,\sigma^2) \quad \propto \quad p(y|X,\theta,\sigma^2)\pi(\theta|\lambda^2,\sigma^2)\ . \tag{3.12}$$

The left hand side of (3.12) is a proper distribution. Consequently, the integration of the right hand side of (3.12) yields $L(\lambda^2,\sigma^2)$, the likelihood for the unknown parameters λ^2 and σ^2,

$$L(\lambda^2,\sigma^2) \;=\; \int_{-\infty}^{\infty} p(y|X,\theta,\sigma^2)\pi(\theta|\lambda^2,\sigma^2)d\theta. \tag{3.13}$$

I. J. Good (1965) referred to the maximization of (3.13) as a type II maximum likelihood method. A critically important result in Akaike (1980a) is that, since $\pi(\theta|y,\lambda^2,\sigma^2)$ is normally distributed, (3.13) can be expressed in the closed form,

$$L(\lambda^2,\sigma^2) \;=\; (2\pi\sigma^2)^{-\frac{N}{2}}|\lambda^2 D^T D|^{\frac{1}{2}}|X^T X+\lambda^2 D^T D|^{-\frac{1}{2}}\exp\left\{-\frac{1}{2\sigma^2}\mathrm{SSE}(\hat{\theta},\lambda^2)\right\}. \tag{3.14}$$

The maximum likelihood estimator of σ^2 is

$$\hat{\sigma}^2 \;=\; \frac{1}{N}\mathrm{SSE}(\hat{\theta},\lambda^2). \tag{3.15}$$

It is convenient to work with -2 log likelihood. Using (3.15) in (3.14) yields

$$-2\log L(\lambda^2,\hat{\sigma}^2) \;=\; N\log 2\pi\hat{\sigma}^2 - \log|\lambda^2 D^T D| + \log|X^T X+\lambda^2 D^T D| + \ N. \tag{3.16}$$

A practical way to determine the value of λ^2 for which the -2 log-likelihood is minimized, is to compute the likelihood for discrete values of λ^2 and search the discrete -2 log likelihood-hyperparameter space for the minimum. A precise estimate of λ^2 may be obtained by a numerical linear search algorithm.

Akaike (1980a) and Good and Gaskins (1980) are very likely the first practical uses of the likelihood of the Bayesian model and the likelihood of the hyperparameters as a measure of the goodness-of-fit of a model to data.

4

Scalar Least Squares Modeling

In this chapter we review some of the applications of smoothness priors modeling of time series that can be done by least squares, or equivalently as linear Gaussian modeling. Smoothness priors trend estimation for scalar time series is treated in Section 4.1. There, the smoothness priors constraint is expressed as a k-th order random walk with a normally distributed zero-mean, unknown variance perturbation. The (normalized) variance is a hyperparameter of the prior distribution. This constraint is a time domain constraint on the priors. The concept of frequency domain priors is introduced and exploited in Sections 4.2 and 4.3. There, the fitting of a smoothness priors long AR model for the spectrum estimation of a scalar stationary time series and a smoothness priors model for transfer function estimation between two simultaneously observed time series, are respectively shown. In addition in Section 4.2, the superiority of the smoothness priors long AR model as compared to ordinary AIC AR model order determined spectral analysis, is demonstrated by a Monte Carlo computation of entropy.

4.1 Estimating a Trend

Here we return to the original problem posed in Whittaker (1923). In that problem the observations y_n, $n = 1, \ldots, N$ are given. They are assumed to consist of the sum of a "smooth" function t_n, $n = 1, \ldots, N$ and observation noise or,

$$y_n = t_n + \varepsilon_n, \quad n = 1, \ldots, N \tag{4.1}$$

where for convenience, we assume $\varepsilon_n \sim N(0, \sigma^2)$. The problem is to estimate the unknown mean value or trend function, t_n, $n = 1, \ldots, N$.

Following the development in Chapter 3, the estimated trend solution balances a tradeoff of goodness-of-fit to the data and goodness-of-fit to a smoothness criterion. That is,

$$\sum_{n=1}^{N} (y_n - t_n)^2 + \lambda^2 \sum_{n=k+1}^{N} \left(\nabla^k t_n\right)^2 \tag{4.2}$$

is to be minimized for some appropriately chosen smoothness tradeoff parameter λ^2. In (4.2), $\nabla^k t_n$ expresses a k-th order difference constraint on the solution, with $\nabla t_n = t_n - t_{n-1}$, $\nabla^2 t_n = \nabla(\nabla t_n)$, etc. Following Akaike (1980a), from the Bayesian point of view, the difference equation constraints in this problem are stochastic. That is, $\nabla^k t_n = w_n$, with w_n assumed to be a normally distributed zero-mean sequence with unknown variance $\tau^2 = \sigma^2/\lambda^2$.

In the context of the least squares computational approach to smoothness priors in Chapter 3, the mean of the posterior normal distribution of the parameter vector $t = (t_1, \ldots, t_N)^T$ minimizes

$$||y - It||^2 + \lambda^2||Dt||^2, \tag{4.3}$$

where D is the $N \times N$ matrix of linear constraints on the "trend" vector t and I is the $N \times N$ identity matrix.

On the assumption that λ^2 were known, the least squares solution for $\hat{t}$, the posterior mean, is the minimizer of

$$\left\| \begin{bmatrix} y \\ 0 \end{bmatrix} - \begin{bmatrix} I \\ \lambda D \end{bmatrix} t \right\|^2, \tag{4.4}$$

with solution

$$\hat{t} = \left(I + \lambda^2 D^T D\right)^{-1} y \tag{4.5}$$

and residual sum of squares,

$$\mathrm{SSE}(\hat{t}, \lambda^2) = y^T y - \hat{t}^T \left(I + \lambda^2 D^T D\right) \hat{t} \tag{4.6}$$

and $\hat{\sigma}^2$ the maximum likelihood estimator of σ^2 is

$$\hat{\sigma}^2 = \frac{1}{N}\mathrm{SSE}(\hat{t}, \lambda^2). \tag{4.7}$$

Under the k-th order difference constraint, $\nabla^k t_n$ on the solution t, with $\nabla t_n = t_n - t_{n-1} = w_n$, $\nabla^2 t_n = \nabla(\nabla t_n)$, $t_n = 2t_{n-1} - t_{n-2} + \omega_n$. Then for $k = 1$ and $k = 2$ those smoothness constraints expressed in terms of the following $N \times N$ constraint matrices are:

$$D_1 = \begin{bmatrix} \alpha & & & & \\ -1 & 1 & & & \\ & -1 & 1 & & \\ & & \ddots & \ddots & \\ & & & -1 & 1 \end{bmatrix}, \quad D_2 = \begin{bmatrix} \alpha & & & & & \\ -\beta & \beta & & & & \\ 1 & -2 & 1 & & & \\ & 1 & -2 & 1 & & \\ & & \ddots & \ddots & \ddots & \\ & & & 1 & -2 & 1 \end{bmatrix}. \tag{4.8}$$

In (4.8) α and β are small numbers that are chosen to satisfy initial conditions. It is interesting to note that the $k = 2$ constraint was used in a non-stochastic framework with an ad-hoc selection of the tradeoff parameter by Phillips (1962) in smoothing the inherent ill-conditioning in numerical solutions to integral equations of the first kind, and subsequently by Hunt (1970, 1973), in inverse problem and image restoration applications.

We introduce the notation $t_{n|N} = t_{n|y_1,\ldots,y_N}$ to signify the estimate of the trend given all the data $\{y_1, \ldots, y_N\}$. That is, the smoothing problem expression of the solution vector is expressed as: $\hat{t} = (t_{1|N}, \ldots, t_{N|N})^T$. The least squares solution (4.5), may be obtained either for discrete values of λ, with the discrete -2 log likelihood-hyperparameter space searched for a minimum, or repeatedly for successive values of λ as determined by a numerical optimizing algorithm. From Chapter 3, the minimized value of -2 log likelihood for this problem is:

$$-2\log L(\hat{\lambda}^2, \hat{\sigma}^2) = N\log 2\pi + N\log\hat{\sigma}^2 - \log|\hat{\lambda}^2 D^T D| + \log|X^T X + \lambda^2 D^T D| + N \quad (4.9)$$

The numerical values of SSE($\hat{t}, \lambda^2$) and of the determinants in (4.9) may be computed as the product of the diagonal terms of a triangular matrix in a Householder transformation least squares algorithm analysis, (Section 2.2). Since $\lambda^2 = \sigma^2/\tau^2$, λ^2 has a noise-to-signal ratio interpretation, larger values of λ correspond to smoother trends.

An Example of Trend Estimation

We consider a $N = 200$ simulated data example in which the smooth trend has the functional form of an asymmetrically truncated Gaussian function. The "true" smooth function t_n, $n = 1, \ldots, N$ and the smooth function with the superposition of t_n and the additive noise is shown in Figure 4.1A. The problem is: Given the noisy observations $\{y_n, n = 1, \ldots, N\}$, estimate the unknown smooth function that is in the noise, i.e. specify $\hat{t}_{n|N}$, $n = 1, \ldots, N$.

The least squares computational problem for order $k = 2$ trend estimation was solved using the Householder transformation method, and a search in the discrete hyperparameter-likelihood space to locate the minimum of -2 log likelihood of the hyperparameter model. The critical role of the hyperparameter is apparent in this example. Figures 4.1B, C, D show the superimposed noisey data, estimated trend and true trend for values of the hyperparameters that are too small, ($\lambda^2 = 1.00$, -2 log likelihood $= 749.929$, $\hat{\sigma}^2 = 0.581$) and too large ($\lambda^2 = 1000.0$, -2 log likelihood $= 666.032$, $\hat{\sigma}^2 = 1.475$) as well as the hyperparameter for which -2 log likelihood is minimized, ($\lambda^2 = 103.9$, -2 log likelihood $= 574.829$, $\hat{\sigma}^2 = 0.871$). As anticipated, with the hyperparameter as defined above, the estimated trend for a too large value of the hyperparameter is too smooth and the estimated trend for a too small value of the hyperparameter is too bumpy. As borne out by the values in Table 4.1.1, the -2 log likelihood vs hyperparameter function is quite flat.

Figure 4.1E is an illustration of -2 log-likelihood versus hyperparameter value function for this problem. We note that this is a smooth function so it is quite easy to realize a reliable minimization of the log-likelihood. It is also important to note that in this example, the trend is estimated with the "incorrect" model $\nabla^2 t_n = w_n$. That is, the true trend function is not a solution of the the stochastically perturbed second order difference equation. The point is that a priori, we do not know a correct

Table 4.1.1: Hyperparameter values and -2 log likelihoods, Trend model

λ^2	-2 log likelihood
30.0	588.617
60.0	577.961
100.0	574.847
150.0	576.672
200.0	581.283
250.0	587.965

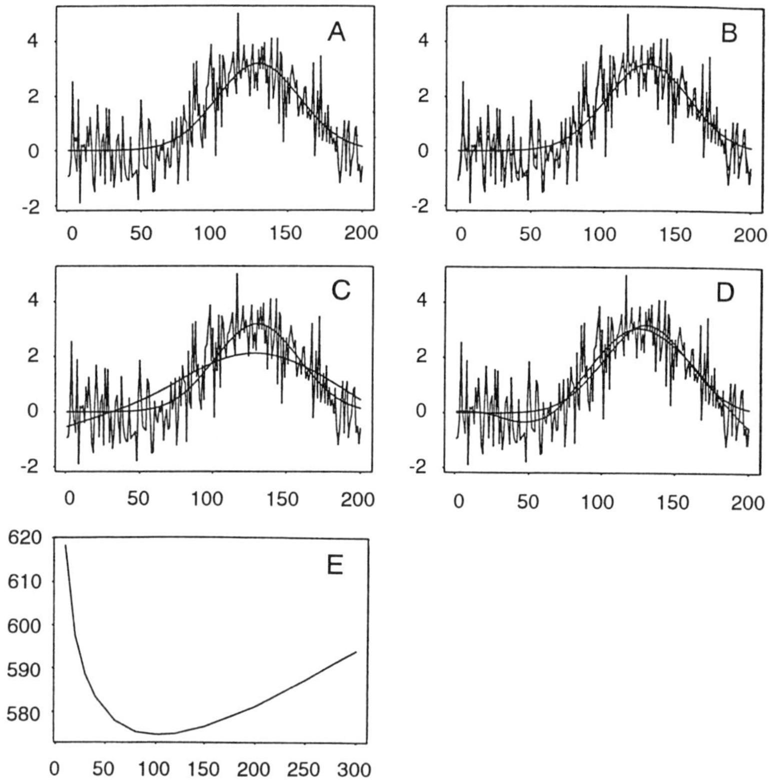

Figure 4.1 A: "True" smooth function and observed data. B: The estimated trend for a too small hyperparameter. C: The estimated trend for a too large hyperparameter. D: The estimated trend for the optimal hyperparameter. E: -2 log-likelihood versus the hyperparameter .

expression for the underlying true smooth functions. Different hyperparameter values result in very different looking solutions. The best of those solutions is a very good approximation to the original unknown smooth function. This key observation was referred to by Shiller, (1973), as the "flexible ruler approach".

Frequency domain priors are introduced in the next section in the context of a smoothness priors long-AR model for spectrum estimation.

4.2 The Long AR Model

Background

A smoothness priors-long autoregressive (AR) model approach is used here for the estimation of the spectral distribution of a stationary time series. That is, as shall be justified later in this section, the AR model order used in our smoothness priors spectrum estimation may be considerably longer than that used in the conventional AIC-AR modelling of the same data. The problem of spectral estimation is a fundamental problem in engineering and in statistics. If the spectral distribution is absolutely continuous, its density function can be estimated by a variety of methods. The two most popular being window and AR estimates. Window estimates are obtained by smoothing the periodogram using "window" functions, while AR estimates are obtained by fitting parametric AR models using "automatic" model selection procedures based on the AIC or the BIC, (Priestly 1981, Chapters 6 and 7, also Brillinger 1993a).

The classical windowed periodogram method of spectral estimation is satisfactory for spectral analysis when the data set is "long". The periodogram is known not to be a consistent estimate of the spectral density function but consistency can be achieved by smoothing the periodogram ordinates. The degree of smoothing is controlled by the window width. Larger window widths smooth out the noise, while smaller window widths tend to yield estimates with spurious features. Daniell (1946), is an early mention of windowed periodogram spectrum analysis. The theory and practice of spectrum estimation developed rapidly with major contributions by Grenander and Rosenblatt (1953, 1957), Parzen (1957a, b) and Blackman and Tukey (1959) among others. Also, see Brillinger (1974, 1975). Recent approaches to the window width selection problem involves the application of cross-validation, (Beltrão and Bloomfield 1987, Hurvich and Beltrão 1990), the bootstrap, (Swanepoel and van Wyk 1986, Politis and Romano 1992) and regression methods, (Wahba 1980, Pawitan and Gangopadhyay 1991).

In doing spectral estimation by the AR modeling approach the estimated spectral density has a specified parametric form. This method has become popular for the fitting of moderate data length spans since the publication of Akaike's FPE and AIC criteria for AR model order selection, (Akaike 1969, 1974a), also see Gersch and Sharpe (1973). However, when the data span is relatively short, three facts render parametric AR and AR-MA modeling methods of spectral estimation statistically unreliable. One is the instability or small sample variability of whatever statistic is used for determining the best order of parametric model fitted to the data. The second is that usually the "parsimonious" parametric model may not be a very good replica of the system that generated the data. The third is that the spectral density of the fitted parametric model can not possibly be correct. Independent of which parametric model order is selected, there is information in the data to select models of different orders. A Bayesian estimate of power spectral density requires that the spectral density of parametric models of different model orders be weighted in accordance with the likelihood and the prior of the model order of different models.

The smoothness priors long AR model of spectral estimation alleviates this problem. A particular class of frequency domain smoothness priors is assumed for the coefficients of AR model of order M, with M relatively large. The likelihood of the hyperparameters that characterize the class of smoothness priors is maximized to yield the best AR model of order M with the best data dependent priors.

The Smoothness Priors Long AR Model

Consider the autoregressive model of order M,

$$y_n = \sum_{m=1}^{M} a_m y_{n-m} + \varepsilon_n. \tag{4.10}$$

In (4.10) ε_n is a Gaussian white noise with mean zero and variance σ^2. The theoretical spectrum of the AR model process is given by, (Grenander and Rosenblatt 1957a, 1957b, Hannan 1961),

$$P(f) = \frac{\sigma^2}{|1 - \sum_{m=1}^{M} a_m \exp\left(-2\pi imf\right)|^2}; \qquad -\frac{1}{2} \le f \le \frac{1}{2}. \tag{4.11}$$

A least squares fit of the AR model to the data, $y_1, \ldots, y_N$, leads to the minimization of

$$\sum_{n=M+1}^{N} \left(y_n - \sum_{m=1}^{M} a_m y_{n-m} \right)^2 . \tag{4.12}$$

If M is comparable to N, the result of the least squares computation can be meaningless. Frequency domain smoothness priors constraints on the distribution of the AR model parameters mitigate this difficulty. Those constraints permit the problem to be cast into a meaningful linear least squares computational framework.

The frequency response function of the whitening filter of the AR process, needed here, is given by

$$A(f) = I - \sum_{m=1}^{M} a_m \exp(-2\pi imf). \tag{4.13}$$

Let a measure of the smoothness, (roughness actually), of the frequency response function be

$$R_k = \int_{-1/2}^{1/2} \left| \frac{d^k A(f)}{df^k} \right|^2 df = (2\pi)^{2k} \sum_{m=1}^{M} m^{2k} a_m^2. \tag{4.14}$$

From the definition in (4.14), a large value of R_k means an unsmooth (in the sense of differential) frequency response function. We also use the zero derivative smoothness constraint,

$$R_0 = \int_{-1/2}^{1/2} |A(f)|^2 df = 1 + \sum_{m=1}^{M} a_m^2 \tag{4.15}$$

as a penalty on the whitening filter.

With these constraints, and with λ^2 and ν^2 fixed, the AR model coefficients a_m, $m = 1, \ldots, M$, minimize

$$\sum_{n=M+1}^{N} \left(y_n - \sum_{m=1}^{M} a_m y_{n-m} \right)^2 + \lambda^2 \sum_{m=1}^{M} m^{2k} a_m^2 + \nu^2 \sum_{m=1}^{M} a_m^2. \tag{4.16}$$

In (4.16), λ^2 and ν^2 are the hyperparameters. By a proper choice of these parameters, our estimates of the AR model coefficients balance the tradeoff between the infidelity of the model to the data and the infidelity of the model to the frequency domain smoothness constraints. For completeness, to within a constant, the Gaussian priors on the AR model coefficients corresponding to the R_0 and R_k constraints are

$$\exp\left\{ -\frac{\lambda^2}{2\sigma^2} \sum_{m=1}^{M} m^{2k} a_m^2 \right\} \exp\left\{ -\frac{\nu^2}{2\sigma^2} \sum_{m=1}^{M} a_m^2 \right\}. \tag{4.17}$$

Following our earlier discussion, define the matrices D, X, y and a by

$$D = \begin{bmatrix} (\nu^2 + \lambda^2)^{1/2} & & & \\ & (\nu^2 + 2^{2k}\lambda^2)^{1/2} & & \\ & & \ddots & \\ & & & (\nu^2 + M^{2k}\lambda^2)^{1/2} \end{bmatrix},$$

$$X = \begin{bmatrix} y_M & \cdots & y_1 \\ y_{M+1} & \cdots & y_2 \\ \vdots & \ddots & \vdots \\ y_{N-1} & \cdots & y_{N-M} \end{bmatrix}, \quad y = \begin{bmatrix} y_{M+1} \\ y_{M+2} \\ \vdots \\ y_N \end{bmatrix}, \quad a = \begin{bmatrix} a_1 \\ a_2 \\ \vdots \\ a_M \end{bmatrix}. \tag{4.18}$$

Then, from the solution of the constrained least squares problem in (4.16), the AR model coefficients satisfy

$$\hat{a} = (X^T X + D^T D)^{-1} X^T y, \tag{4.19}$$

with the residual sum of squares,

$$S(\lambda^2, \nu^2) = y^T y - \hat{a}^T (X^T X + D^T D) \hat{a}. \tag{4.20}$$

The likelihood of the hyperparameter of the model is

$$L(\lambda^2, \nu^2, \sigma^2) = \left(\frac{1}{2\pi\sigma^2} \right)^{N/2} |D^T D|^{1/2} |X^T X + D^T D|^{-1/2} \exp\left\{ -\frac{1}{2\sigma^2} S(\lambda^2, \nu^2) \right\} \tag{4.21}$$

Table 4.2.1: Hyperparameter values, -2 log likelihoods in modeling the Canadian lynx data.

k	ν	λ	-2 log likelihood
1	0.000	0.173	-12.77
2	0.346	0.022	-10.01
3	0.346	0.014	-7.51
4	0.346	0.0002	-6.22

Given λ^2 and ν^2, the maximum likelihood estimate of σ^2 is $\hat{\sigma}^2 = S(\lambda^2, \nu^2)/N$. The maximum likelihood estimates of λ^2 and ν^2 are obtained by minimizing

$$-2\ \log L(y|\lambda^2, \nu^2, \hat{\sigma}^2) = N \log 2\pi\hat{\sigma}^2 - \log |D^T D| + \log |X^T X + D^T D| + N. \quad (4.22)$$

with respect to λ^2 and ν^2. One way to estimate the hyperparameters is to compute the likelihood over a discrete k, λ^2, ν^2 parameter grid and search over the resulting discrete likelihood-hyperparameter space for the minimum of -2 log likelihood. In fact we used the Davidon-Fletcher-Powell type quasi-Newton optimization algorithm described in Section 2.4 to minimize the expression in (4.22).

The frequency domain smoothness priors constraint used here has an interpretation as a constraint on the smoothness of the whitening filter of the AR model. (The 0-th derivative has an energy constraint interpretation.) An important feature of our computations is that they are computationally tractable which allows us to remain within the framework of the general linear model.

An Example, Analysis of Canadian Lynx Data

The data example discussed here is the analysis of the Canadian Lynx data ($n = 114$). This data set is one of several sets of data that have attracted considerable and special attention. (It is also discussed in Chapter 14.) The data, the annual record of the number of Canadian lynx trapped in the Mackenzie River district of Northwest Canada for the period 1821-1924, originally appeared in Elton and Nicholson (1942). Discussion of the earliest statistical analysis (Moran 1953a, 1953b) and reviews of subsequent and at that time contemporary analyses appear in Campbell and Walker (1977).

In our analysis, AR models of order up to 20 were first fitted by the least squares algorithm-AIC criterion method. The AIC best model order was 11. Smoothness priors AR model orders up to 20 and up to the fourth order ($k = 4$), smoothness prior constraint were tried. The hyperparameters λ and ν which minimize of -2 log-likelihood were determined by a version of the Davidon-Fletcher-Powell optimization algorithm for each value of the differentiation order k. Numerical details are in Table 4.2.1.

The overall AIC best smoothness priors model was model order $M = 14$, $k = 1$, $\nu = 0$, $\lambda = 0.173$. The smoothness priors AR model estimate of the spectrum

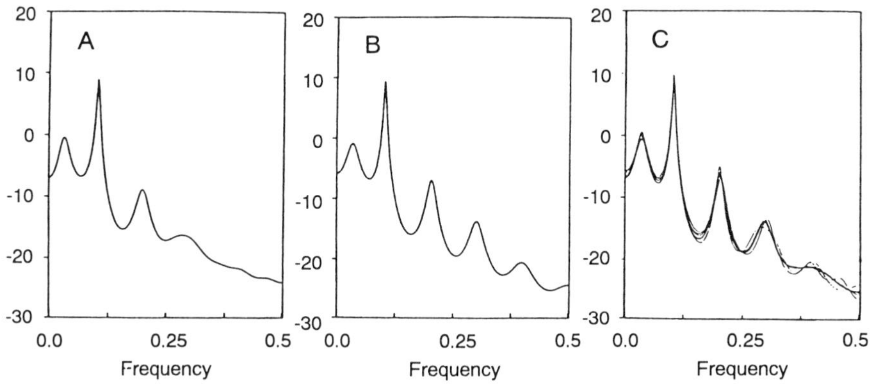

Figure 4.2 A: Smoothness priors spectrum estimate. B: AIC best spectrum estimate. C: Superimposed estimated spectra from different AR model orders.

is shown in Figure 4.2A. The AIC criterion-AR modeled spectrum is shown in Figure 4.2B. In the Bayesian model spectral estimate, the peaks at the high frequencies are significantly reduced compared to the conventional AR model spectrum estimate, while the ones in the lower frequencies remain unchanged. Figure 4.2C shows the superimposed estimated spectra obtained from AR models with different model orders. The shape of the two rightmost peaks in Figure 4.2C vary considerably with model order.

This situation is typical of the problem of estimating a spectral density by fixed order parametric models. If the model fitted to the data is not in the class of models which generated the data, the model fitting is only approximate. The selection of the best non Bayesian parametric model ignores the evidence for other parametric models, when in fact it should be taken into account. The suppression of those peaks by the smoothness priors-long AR model method, shown in Figure 4.2A, therefore seems quite reasonable and we conclude that spectrum estimation by fixed order models is not statistically reliable.

A Second Example, The Simulated Sum of Sinusoids

We considered another example in order to illustrate and contrast the properties of the smoothness priors long AR modeling method with those of the minimum AIC least squares AR modeling. Both model types were fitted to the simulated data corresponding to the sum of two sines waves observed in the presence of added Gaussian noise. This then is a case of the estimation of a line spectrum. Some of the subtlety in the case of estimating line spectra or mixed spectra has been treated in Priestly (1962a, 1962b), Stoica *et al.* (1991) and Kooperberg *et al.* (1995).

The data for the second example was generated by

$$y_n = \sin\frac{2\pi n}{25} + \sin\frac{2\pi n}{16} + \varepsilon_n, \tag{4.23}$$

where ε_n is a Gaussian pseudorandom sequence with variance 0.09. The data length was 140 and we fitted an $M = 40$-th order model. The hyperparameter best models for each difference order $k = 1, \ldots, 5$, were again determined by the minimization of -2 log likelihood. Table 4.2.2 shows the estimated parameters. For $k \geq 2$, $\lambda = 0$ was selected as the best hyperparameter. Thus the coefficients are identical for $k \geq 2$. The overall best model was $k = 1$, $\nu = 0.887$ and $\lambda = 0.089$. The estimated spectrum is shown in Figure 4.3A. Figure 4.3B shows the spectrum obtained by the AIC best order (order = 30) AR model. Comparison of Figures 4.3A and 4.3B reveals that the Bayesian procedure reproduces the sharp spectral peaks of the signal and yields a much smoother spectral estimate than that obtained by the AIC-AR model. The model in (4.23) is reasonably represented as an ARMA(4,4) process and the equivalent AR model is of infinite order. Ordinary long AR models fitted to these data tend to exhibit the behavior illustrated in Figure 4.3A. The spectral peaks are clearly distinguishable and the remaining spectrum that should be flat, is quite bumpy.

Table 4.2.2: Hyperparameter values, -2 log likelihoods in modeling the simulated two sine wave data.

k	ν	λ	-2 log likelihood
1	0.887	0.022	-115.17
2	1.065	0.000	-114.87
3	1.065	0.000	-114.87
4	1.065	0.000	-114.87
5	1.065	0.000	-114.87

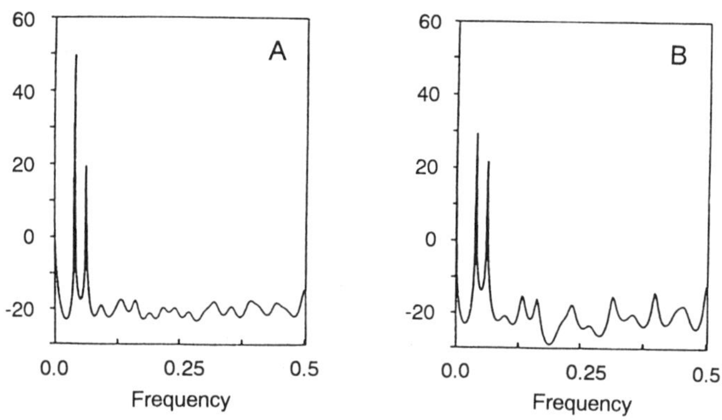

Figure 4.3 A: Smoothness priors spectrum estimate. B: AIC best spectrum estimate.

A Monte Carlo Entropy Computation Study of the Bayesian Procedure

The statistical performance of the smoothness priors long AR model method and the minimum AIC-AR model methods of spectral analysis were compared using a Monte-Carlo simulation study of the goodness-of-approximation of the estimated probability distributions to the true distribution.

A. Motivation and Background

Let $y = (y_1, \ldots, y_n)$ be the observed data. Let $f(x|y)$ be the probability density function of the fitted model given the data y and let $g(x)$ be the probability density function of the true model. Then, the goodness-of-approximation of f with respect to g measured by the entropy of the true distribution with respect to the model is defined by, (Akaike 1977a),

$$B(g, f) = \int g(x) \log \left\{ \frac{f(x|y)}{g(x)} \right\} dx. \tag{4.24}$$

The quantity $B(g, f)$ is the negative of the Kullback-Leibler number. That is, $B(g, f) = -I(g, f)$. Following from Section 2.1.1, $B(g, f) \leq 0$ and $B(g, f) \equiv 0$ if and only if $f = g$ almost everywhere. Different distributions, f_i can be compared to a known true distribution g. The model for which $B(g, f_i)$, $i = 1, 2, \ldots, L$ is a maximum is the maximum entropy best model.

Motivations for the use of an entropy measures of dissimilarity for the comparison of models may be found in Jaynes (1957), Kullback (1958), Akaike (1977a), Kitagawa and Akaike(1981), and Johnson and Shore (1979) for example. There is a separate literature on the use of the Kullback-Leibler or equivalently the Itakura-Saito type (neg-entropy) measures in speech, see De Sousa and Thomson (1982), Gray *et al.* (1980) and, Makhoul (1975) for example and the references therein.

A practical computational procedure for a maximum entropy comparison of alternative models follows. Let y be the data vector from which the i-th fitted model $f_i(x|y)$ is computed with $g(x)$ assumed known. The alternative models can be compared by computing the average entropy $E_Y[B(g(\cdot), f_i(\cdot|Y))]$ for each model where the average is taken with respect to the data distributions. The maximum expected entropy best model is the model for which the average entropy is maximum. A Monte Carlo experiment permits a law of large numbers approximation to the average entropy computations.

We computed the average entropy from the simulated data for both the smoothness priors long AR model and for the minimum AIC-AR model method in two different situations of interest. AR(2) and ARMA(4,2) models were chosen as simulation (true) models. Thus, the choice of simulations and fitted models span the situations in which the minimum AIC-AR of fitted parametric model is within the class of models that generated the data (the AR(2) simulation model), and when the fitted model does not include the class of models that generated the data, (the ARMA(4,2) model, since it is the equivalent of an infinite AR order model.) The latter situation, in which both the minimum AIC-AR model and the smoothness priors model are not in the class of models that generated the data is the situation of greatest interest in evaluating Bayesian smoothness priors models. ARMA models were not fitted to simulated data for several reasons. First of all, both the AR(2) and ARMA(4,2) simulation models would be in the class of fitted ARMA models thereby depriving us of the most important situation for study. Also, fitting AR models requires a one-dimensional search for the model order whereas fitting ARMA models requires a model order search in the plane as well as numerical optimization in the parameter space. That is a time consuming task and should be avoided whenever possible.

As a result of some preliminary experimentation, which included the fitting of longer AR models, a smoothness priors long AR model order 20 was chosen. We anticipated and discovered that AR(20) was sufficiently long so that the minimum AIC procedure did not select ordinary AR models larger than AR(20) for the simulated data. It is important to note that if we allowed smoothness priors with different values of M, we would also be required to establish priors for those different M's. Although that is a subject of considerable interest, it is quite secondary to our principal purposes here.

B. Some Details and Discussion of Results

100 sets of 120 observations were generated by the AR model

$$y_n = 0.9y_{n-1} - 0.81y_{n-2} + \varepsilon_n \tag{4.25}$$

and separately by the ARMA model

$$\begin{aligned} y_n \; = \; & 0.12y_{n-1} - 0.7504y_{n-2} + 0.0157y_{n-3} - 0.5417y_{n-4} \\ & + \varepsilon_n + 0.9\varepsilon_{n-1} + 0.081\varepsilon_{n-2} \end{aligned} \tag{4.26}$$

where ε_n is a Gaussian pseudorandom number with mean zero and variance one.

Smoothness priors models of order 20 with the k-th order ($k = 1, \ldots, 7$) smoothness prior were fitted to each data set. The hyperparameters were searched over the discrete values $\nu = i^2\sigma_0,\ i = 0, 1, \ldots, 4$ and $\lambda = 2^{j-3k},\ j = 1, 2, \ldots, 10$. The overall Bayesian maximum a posteriori model is the one with the k which attains the overall maximum of the log-likelihood.

The Bayesian and the minimum AIC-AR models for spectral analysis were compared using three related entropy of distribution measures: two types of entropies of the probability density function and an integrated square of the difference of log spectral density measure. Discussion of the alternative dissimilarity measures is beyond the scope of our interests here. Therefore only the formulas for those measures and a discussion of the computational results obtained are given here. Complete numerical and tabulated results follow after the description of the results obtained.

The first dissimilarity measure is the entropy of the true distribution with respect to the postulated distribution, defined by

$$W_1 = B(g, f) = \int_{-\infty}^{\infty} \log\left\{\frac{f(x)}{g(x)}\right\} g(x)dx \tag{4.27}$$

where $g(x)$ and $f(x)$ are the probability density functions of the true and estimated models, respectively.

In the cases of the AR order the entropy is specifically given by

$$\frac{1}{2}\left[1-\log\left(\frac{\hat{\sigma}^2}{\sigma^2}\right)-\frac{1}{\hat{\sigma}^2}\left\{\gamma_0-2\sum_{m=1}^{M}\hat{a}_m\gamma_m+\sum_{m=1}^{M}\sum_{l=1}^{M}\hat{a}_m\hat{a}_l\gamma_{m-l}\right\}\right] \tag{4.28}$$

where γ_m is the m-lag theoretical covariance of the true process, $\hat{a}_m$ is the m-th AR coefficient of the estimated model, and σ^2 and $\hat{\sigma}^2$ are the innovations variances of the true and estimated models, respectively. (This result is derived in Kitagawa and Akaike 1978, Gray *et al.* 1980 and Gersch 1981, for example.)

W_2, a variation of W_1, is taken to be the normalization with respect to the unknown variances σ^2 of the Bayesian and minimum AIC-AR models. That is, W_1 is a dissimilarity measure in which both the dissimilarity in energy levels and the dissimilarity in the shapes of the distribution are relevant. W_2 is a measure only of the shape dissimilarity. That is achieved using the formula for W_1 with $\hat{\sigma}^2 = \sigma^2$. (Gersch *et al.* 1980 offers a motivation for this normalization in a classification of time series context.) W_2 is very likely the most informative measure of dissimilarity.

The third measure,

$$W_3 = \int_{-\frac{1}{2}}^{\frac{1}{2}} [\log p(f) - \log q(f)]^2 df, \tag{4.29}$$

is a measure of the dissimilarity of the true and estimated log spectral densities. This measure was advocated for example by Makhoul (1975). For $p(f) \approx q(f)$, this measure can be seen to be proportional to the negentropy.

The computational results are shown in Table 4.2.3–4.2.5. First, in fitting the smoothness priors long AR model and the minimum AIC-AR model to the AR(2) data in Table 4.2.3, we see evidence of the now well known risk that occasionally excessively large order models are fitted in minimum AIC-AR modeling. We also observed that the selection of the differential order in the long AR Bayesian model was quite stable.

Table 4.2.3: Monte Carlo study, Model order occurrence.

Smoothness Prior Order K	1	2	3	4	5	6	7
AR(2)	2	2	6	4	85	1	0
ARMA(4,2)	0	4	9	9	7	7	64

AR order M	0	1	2	3	4	5	6	7	8	9	
AR(2)	0	75	9	4	2	6	0	0	0	0	
ARMA(4,2)	0	0	0	0	0	1	9	9	17	9	
M	10	11	12	13	14	15	16	17	18	19	20
AR(2)	0	3	0	0	0	0	0	0	0	1	0
ARMA(4,2)	15	7	8	3	7	1	7	1	3	2	1

Also from Table 4.2.3, in fitting the minimum AIC-AR model to the simulated ARMA(4,2) model data we again observe broad variations in the minimum AIC-AR order models. This evidence suggests the relative stability in model selection achieved by the smoothness priors model as compared with that achieved by the ordinary AIC-AR modeling. It also suggests that ordinary parametric model order selection can be delicate.

Table 4.2.4: Entropy and other dissimilarity measures of Bayesian and least squares AR model fitting performance, AR process.

	order	W_1	W_2	W_3
Bayesian	1	-92.1	-4.1	44.3
	2	-90.4	-2.4	34.6
	3	-83.9	-2.0	31.8
	4	-82.2	-1.9	30.3
	5	-81.8	-1.7	28.8
	6	-84.0	-1.9	30.3
	7	-85.1	-2.1	31.6
	max. lk	-83.3	-2.0	30.0
Least Squares	0	-145.5	-143.1	885.6
	1	-131.7	-95.6	554.0
	2	-86.1	-1.2	25.1
	3	-89.5	-1.7	28.9
	4	-92.9	-2.1	32.5
	5	-96.2	-2.6	36.0
	6	-100.8	-3.2	40.8
	7	-104.1	-3.7	44.5
	8	-108.5	-4.2	48.9
	9	-113.3	-4.8	53.5
	10	-117.3	-5.3	57.3
	11	-123.0	-6.1	63.0
	12	-127.0	-6.6	66.9
	13	-130.8	-7.0	70.7
	14	-133.9	-7.4	74.0
	15	-139.5	-8.0	79.1
	16	-145.4	-8.8	84.8
	17	-150.8	-9.5	89.9
	18	-155.4	-10.0	94.0
	19	-161.7	-10.6	99.5
	20	-167.4	-11.2	105.2
	AIC best	-96.4	-2.5	35.4

Table 4.2.5: Entropy and other dissimilarity measures of Bayesian and least squares AR model fitting performance, ARMA process.

	order	W_1	W_2	W_3
Bayesian	1	-34.0	-9.6	91.8
	2	-32.0	-8.6	83.4
	3	-31.6	-8.5	82.3
	4	-31.5	-8.5	82.3
	5	-31.5	-8.6	83.0
	6	-31.5	-8.7	84.5
	7	-31.5	-8.7	85.1
	max. lk	-31.9	-8.7	84.4
Least Squares	0	-51.7	-50.8	377.1
	1	-49.7	-46.0	363.5
	2	-48.9	-39.8	386.3
	3	-44.5	-30.4	290.3
	4	-37.5	-17.0	173.2
	5	-38.3	-15.6	152.2
	6	-34.5	-10.8	106.0
	7	-36.4	-10.6	100.3
	8	-36.9	-9.1	83.9
	9	-39.0	-9.4	83.2
	10	-39.6	-8.6	77.2
	11	-41.4	-8.8	78.4
	12	-42.1	-8.5	75.5
	13	-43.9	-8.8	78.1
	14	-45.0	-8.8	78.7
	15	-47.0	-9.1	81.7
	16	-49.0	-9.6	84.8
	17	-51.5	-10.0	88.9
	18	-54.0	-10.4	92.8
	19	-56.5	-10.9	96.8
	20	-59.2	-11.5	101.6
	AIC best	-47.3	-9.9	91.3

With respect to both the AR model and ARMA model simulation experiments, examination of the mean values of W_1, W_2 and W_3, (in Tables 4.2.4 and 4.2.5), of the fitted Bayesian and minimum AIC-AR models it is evident that variation of these values for the Bayesian models with prior differentiation order k is rather small compared to the variation of those values with fixed AR model order. This evidence again suggests that our Bayesian procedure is relatively stable and is not very sensitive to the selection of differentiation order and that the Bayesian procedure reduces the risk incurred with potential large model order selection.

With respect to the more important ARMA model simulation experiment results, for each of the dissimilarity measures the essential result is that the Bayesian long AR model procedure yielded superior model fits as compared to the fitted AR models.

The evidence obtained supports our earlier remark that the statistical variability of model order selection procedures used for short data span conditions tends to make parametric model spectral analysis procedures statistically unreliable. That evidence also illustrates the importance of obtaining statistical appraisals of model fitting and spectral estimation procedures for the usual practical situation when the class of models fitted to the data does not include the class of models that generated the data.

4.3 Transfer Function Estimation

Here we consider the relationship between the input and output time series of a process that is referred to as the "transfer function". In the transfer function estimation problem, jointly stationary input and output data is assumed to be observed in the presence of additive (colored) noise. The smoothness priors method is particularly relevant when only short spans of data are available, when the impulse response is relatively long-tailed and when the low order polynomial ARMAX type model can not capture the true model structure. The method is illustrated by the analysis of the Box-Jenkins Series J data.

The models in Åstrom and Bohlin (1965) and Box and Jenkins (1970), are the classical parametric time domain transfer function models. Hannan (1987), is an important review of the subject. ARMAX type models characterized by polynomial operators on the input, the output and the observation noise are fitted to the observed input and output data. (The observation noise in the Åstrom-Bohlin model is MA noise. It is AR noise in the Box-Jenkins model.) Those methods require the specification of three polynomial operator orders, one each for the input, output and noise polynomials and the estimation of the unknown polynomials coefficients via the minimization of a performance functional. Typically that computation is achieved by a computationally costly nonlinear optimization procedure. In such procedures with only short spans of data, it is only feasible to search for solutions over low polynomial orders.

Despite the fact that conventional transfer estimation methods have been extensively used, the influence of the sampling variability in the polynomial model order selection on the transfer function estimation performance remains to be explored. Another objection to the use of low order polynomial ARMAX models is that the parsimonious parametric model may not be a good characterization of the system that generated the data. Here too, from a Bayesian point of view, the conventional parametric modeling methods can not yield "correct" models. (See the discussion in "Background" in Section 4.2.)

In our method, an M-th order impulse response between input and output plus an M-th order autoregressive (AR) model for the additive noise is assumed. M can be quite large. This model is equivalent to an ARMAX plus white noise model. Here, as in the smoothness priors long AR model problem, frequency domain priors are used. We assume integrated square zero-th and k-th order derivative frequency domain smoothness constraints on the polynomial operators.

4.3.1 Analysis

Assume that input/output jointly stationary time series data x_n, y_n, $n = 1, \ldots, N$ is observed and that the output y_n is observed in the presence of additive colored noise, w_n. Consider a representation of the input/output plus noise in the causal or physically realizable impulse response plus colored noise form,

$$y_n = \sum_{m=1}^{\infty} b_m x_{n-m} + w_n, \qquad w_n = \sum_{m=1}^{\infty} a_m w_{n-m} + u_n. \tag{4.30}$$

In (4.30) u_n is assumed to be a Gaussian zero-mean uncorrelated sequence with unknown variance σ_u^2. Also, b_m is an impulse response sequence and w_n is in AR model form, (Kavalieris 1993 also uses this model).

Using the assumed stationarity, substituting $w_{n-j} = y_{n-j} - \sum_{i=1}^{\infty} b_i x_{n-i-j}$ into the second equation of (4.30) yields the infinite order transfer function uncorrelated noise model

$$y_n = \sum_{m=1}^{\infty} c_m y_{n-m} + \sum_{m=1}^{\infty} d_m x_{n-m} + u_n \tag{4.31}$$

with

$$\begin{aligned} c_m &= a_m, \quad m = 1, 2, \cdots \\ d_m &= b_m - \sum_{i=1}^{m-1} b_i a_{m-i}, \quad m = 1, 2, \cdots \end{aligned} \tag{4.32}$$

Equation (4.31) is an ARMAX model with additive white noise u_n. Because of the finite data observation span, the infinite order transfer function model in (4.31) is approximated by the finite order transfer function model

$$y_n = \sum_{m=1}^{M} c_m y_{n-m} + \sum_{m=1}^{M} d_m x_{n-m} + u_n, \tag{4.33}$$

with M assumed to be "large". (The choice of M may be determined by the maximization of a likelihood and the minimization of Akaike's AIC.) The coefficients c_m, d_m, $m = 1, \ldots, M$ are estimated by the Bayesian smoothness priors procedure described below. The estimates of the coefficients of the model in (4.33) are subsequently obtained from (4.32) by the formulas

$$\begin{aligned} a_m &= c_m, \quad m = 1, \ldots, M; \\ a_m &= 0, \quad m = M+1, \ldots \end{aligned} \tag{4.34}$$

$$\begin{aligned} b_m &= d_m + \sum_{i=1}^{m-1} a_i b_{m-i}, \quad m = 1, \ldots, M; \\ b_m &= \sum_{i=1}^{M} a_i b_{m-i}, \quad m = M+1, M+2, \ldots. \end{aligned} \tag{4.35}$$

The power spectrum of the noise w_n is given by

$$S(f) = \frac{\sigma^2}{|1 - \sum_{m=1}^{M} a_m \exp(-2\pi i f m)|^2}, \tag{4.36}$$

where σ^2 is the innovations variance of the estimated model in (4.33).

For our frequency domain smoothness priors constraints on the c_m, d_m parameters, we require the frequency domain quantities $C(f)$ and $D(f)$,

$$C(f) = \sum_{m=1}^{M} c_m \exp(-2\pi imf), \quad D(f) = \sum_{m=1}^{M} d_m \exp(-2\pi ifm). \tag{4.37}$$

Then, a convenient form for the frequency response function is

$$h(f) = \frac{D(f)}{1 - C(f)} = \frac{\sum_{m=1}^{M} d_m \exp(-2\pi ifm)}{1 - \sum_{m=1}^{M} c_m \exp(-2\pi ifm)}. \tag{4.38}$$

Let R_k and Q_k respectively, measures of the roughness of the $C(f)$ and $D(f)$ polynomials, be characterized by the integrated square k-th derivative of those operators,

$$\begin{aligned} R_k &= \int_{-\frac{1}{2}}^{\frac{1}{2}} \left|\frac{d^k C(f)}{df^k}\right|^2 df = (2\pi)^{2k} \sum_{m=1}^{M} m^{2k} c_m^2, \\ Q_k &= \int_{-\frac{1}{2}}^{\frac{1}{2}} \left|\frac{d^k D(f)}{df^k}\right|^2 df = (2\pi)^{2k} \sum_{m=1}^{M} m^{2k} d_m^2. \end{aligned} \tag{4.39}$$

From the definitions in (4.33), large values of R_k and Q_k, respectively mean an unsmooth, in the sense of k-th differential, frequency domain measure of the c and d polynomials. We also introduce the 0-th derivative smoothness constraints

$$R_0 = \int_{-\frac{1}{2}}^{\frac{1}{2}} |C(f)|^2 df = \sum_{m=1}^{M} c_m^2, \quad Q_0 = \int_{-\frac{1}{2}}^{\frac{1}{2}} |D(f)|^2 df = \sum_{m=1}^{M} d_m^2. \tag{4.40}$$

Let the differential orders for the numerator and denominator polynomials to be k_1 and k_2, respectively. With these frequency domain priors the constrained least squares problem for fixed values of k_1, k_2 and τ_j^2, $j = 1, \ldots, 4$, determines the c_m, d_m, $m = 1, \ldots, M$ that minimizes

$$\begin{aligned} \sum_{n=1}^{N} \left(y_n - \sum_{m=1}^{M} c_m y_{n-m} - \sum_{m=1}^{M} d_m x_{n-m} \right)^2 &+ \sum_{m=1}^{M} [\tau_1^2 c_m^2 + \tau_2^2 m_1^{2k} c_m^2] \\ &+ \sum_{m=1}^{M} [\tau_3^2 d_m^2 + \tau_4^2 m_2^{2k} d_m^2]. \end{aligned} \tag{4.41}$$

In (4.41) τ_j^2, $j = 1, \ldots .4$ are the tradeoff hyperparameters. By a proper choice of the tradeoff parameters, our estimate of the model parameters, c_m, d_m, $m = 1, \ldots, M$, balance a tradeoff between infidelity of the transfer function solution to the data and infidelity to the smoothness constraints.

In terms of the linear model formula in Chapter 3, identify the vector y and the matrices D, θ, and X

$$D = \begin{bmatrix} (\tau_1^2 + \tau_2^2)^{1/2} & & & & \\ & (\tau_3^2 + \tau_4^2)^{1/2} & & & \\ & & (\tau_1^2 + 2^{2k_1}\tau_2^2)^{1/2} & & \\ & & & \ddots & \\ & & & & (\tau_3^2 + M^{2k_2}\tau_4^2)^{1/2} \end{bmatrix},$$

$$X = \begin{bmatrix} y_0 & x_0 & \cdots & y_{1-M} & x_{1-M} \\ y_1 & x_1 & \cdots & y_{2-M} & x_{2-M} \\ \vdots & \vdots & & \vdots & \vdots \\ y_{N-1} & x_{N-1} & \cdots & y_{N-M} & x_{N-M} \end{bmatrix}, \quad y = \begin{bmatrix} y_1 \\ y_2 \\ \vdots \\ y_N \end{bmatrix}, \quad \theta = \begin{bmatrix} c_1 \\ d_1 \\ \vdots \\ c_M \\ d_M \end{bmatrix}. \tag{4.42}$$

The -2 log-likelihood for the hyperparameters is minimized via a Davidon-Fletcher-Powell quasi-Newton algorithm, (Section 2.4). That algorithm is exterior to a Householder transformation least squares solution of the constrained least squares problem.

Here again we use the AIC statistic,

$$\text{AIC} = -2\log L(\hat{\tau}^2, \hat{\sigma}^2) + 2(\text{number of parameters estimated}) \tag{4.43}$$

to determine the order M for the transfer function.

4.3.2 A Transfer Function Analysis Example

Here the transfer function analysis of the Box-Jenkins Series J gas furnace data by the smoothness priors, (SP), and Box-Jenkins, (BJ), methods are described and compared. Extensive Monte Carlo studies of the statistical performance of the SP method and an asymptotically maximum likelihood method, too lengthy to be included here, are in Gersch (1987).

The generic Box-Jenkins transfer function model is

$$\begin{aligned} y_n &= a_1 y_{n-1} + \cdots + a_p y_{n-p} + b_1 x_{n-1-d} + \cdots + b_q x_{n-q-d} + w_n \\ w_n &= c_1 w_{n-1} + \cdots + c_r w_{n-r} + u_n. \end{aligned} \tag{4.44}$$

In (4.44), x_n, y_n, w_n, $n = 1, \ldots, N$ are respectively, the observed input and output and the unobserved added noise and u_n, $n = 1, \ldots, N$ is a normal zero mean i.i.d. random variable with variance σ_u^2 and d is a delay between input and output time series. The magnitudes of the BJ model parameters are $d = 2$, $p = 2$, $q = 3$, $r = 2$ with model coefficients: $a = (0.57, 0.01)$; $b = (0.53, -0.37, -0.51)$; $c = (1.53, -0.63)$, $\sigma_u^2 = 0.051$, (Box-Jenkins, Section 11.4). For the minimum AIC SP model, $k_1 = 4$, $k_2 = 2$ and the dimensional parameters are $p = q = r = 4$. The $d = 0$ model is the AIC best shift parameter model. (In this data example, the higher order SP model automatically accounts for the delay between input and output data, without requiring an additional non-zero d parameter.) The a, b polynomial coefficients of the SP model are: $a = (1.588, -0.705, -0.132, 0.149)$, $b = (0.171, -0.438, -0.175, 0.086)$. Table 4.3.1 shows the values of the AIC for the differential orders k_1, $k_2 = 1, \ldots, 5$ for the order $M = 4$ model. The similarity of the AIC values for the range of differential order models in Table 4.3.1 indicate that the SP models for this data are not very sensitive to the differential orders k_1, k_2.

The variances of the tracking error for the BJ and SP models (the sums of squares of the residuals, SSE) were 0.70187 and 0.68662, respectively.

The impulse response, transfer function amplitude response and phase response, (in radians), of the BJ and SP models are shown in Figure 4.4A. (Compare the windowed periodogram analysis of the B-J data, Jenkins and Watts, p446.) The zero initial impulse response of the BJ model due to the fact that $d = 2$ in the

Table 4.3.1: AIC's of SP Model, Parametric in k_1 and k_2, d=0, M=4.

k_1	$k_1 = 1$	$k_1 = 2$	$k_1 = 3$	$k_1 = 4$	$k_1 = 5$
1	46.89	45.70	45.66	46.35	47.38
2	46.44	46.19	45.32	46.00	46.63
3	46.27	45.03	45.12	45.79	46.44
4	46.67	45.01	45.10	45.77	46.41
5	46.41	45.07	45.07	45.72	46.39

BJ model. The optimal SP model delay parameter d is zero, correspondingly the initial SP impulse response is not zero. The BJ modeled transfer function and phase function versus frequency each have some relatively abrupt kinks in their responses as compared to those for the SP modeled results. The power spectral densities of the residuals from the SP and BJ models computed from fitted AR models were very similar to each other.

The impulse response, noise spectrum, transfer function and phase function for the $k_1 = 1, k_2 = 1$ SP models for the B-J Series J data, and for the data truncated at $n = 75$ are shown superimposed on each other in Figure 4.4B. The similarity of the full data and $n = 75$ data model properties support the conjecture that SP modeling method might be reasonable for relatively short length data spans. This property, a consequence of the assumption of priors on the model parameters is equivalent to the observation of additional data in the Bayesian paradigm.

The effects of the choice of model order M, on the impulse response, the noise spectrum and the transfer function amplitude and phase of the SP model are shown superimposed in Figure 4.4C for the optimal SP $M = 4$ model and the likelihood best SP model of order $M = 30$. The results computed for the $M = 30$ SP model wiggle only slightly around those for the order $M = 4$ model. Those results indicate that, on the provision that the model order, M, is sufficiently large, the specification of the order of the SP model does not very critically influence the transfer function characteristics.

The values of the a and b polynomial coefficients of the $M = 10$ optimal SP model are:

$$a = (1.579, -0.702, -0.024, 0.062, 0.011, -0.000, -0.002, -0.000, 0.000, 0.001)$$
$$b = (0.151, -0.384, -0.177, -0.007, 0.030, 0.086, 0.042, 0.006, -0.017, -0.007).$$

The pattern of a, b polynomial coefficients is similar for the larger order SP models. The tapering toward zero values effect of the smoothness priors constraints on the model parameters, particularly on the higher order a polynomial parameters and the relatively long tail b model parameters helps explain the similarity of the $M = 4$ and $M = 30$ model properties.

Figure 4.4D shows the results of ordinary least squares, (OLS), models of orders $M = 5$ and $M = 10$ fitted to the BJ series J data. The $M = 5$ OLS model properties are quite similar to the optimal SP model properties. (The OLS $M = 5$ model is actually a sum of squares tracking error superior model of the BJ series J data than

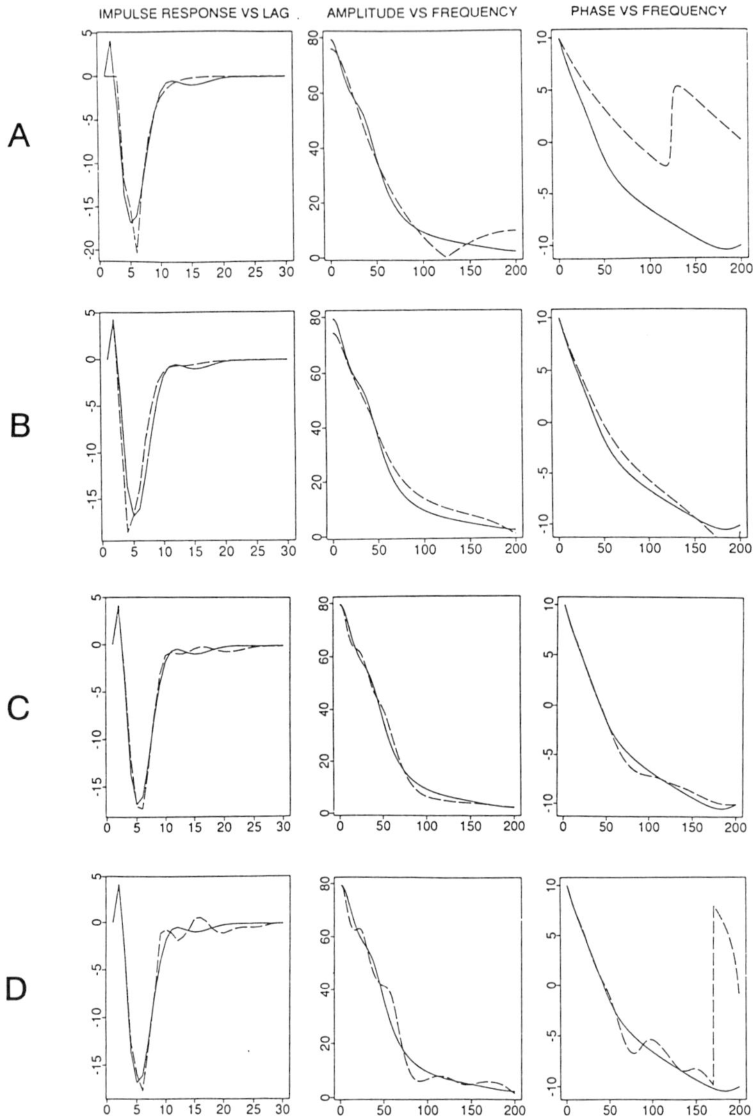

Figure 4.4 Impulse response, amplitude response and phase response. A: The Box-Jenkins and smoothness priors models. B: The smoothness priors models for the full $n = 296$ and $n = 75$ truncated data. C: The smoothness priors models, M = 4 and M = 30. D: The smoothness priors model, M = 4 and OLS model, M = 10.

the original BJ model.) The computed properties of the OLS $M = 10$ model indicate a larger variance, (wiggle a lot more), than the SP $M = 30$ model. The wiggles in the OLS $M = 20$ model properties are much more severe than those in the OLS $M = 10$ model. This is evidence that the SP model properties are relatively insensitive to model order in comparison with other transfer function modeling methods.

The effects of increasing differential orders k_1, k_2 on transfer function properties were also computed. The priors are frequency domain measures of roughness whose values increase with increasing differential orders. As expected, the amplitude response, noise spectrum and phase response of the higher order smoothness constraint differential models were smoother than those for the lower differential order model.

5

Linear Gaussian State Space Modeling

Linear Gaussian state space modeling is treated in this chapter. The prediction, filtering and smoothing formulas in the standard Kalman filter are shown. Model identification or, computation of the likelihood of the model is also treated. Some of the well known state space models that are used in this book as well as state space modeling of missing observations and a state space model for unequally spaced time series are shown. The final section is a discussion of the information square root filter/smoother, that we use in linear Gaussian state space seasonal decomposition modeling in Chapter 9. Not necessarily linear - not necessarily Gaussian state space modeling is treated in Chapter 6. A variety of illustrative examples of linear state space modeling is shown in Chapter 7.

5.1 Introduction

The concept of *state* is one of the fundamental concepts of system theory, (Zadeh and Desoer 1963, Kailath 1980, Caines 1988). The state of a system has an interpretation as the summary of the past behavior of the system. The state taken together with the future system inputs determines all future states and system outputs. Furthermore, the current output is a function of the current state and current input values only. The importance of state space systems for time series analysis is a consequence of the Markov process property of the state which facilitates computation of the likelihood of a state space model of observed data, (Schweppe 1965).

Let a generic linear state space model be given by

$$\begin{aligned} x_n &= F_n x_{n-1} + G_n w_n && (5.1)\\ y_n &= H_n x_n + \varepsilon_n, && (5.2) \end{aligned}$$

where x_n is the state at time n, the state input process, (process or system noise) and the observation noise process respectively are, $w_n \sim P_w$ and $\varepsilon_n \sim P_\varepsilon$. That is, in general it conceptually assumed that the system noise and the observation noise can have arbitrary distributions. Also, for convenience we assume that $E(w_n, \varepsilon_m) = 0$, for all n and m. Equation (5.1) is referred to as the evolutionary (or system transition) equation, while (5.2) is referred to as the observation equation.

The state space is the set of all possible values of x_n, $n = 1, 2, \cdots$, and F_n is the state transition matrix function at time n. The input process w_n, $n = 1, \ldots, N$ and the observation error process ε_n, $n = 1, \ldots, N$ are orthogonal and orthogonal to each other. The input process is orthogonal to the initial state x_0 which is distributed in general, in accordance with an arbitrary distribution $p(x_0)$. These latter properties account for the Markovian property of the state process, (Caines 1988, Theorem 4.2).

Kalman (1960) and Kalman and Bucy (1961) introduced stochastic state space methods to the system theory community. Several different derivations of the Kalman algorithm are well known. Ho and Lee (1964) observed that if the initial state x_0 is Gaussian distributed, and the conditional densities are Gaussian, their means and variances can be computed by algebraic formulas and propagate according to the now well-known Kalman filter equations and that the Kalman filter is in fact the Bayes solution for this model. The popularity of the state space modeling of time series is a consequence of the simplicity of the algebraic computations of the evolution of the state means and state covariances and the simplicity of the computation of the likelihood of the model under Gaussian disturbances assumptions. Schweppe (1965) showed how state space representations and the Kalman filter could be used to calculate exact likelihoods for given values of the unknown parameters again assuming Gaussian disturbances. Duncan and Horn (1972) proved that the (filtered) Kalman state estimator is the minimum mean square linear estimator and minimum variance unbiased estimator of the state and that the Kalman filter could handle random effects by including the random parameters in the state vector. The unknown variances of the random parameters can be estimated by nonlinear optimization computations that are external to the Kalman filter. Harvey and Phillips (1979) extended the use of state space models to regression with ARMA errors. (Among the earliest and most ubiquitous time series state space models were those for AR and ARMA time series, see Section 5.3). Jones (1981) used state space methods to fit continuous time linear differential equations to unequally spaced data when there is observational error. Wecker and Ansley (1983) used state space methods for modeling growth curves and showed the relationship of these methods to splines. Meinhold and Singpurwalla (1983) is an expository derivation of the Kalman filter from the Bayesian viewpoint. It is known that essentially every type of stationary and nonstationary linear and nonlinear time series model can be cast in a state space form. Also, because of the recursive computational properties associated with the Markovianness of the state process, (independent of optimization procedures external to state space computations), the likelihood of a time series model can be computed in $O(N)$ time. That is in contrast with ordinary least squares computations which are $O(N^3)$. Of the large number of books which include the Kalman filter we mention only Jazwinski (1970), Sage and Melsa (1971), Anderson and Moore (1979), Kailath (1980), Caines (1988), Brockwell and Davis (1991) and Harvey (1989). A collection of articles edited by Sorenson (1985), covers the history and numerous applications of the method.

Our own interest in state space modeling of time series was propelled by the realization of state space smoothness priors models for nonstationary mean time series, Brotherton and Gersch (1981), Kitagawa (1981). Those initial linear model, Gaussian disturbances papers were stimulated by and were extensions of Akaike (1980a). Some additional smoothness priors state space linear Gaussian models are Gersch and Kitagawa (1983), Kitagawa and Gersch (1984, 1985a, 1985b). Kitagawa (1987) and in several subsequent papers, demonstrated state space modeling for not necessarily linear not necessarily Gaussian time series. (This latter topic is treated in detail in Chapter 6).

The concern of most contemporary texts' treatment of state space modeling is with the Gaussian disturbances linear state space model. Such modeling invariably exploit the Kalman filter algorithm. It is shown here in Chapter 5. It is applied to the smoothness priors Gaussian disturbances linear state space modeling of nonstationary mean and nonstationary covariance time series in subsequent chapters. A derivation of the Kalman filter, starting from the general state space model is included in Chapter 6.

5.2 Standard State Space Modeling

By the term "standard state space" we mean the linear Gaussian state space model. Let the linear Gaussian state space model be given by

$$\begin{aligned} x_n &= F_n x_{n-1} + G_n w_n \\ y_n &= H_n x_n + \varepsilon_n \\ \begin{bmatrix} w_n \\ \varepsilon_n \end{bmatrix} &\sim N\left(\begin{bmatrix} 0 \\ 0 \end{bmatrix}, \begin{bmatrix} Q_n & 0 \\ 0 & R_n \end{bmatrix}\right), \end{aligned} \tag{5.3}$$

where y_n is an ℓ-dimensional time series, x_n is m-dimensional state vector. w_n and ε_n are, k- and ℓ-dimensional white noise sequences. F_n, G_n and H_n are $m \times m$, $m \times k$ and $\ell \times m$ matrices, respectively. Also for convenience, (and as is conventional), we assume that $E(w_n, \varepsilon_m) = 0$, for all n and m.

State Estimation and Identification

A large variety of problems of time series analysis can be solved by estimating the state vector. The problem of state estimation is to evaluate $p(x_n|Y_m)$, the distribution of x_n given the observations $Y_m \equiv (y_1, \ldots, y_m)$ and the initial values $x_{0|0}$ and $V_{0|0}$. The state prediction, filtering and smoothing problems respectively refer to the situations in which, $m < n$, $m = n$, and $m > n$.

The sequence of events in Kalman state space filtering is as follows: Starting with the initial conditions $x_{0|0}$, $V_{0|0}$, the initial operation is to predict the next state x_1. In standard notation, the quantity $x_{1|0}$ is computed. Following that the observation y_1 is used in a filtering stage to compute the value of the state at time $n = 1$. That is represented by $x_{1|1}$. The process is recursive, prediction is the next operation and $x_{2|1}$ is computed followed by the filtering stage during which $x_{2|2}$ is computed. The

Kalman filter recursive computational process for the assumed set of observations, y_1, y_2, y_3, y_4 is diagramed below.

$$
\begin{array}{cccccccccc}
x_{1|0} & \rightarrow & x_{2|0} & \rightarrow & x_{3|0} & \rightarrow & x_{4|0} & \rightarrow & x_{5|0} & \rightarrow \\
\Downarrow & & & & & & & & & \\
x_{1|1} & \Rightarrow & x_{2|1} & \rightarrow & x_{3|1} & \rightarrow & x_{4|1} & \rightarrow & x_{5|1} & \rightarrow \\
& & \Downarrow & & & & & & & \\
x_{1|2} & \leftarrow & x_{2|2} & \Rightarrow & x_{3|2} & \rightarrow & x_{4|2} & \rightarrow & x_{5|2} & \rightarrow \\
& & & & \Downarrow & & & & & \\
x_{1|3} & \leftarrow & x_{2|3} & \leftarrow & x_{3|3} & \Rightarrow & x_{4|3} & \rightarrow & x_{5|3} & \rightarrow \\
& & & & & & \Downarrow & & & \\
x_{1|4} & \leftarrow & x_{2|4} & \leftarrow & x_{3|4} & \leftarrow & x_{4|4} & \Rightarrow & x_{5|4} & \rightarrow \\
& & & & & & & & \Downarrow &
\end{array}
$$

In this diagram, the double arrows pointing right refer to the time-update or prediction step. The double arrows pointing downward refer to the measurement-update or filtering step that immediately follows an observation. The single arrows pointing right identify a multi-step-ahead prediction step which is not immediately preceded by a filtering step. Finally, the single arrows pointing left refer to the smoothing operation.

The specific computations are given below by the well known Kalman filter formulas, (e.g. Anderson and Moore 1979).

Prediction (Time-Update)

$$
\begin{aligned}
x_{n|n-1} &= F_n x_{n-1|n-1} \\
V_{n|n-1} &= F_n V_{n-1|n-1} F_n^T + G_n Q_n G_n^T.
\end{aligned} \tag{5.4}
$$

We assume that the initial conditions $x_{0|0}$ and $V_{0|0}$ are given.

Filter (Measurement-Update)

$$
\begin{aligned}
K_n &= V_{n|n-1} H_n^T (H_n V_{n|n-1} H_n^T + R_n)^{-1} \\
x_{n|n} &= x_{n|n-1} + K_n (y_n - H_n x_{n|n-1}) \\
V_{n|n} &= (I - K_n H_n) V_{n|n-1}.
\end{aligned} \tag{5.5}
$$

Using the outputs of the Kalman filter, the smoothed state x_n given the entire set of observations Y_N is given by the fixed interval smoother:

Smoothing

$$
\begin{aligned}
A_n &= V_{n|n} F_{n+1}^T V_{n+1|n}^{-1} \\
x_{n|N} &= x_{n|n} + A_n (x_{n+1|N} - x_{n+1|n}) \\
V_{n|N} &= V_{n|n} + A_n (V_{n+1|N} - V_{n+1|n}) A_n^T
\end{aligned} \tag{5.6}
$$

Identification of the Model

Recursive methods based on state space representations are known to be very efficient for calculating likelihood functions of discrete time Gaussian processes, (Mehra 1971, Caines and Rissanen 1974, Akaike 1978, Harvey and Phillips 1979). In fact, the state space model and the Kalman filter do yield an efficient method for the computation of the likelihood of the time series models. The likelihood can be expressed by using the conditional distributions as follows:

$$\begin{aligned} L(\theta) &= f(y_1, \ldots, y_N|\theta) \\ &= \prod_{n=1}^{N} f(y_n|y_1, \ldots, y_{n-1}) \\ &= \prod_{n=1}^{N} f(y_n|Y_{n-1}). \end{aligned} \tag{5.7}$$

For scalar time series, $\ell = 1$, the individual terms in the last expression are given by

$$f(y_n|Y_{n-1}) = \frac{1}{\sqrt{2\pi r_n}} \exp\left\{-\frac{(y_n - H_n x_{n|n-1})^2}{2r_n}\right\}, \tag{5.8}$$

with $r_n = H_n V_{n|n-1} H_n^T + R_n$. Therefore, the log-likelihood of the model is given by

$$\ell(\theta) = \log L(\theta) = -\frac{1}{2}\left\{N \log 2\pi + \sum_{n=1}^{N} \log r_n + \sum_{n=1}^{N} \frac{(y_n - H_n x_{n|n-1})^2}{r_n}\right\}. \tag{5.9}$$

The maximum likelihood estimate of the parameter, $\hat{\theta}$, is obtained by maximizing (5.9) with respect to those parameters.

Estimation of the variance

In the case of univariate time series, the maximum likelihood estimate of σ^2, the unknown observation variance, is easily obtained. The log-likelihood is given by (5.9). Put $\theta = (\sigma^2, \theta^*)$, $Q_n = \sigma^2 Q_n^*$ and $V_{0|0} = \sigma^2 V_{0|0}^*$, then $V_{n|n-1}$ can be expressed as $V_{n|n-1} = \sigma^2 V_{n|n-1}^*$, where $V_{n|n-1}^*$ is obtained by

$$\begin{aligned} V_{n|n-1}^* &= F_n V_{n-1|n-1}^* F_n^T + G_n Q_n^* G_n^T \\ V_{n|n}^* &= (I - K_n H_n) V_{n|n-1}^* \\ K_n &= V_{n|n-1}^* (V_{n|n-1}^* + 1)^{-1} \end{aligned} \tag{5.10}$$

Then the log-likelihood is given by

$$\log p(y_1, \cdots, y_N|\sigma^2, \theta^*) = -\frac{1}{2}\sum_{n=1}^{N}\left\{\log 2\pi\sigma^2 r_n^* + \frac{(y_n - H_n x_{n|n-1})^2}{\sigma^2 r_n^*}\right\}, \tag{5.11}$$

where $r_n^* = H_n V_{n|n-1}^* H_n^T + 1$. Therefore the maximum likelihood estimate of σ^2 is obtained by

$$\hat{\sigma}^2 = \frac{1}{N}\sum_{n=1}^{N} \frac{(y_n - H_n x_{n|n-1})^2}{r_n^*}. \tag{5.12}$$

Substitution of $\hat{\sigma}^2$ into (5.11) yields

$$\log p(y_1, \cdots, y_N|\hat{\sigma}^2, \theta^*) = -\frac{N}{2}(\log 2\pi\hat{\sigma}^2 + 1) + \sum_{n=1}^{N}\left\{\log r_n^* + \frac{(y_n - H_n x_{n|n-1})^2}{r_n^*}\right\}. \tag{5.13}$$

The maximum likelihood estimate of θ^* can be obtained by numerically maximizing this log-likelihood function.

The AIC is then defined by

$$\begin{aligned} \text{AIC} &= -2(\text{maximum log likelihood}) + 2(\text{number of parameters}) \\ &= -2\ell(\hat{\theta}) + 2|\theta| \end{aligned} \tag{5.14}$$

where $|\theta|$ is the number of parameters estimated. The AIC provides a unified measure for the selection of the best of alternative time series models.

5.3 Some State Space Models

In this section state space models for the fitting of ARMA and AR models to stationary time series data are shown.

There are four possible canonical ARMA model forms. Their structure depends on where the coefficients in the AR part of the model are placed. This system theoretic topic is discussed for example in Kailath (1980). An ARMA model of AR order m and MA order ℓ is,

$$y_n = \sum_{j=1}^{m} a_j y_{n-j} + \varepsilon_n - \sum_{j=1}^{\ell} b_j \varepsilon_{n-j}. \tag{5.15}$$

A particular state space model realization of the ARMA model that we use is,

$$\begin{aligned} F_n &= \begin{bmatrix} a_1 & 1 & & \\ a_2 & & \ddots & \\ \vdots & & & 1 \\ a_k & & & \end{bmatrix}, \quad G_n = \begin{bmatrix} 1 \\ -b_1 \\ \vdots \\ -b_{k-1} \end{bmatrix}, \\ H_n &= [\, 1 \ 0 \ \cdots \ 0 \,], \end{aligned} \tag{5.16}$$

where $k = \max(m, \ell + 1)$ and $a_j = 0$ for $j > m$ and $b_j = 0$ for $j > \ell$. The Kalman filter facilitates the exact maximum likelihood estimation of the ARMA model and there have been numerous publications on this topic, (see for example Akaike 1978, Gardner *et al.*, 1980 and Jones 1981, 1993).

An AR model of order m is,

$$y_n = \sum_{j=1}^{m} a_j y_{n-j} + \varepsilon_n. \tag{5.17}$$

A state space representation of an AR model can be simply realized as a special case of the state space ARMA model by setting the MA coefficients, $b_1, \ldots, b_{k-1}$, in the matrix G_n in (5.16) to zero. On the other hand, when we use AR models as a component among other time series components in time series decomposition problems, we usually use the state space representation,

$$F_n = \begin{bmatrix} a_1 & a_2 & \cdots & a_m \\ 1 & & & \\ & \ddots & & \\ & & 1 & \end{bmatrix}, \quad G_n = \begin{bmatrix} 1 \\ 0 \\ \vdots \\ 0 \end{bmatrix}, \tag{5.18}$$
$$H_n = [\, 1 \; 0 \; \cdots \; 0 \,].$$

5.4 Modeling With Missing Observations

Consider a state space model

$$\begin{aligned} x_n &= F_n x_{n-1} + G_n w_n \\ y_n &= H_n x_n + \varepsilon_n \end{aligned} \tag{5.19}$$

where $w_n \sim N(0, \tau^2)$ and $\varepsilon_n \sim N(0, \sigma^2)$. Let $x_{n|m}$ and $V_{n|m}$ denote the mean and the variance of x_n given the observations up to time m, $Y_m \equiv \{y_1, \cdots, y_m\}$. The one step ahead prediction $x_{n|n-1}$ and the filter $x_{n|n}$ with their variances can be obtained by the Kalman filter (5.4) and (5.5).

When there are missing observations the likelihood is evaluated efficiently using the Kalman filter as in Jones (1981,1993), Kitagawa and Akaike (1981) and Kohn and Ansley (1986). This is an important property of the Kalman filter. When the observation y_n is missing, $Y_n = Y_{n-1}$ holds and therefore we have

$$x_{n|n} = x_{n|n-1} \qquad V_{n|n} = V_{n|n-1}. \tag{5.20}$$

This means that if the observation y_n is missing, it is only necessary to put $K_m = 0$ and skip filtering at the m-th step.

Likelihood of the model and parameter estimation

Let $I(n)$ be the set of indices of time instants for which the observation is actually obtained. If there are no missing observations, we have that $I(n) = \{1, \cdots, n\}$. Then defining Y_n by $Y_n \equiv \{y_i | \ i \in I(n)\}$ generalizes the definition of Y_n and the log-likelihood of time series is obtained by

$$\ell(\theta) = \log p(Y_N|\theta) = \sum_{n \in I(N)} \log p(y_n | Y_{n-1}, \theta). \tag{5.21}$$

Since the mean and the covariance of the one step ahead predictor are available even with observations missing, we compute the log-likelihood by

$$\begin{aligned} \ell(\theta) = & -\frac{1}{2} \sum_{n \in I(N)} \Big\{ \ell \log 2\pi + \log |V_{n|n-1}| \\ & + (y_n - H_n x_{n|n-1})^T V_{n|n-1} (y_n - H_n x_{n|n-1}) \Big\} \end{aligned} \tag{5.22}$$

where the summation is taken with respect to the available observations.

Using this method, it is possible to estimate the parameters of the model, even when there are many missing observations. If necessary, it is also possible to estimate the missed observations by applying the smoothing algorithm. Examples that illustrate these facts are shown in Chapter 7.

5.5 Unequally Spaced Observations

The analysis of irregularly observed time series, or time series with missing data, is one of the most important problems faced by applied researchers whose data arises in the form of time series. Here, motivated by Jones (1980), we approach the modeling of irregularly spaced data by the fitting of continuous time autoregressive models via a state space representation and then diverging from Jones' approach, convert that representation to a discrete time state space model. The exact likelihood for Gaussian processes can then be calculated using Kalman recursive estimation and a non-linear optimization program to calculate the maximum likelihood estimates of the unknown system parameters.

Shapiro and Silverman (1960), Beutler (1970), and Brillinger (1972), are of the earliest efforts in modeling with unequally spaced data. Their objectives, to achieve alias free unequally spaced sampling methods, were primarily motivated by spectrum estimation. Robinson (1977) and Jones (1977) both considered a parametric modeling approach that gave the exact maximum likelihood estimate of a first order continuous time autoregression (Markov process). Jones (1980) used the Kalman recursive calculation techniques for the fitting of ARMA models with unequally spaced observations.

Here we show a continuous time-discrete time state space modeling variation of Jones' approach. Maximum likelihood estimation of the single unknown model parameter in our model can be obtained by using a nonlinear optimization program. Akaike's AIC criterion is used for model identification.

Let the observations of a continuous-time time series $y(t)$ be expressed by

$$y(t) = f(t) + \varepsilon(t) \tag{5.23}$$

where $\varepsilon(t)$ is an observation error and $f(t)$ is a continuous process generated by the k-th order continuous time AR process

$$f^{(k)}(t) = a(1)f^{(k-1)}(t) + \cdots + a(k)f(t) + w(t). \tag{5.24}$$

In (5.24), $f^{(k)}(t)$ is the k-th derivative of $f(t)$, $w(t)$ is the formal derivative of the Wiener process $W(t)$ with dispersion parameter τ^2, (a continuous time "white noise" with spectral density τ^2). The Weiner or Brownian motion process is treated for example in Doob (1953) and Caines (1988). Jones (1981) showed a state space representation of this system and applied a continuous time Kalman filter algorithm to fit this model to data.

Here we develop a discrete time series model that expresses the relation between successive observations. This enables us to apply the same filtering and smoothing methodology as we do in the discrete time problem. Let $x(t)$ be defined by the succession of increasing order function derivatives $x(t) = (f(t), f^{(1)}(t), \cdots, f^{(k-1)}(t))^T$. Then we have a continuous state space representation, (Jones 1981),

$$\begin{aligned} dx(t) &= Ax(t)dt + BdW(t) \\ y(t) &= Cx(t) + \varepsilon(t) \end{aligned} \tag{5.25}$$

with

$$A = \begin{bmatrix} 0 & 1 & & \\ & & \ddots & \\ & & & 1 \\ a(k) & a(k-1) & \cdots & a(1) \end{bmatrix}, \quad B = \begin{bmatrix} 0 \\ \vdots \\ 0 \\ 1 \end{bmatrix}, \quad C^T = \begin{bmatrix} 1 \\ 0 \\ \vdots \\ 0 \end{bmatrix}. \tag{5.26}$$

The state $x(t)$ is then expressible as

$$x(t) = F(t-s)x(s) + \int_s^t F(t-u)BdW(u), \tag{5.27}$$

where $F(u)$ is the transition matrix associated with the system of liner differential equations

$$\dot{x}(t) = Ax(t). \tag{5.28}$$

Then the process $y(t)$ has a state space representation that is appropriate for unequally spaced data.

$$\begin{aligned} x(t) &= F(t-s)x(s) + Gw(t-s) \\ y(t) &= Hx(t) + \varepsilon(t) \end{aligned} \tag{5.29}$$

with

$$\begin{aligned} G &= I_k, \quad H = [\,1\ 0\ \cdots\ 0\,] \\ v(t,s) &= \int_s^t F(t-u)BdW(u). \end{aligned} \tag{5.30}$$

In the above representation, the covariance of the system noise $v(t,s)$ is obtained from

$$\mathrm{cov}(t,s) = \tau^2 \int_s^t F(t-u)BB^tF(t-u)^t du. \tag{5.31}$$

As a particular example of this model we obtain a useful model for the smoothing of unequally spaced data, (Kitagawa 1984)

$$f^{(k)}(t)dt = dW(t). \tag{5.32}$$

In this case, the matrices A, B and C in the continuous state space representation (5.25) are defined by

$$A = \begin{bmatrix} 0 & 1 & & \\ & & \ddots & \\ & & & 1 \\ 0 & 0 & \cdots & 0 \end{bmatrix}, \quad B = \begin{bmatrix} 0 \\ \vdots \\ 0 \\ 1 \end{bmatrix}, \quad C^T = \begin{bmatrix} 1 \\ 0 \\ \vdots \\ 0 \end{bmatrix}. \tag{5.33}$$

Since the determinant $|\lambda I - A| = 0$ has k multiple roots at $\lambda = 0$, the transition matrix associated with the differential equation (5.25) is given by

$$F(t) = \begin{bmatrix} 1 & t & \frac{t^2}{2} & \cdots & \frac{t^{k-1}}{(k-1)} \\ & 1 & t & \cdots & \frac{t^{k-2}}{(k-2)} \\ & & 1 & \ddots & \vdots \\ & & & \ddots & t \\ & & & & 1 \end{bmatrix}. \tag{5.34}$$

Since

$$F(t-u)B = \left[\frac{(t-u)^{k-1}}{(k-1)!}, \frac{(t-u)^{k-2}}{(k-2)!}, \cdots, t-u,\ 1\right]^T, \tag{5.35}$$

the covariance of the system noise is governed by

$$\begin{aligned} \operatorname{cov}\left(v_i(t,s), v_j(t,s)\right) &= \tau^2 \int_s^t \frac{(t-u)^{k-i}}{(k-i)!} \frac{(t-u)^{k-j}}{(k-j)!} du \\ &= \frac{\tau^2 (t-u)^{2k-i-j+1}}{(2k-i-j+1)(k-i)!(k-j)!}. \end{aligned} \tag{5.36}$$

Then the log-likelihood of the model is given by

$$\ell = -\frac{1}{2}\left\{ N \log 2\pi + \sum_{i=1}^{N} \log v(t_i) + \sum_{i=1}^{N} \frac{(y(t_i) - Hx(t_i|t_{i-1}))^2}{2v(t_i)} \right\}, \tag{5.37}$$

where $v(t_i)$ is given by $v(t_i) = HV(t_i|t_{i-1})H^T + R$.

5.6 An Information Square-Root Filter/Smoother

Some details of an information square root filter and smoother algorithms that are used in state space seasonal adjustment modeling program DECOMP, Chapter 9, are shown here. The information square root filter/smoother, a square root of the information matrix algorithm with numerically better conditioned properties than the Kalman filter/smother is extensively treated in Bierman (1977) also see Dyer and McReynolds (1969), Kaminski *et al.* (1971), Nishimura (1976), Rauch *et al.* (1965) and Anderson and Moore (1979). This algorithm was adopted for the following reasons:

1. The algorithm can start the state space recursion computations with an infinite covariance matrix since it uses the inverse of the covariance matrix.

2. It can handle delicate problems without incurring numerical instability because of the double precision property of the square root algorithm.

3. The model is easily adaptable for seasonal adjustment modeling, (Chapter 9).

4. The memory required for achieving the computation of $x_{n|N}$, the smoothed state estimate, is considerably less than that required by other conventional filters.

Define the upper triangular matrices R_n, S_n, U_n and W_n by

$$\begin{aligned} V_{n|n-1} &= R_n^T R_n, \\ V_{n|n} &= S_n^T S_n, \\ \mathrm{E}\omega_n\omega_n^T &= Q_n = W_n^T W_n, \\ \mathrm{E}\varepsilon_n\varepsilon_n^T &= R_n = U_n^T U_n. \end{aligned} \tag{5.38}$$

Then, one step prediction $x_{n|n-1}$, the filtered estimate $x_{n|n}$ and their covariance matrices are obtained by the following information square root filter,

Time update

$$\left[\begin{array}{c|c|c} W_{n-1}^{-1} & 0 & 0 \\ \hline -S_{n-1}^{-1}F_{n-1}^{-1}G_{n-1} & S_{n-1}^{-1}F_{n-1}^{-1} & c_{n-1} \end{array}\right] \longrightarrow \left[\begin{array}{c|c|c} P_n & T_n & a_n \\ \hline 0 & R_n^{-1} & b_n \end{array}\right]$$

$$x_{n|n-1} = R_n b_n. \tag{5.39}$$

Measurement update

$$\left[\begin{array}{c|c} R_n^{-1} & b_n \\ \hline U_n^{-1}H_n & U_n^{-1}y_n \end{array}\right] \longrightarrow \left[\begin{array}{c|c} S_n^{-1} & c_n \\ \hline 0 & e_n \end{array}\right]$$

$$x_{n|n} = S_n c_n. \tag{5.40}$$

The notation $\longrightarrow$ means the reduction to an upper triangular matrix by an appropriate Householder transformation. The filter algorithm requires initial values S_0^{-1} and c_0. S_0^{-1} is obtained as the square root of the unconditional variance of the state vector. The initial vector c_0 is then obtained by $c_0 = S_0^{-1}x_{0|0}$ where $x_{0|0}$ is the unconditional mean of the state vector. The smoothed estimate $x_{n|N}$ is obtained by the following backward smoothing algorithm.

Smoothing

$$\begin{aligned} \omega_{n-1|N} &= P_n^{-1}\{a_n - T_n x_{n|N}\}, \\ x_{n-1|N} &= F_{n-1}\{x_{n|N} - G_{n-1}\omega_{n-1|N}\}, \end{aligned} \tag{5.41}$$

where $x_{N|N}$ is obtained by $x_{N|N} = S_N c_N$ and P_n, T_n and a_n are obtained by (5.39).

6

General State Space Modeling

Non-Gaussian state space modeling of time series and the more general, general state space modeling are treated in this chapter. They were introduced because of the need to model time series with abrupt discontinuities, and time series with outliers and to model time series whose state and or observation processes were nonlinear. The general state space model and its recursive formulas for prediction, filtering and smoothing are treated in Section 6.2 after an introduction in Section 6.1. Three alternative computational approaches for realizing the general state space modeling, numerical integration method, a Gaussian mixture-two filter formula method and the most recently developed Monte Carlo "particle-path tracing" method are discussed respectively in Sections 6.3, 6.4 and 6.5. In addition, starting from the general state space model, an alternative derivation of the Kalman filter is shown in Section 6.6. Applications of the non-Gaussian methods appear in Chapters 8 through 16. In those chapters where appropriate, the non-Gaussian and general state space methods performance are contrasted with the linear Gaussian state space analysis methods.

6.1 Introduction

The general state space model is treated in this chapter. Under that topic, we include the generalization of linear Gaussian state space modeling to the non-Gaussian process and observation noise state space modeling. The more general state space modeling which includes nonlinearities in the state and/or the observation process is also treated. Markovian assumptions underlying the general state space modeling process imply recursive formulas for the state's prediction, filtering and smoothing conditional densities. The mean and covariance of those densities propagate according to the well-known Kalman filter equations only if the initial state density and the system and observation noise processes are Gaussian. Otherwise, those densities must be approximated. Sorenson (1985) distinguishes local and global density approximations. The extended Kalman filter, (EKF), is the most widely used of the local approximations. Various series expansions have led to other local approaches,

(Sorenson and Stubberud 1968, Srinivasan 1970, Hecht 1973, McReynolds 1975 and Willman 1981 for example).

There are basically two global approaches: direct approximation of the densities and approximation of the integrals in the recursive formulae. Both approaches are based on sampling the densities at points distributed through the region containing non-negligible probability. In function approximation, the points and the approximate density values at those point are defined on the basis of matching some interpolating function which is chosen to make the integrations tractable. The integration approximation picks the points based on a particular numerical integration scheme.

The simplest of the global function approximation methods is the point-mass method where the densities are approximated by point masses located on a rectangular grid, (Bucy 1969, Bucy and Senne 1971). An alternative approach is to use a weighted sum of Gaussians, each centered on a different point in state space, (Sorenson and Alspach 1971, de Figuieredo and Jan 1971, Alspach 1974). The storage and computational requirements of these techniques strained the limited computer capabilities available in the 1970's. The basic issue was trading off computational complexity against the number of grid points required to achieve a desired accuracy. The revolution in computing capabilities in the 1980's motivated a renewed look at some of the simpler methods, expanded interest in time series modeling, and motivated innovative approaches. West *et al.* (1985), a Bayesian dynamic generalized linear model approach and Kitagawa (1987), a non-Gaussian state space modeling approach are two of the more important innovations.

Kitagawa (1987), is a direct simple spline function numerical integration approach to achieving the non-Gaussian state space recursions. It was motivated by the need to model time series with abrupt discontinuities, outliers, nonlinearities and also to model nonhomogeneous discrete distributions. Kramer and Sorenson (1988) is piecewise constant approximation to the density functions, however, smoothing was not addressed in that paper. Alternative approaches to non-Gaussian, nonlinear state space modeling primarily stimulated by Kitagawa (1987), include step function approximation which does not neglect smoothing (Kitagawa 1988, 1991), Gauss-Hermite polynomial expansion integration (Schnatter 1992), a random placement of knots spline function approximation method, (Tanizaki 1993) and Monte Carlo integration methods, Carlin *et al.* (1992), Frühwirth-Schnatter (1994), and also Carter and Kohn (1993). Hodges and Hale (1993) is a computationally more efficient method than Kitagawa (1987) that is suitable for scalar time series. Authors who have shown a Bayesian approach to the analysis of time series with outliers include West and Harrison (1976), West (1981), Meinhold and Singpurwalla (1989) and Carlin *et al.* (1992). Smoothness priors applications of the numerical integration approach include Kashiwagi (1993), to a nonhomogeneous Poisson process medical data analysis application and Komaki (1993) in an interesting Markov state switching application in endocrinology.

The direct integration methods are actually only applicable to the analysis of low-dimensional state space systems, (a four dimensional state system is large for numerical integration methods). Kitagawa (1994) is a two-filter formula that realizes recursive smoothing in a Gaussian mixture modeling. The motivation was the necessity of modeling the seasonal time series in a non-Gaussian state space systems framework that would permit abrupt discontinuities and outliers. Also, Kitagawa (1994) solves the Gaussian mixture modeling problem introduced in Harrison and Stevens (1976). Harrison and Stevens' idea was that by permitting the non-Gaussian noise to be represented by a sum of Gaussian distributions, the Kalman filter recursive computations could be realized on each of the Gaussian components. Several problems accrue with that approach. One is that the number of states grows exponentially with increasing time. That required some kind of generally ad-hoc procedure to control the growing state dimension. In the treatment of the Gaussian mixture-two filter formula approach in Section 6.4, a Kullback-Leibler measure of dissimilarity is used to control that growth in state dimension. A second problem with the aforementioned approaches is that the denominator in the smoothing computation is not Gaussian and coping with that fact usually imposes an ad-hoc approximation in the smoothing operation. The Gaussian mixture-two filter method is a solution to that problem.

Finally, Kitagawa (1995) is conceptually a relatively simple but computationally intensive Monte Carlo "particle-path tracing" alternative to the numerical integration and Gaussian mixture-two filter formula approaches to the non-Gaussian and general state space methods. In that method the successive prediction, filtering and smoothing conditional probability density functions are approximated by many of their realizations.

6.2 The General State Space Model

Consider a system described by a general state space model

$$\begin{aligned} x_n &\sim q(x_n|x_{n-1}) \\ y_n &\sim r(y_n|x_n), \end{aligned} \tag{6.1}$$

where y_n is the observed time series and x_n is the unknown state vector. q and r are conditional distributions of x_n given x_{n-1} and of y_n given x_n, respectively. The initial state vector x_0 is distributed according to the distribution $p(x_0|Y_0)$. The problem of state estimation is to evaluate $p(y_n|Y_m)$, the distribution of y_n given the observations $Y_m = (y_1, ..., y_m)$ and the initial distribution. This general state space model includes various important time series models. Some of the examples are:

(i) The linear state space model with non-Gaussian white noises w_n and ε_n

$$\begin{aligned} x_n &= Fx_{n-1} + Gw_n \\ y_n &= Hx_n + \varepsilon_n. \end{aligned} \tag{6.2}$$

(ii) The nonlinear model

$$\begin{aligned} x_n &= f(x_{n-1}) + w_n \\ y_n &= h(x_n) + \varepsilon_n. \end{aligned} \tag{6.3}$$

(iii) Discrete Process

$$\begin{aligned} x_n &= Fx_{n-1} + Gw_n \\ y_n &\sim Dist(x_n). \end{aligned} \tag{6.4}$$

We note that for linear Gaussian systems, the conditional densities $p(x_n|Y_{n-1})$, $p(x_n|Y_n)$ and $p(x_n|Y_N)$ are characterized by the mean vectors and the covariance matrices and hence (6.5)–(6.7) are equivalent to the well known Kalman filter and the fixed interval smoothing algorithms, (Kalman 1960, Anderson and Moore 1979).

6.2.1 General Filtering and Smoothing

For state estimation in the general state space model, we need to evaluate $p(x_n|Y_m)$, the conditional distribution of x_n given observations Y_m. Using only the definition of conditional probability, the concept of a marginal distribution and Bayes' law, the recursive formulas for obtaining the one step ahead predictor, the filter and the smoother are as follows (Kitagawa 1987, 1988):

One step ahead prediction:

$$\begin{aligned} p(x_n|Y_{n-1}) &= \int_{-\infty}^{\infty} p(x_n, x_{n-1}|Y_{n-1})dx_{n-1} \\ &= \int_{-\infty}^{\infty} p(x_n|x_{n-1}, Y_{n-1})p(x_{n-1}|Y_{n-1})dx_{n-1} \\ &= \int_{-\infty}^{\infty} q(x_n|x_{n-1})p(x_{n-1}|Y_{n-1})dx_{n-1} \end{aligned} \tag{6.5}$$

Filtering:

$$\begin{aligned} p(x_n|Y_n) &= p(x_n|y_n, Y_{n-1}) \\ &= \frac{p(y_n|x_n, Y_{n-1})p(x_n|Y_{n-1})}{p(y_n|Y_{n-1})} \\ &= \frac{r(y_n|x_n)p(x_n|Y_{n-1})}{p(y_n|Y_{n-1})} \end{aligned} \tag{6.6}$$

where $p(y_n|Y_{n-1})$ is obtained from $\int r(y_n|x_n)p(x_n|Y_{n-1})dx_n$.

Smoothing:

$$\begin{aligned} p(x_n|Y_N) &= \int_{-\infty}^{\infty} p(x_n, x_{n+1}|Y_N)dx_{n+1} \\ &= \int_{-\infty}^{\infty} p(x_{n+1}|Y_N)p(x_n|x_{n+1}, Y_N)dx_{n+1} \\ &= \int_{-\infty}^{\infty} p(x_{n+1}|Y_N)p(x_n|x_{n+1}, Y_n)dx_{n+1} \\ &= p(x_n|Y_n)\int_{-\infty}^{\infty} \frac{p(x_{n+1}|Y_N)p(x_{n+1}|x_n, Y_n)}{p(x_{n+1}|Y_n)}dx_{n+1} \\ &= p(x_n|Y_n)\int_{-\infty}^{\infty} \frac{p(x_{n+1}|Y_N)q(x_{n+1}|x_n)}{p(x_{n+1}|Y_n)}dx_{n+1}. \end{aligned} \tag{6.7}$$

The formulas (6.5), (6.6) and (6.7) exhibit the recursive relations between state distributions. The conditional distribution of the state $p(x_n|Y_m)$ is in general non-Gaussian. A numerical method for the realization of these formulas is shown in Section 6.3.

6.2.2 Model Identification

The non-Gaussian model presented in the previous section generally has some unknown parameters. Our method of choice for parameter estimation is the maximization of the log likelihood by

$$\begin{aligned} \ell(\theta) &= \log p(y_1, ..., y_N) \\ &= \sum_{n=1}^{N} \log p(y_n | y_1, ..., y_{n-1}) \\ &= \sum_{n=1}^{N} \log p(y_n | Y_{n-1}). \end{aligned} \tag{6.8}$$

Note that $p(y_n|Y_{n-1})$ appears as the denominator in (6.6). Therefore the log-likelihood is obtained as the by-product of the non-Gaussian filter.

As before, given several candidate models, each including a specification of the particular system noise and/or observational noise distributions, the model with the minimum value of the Akaike information criterion (AIC),

$$\text{AIC} = -2 \max \ell(\hat{\theta}) + 2(\text{number of free parameters}), \tag{6.9}$$

is selected as the best model. This criterion enables comparison of different models with different distribution assumptions.

6.3 Numerical Synthesis of the Algorithms

In this section we show a numerical realization of the filtering and smoothing formulas identified in the previous section. Only the formulas for a one dimensional system, with the number of states, $k = 1$ and the dimension of the output, $l = 1$ is shown here. In principle, it is not difficult to generalize this formula to a multi-dimensional system. The main reason for focusing on the simplest case is to avoid notational complexity. Even with this simplest case, it is possible to exhibit the potential of non-Gaussian modeling. The trend with abrupt discontinuities example in Chapter 8 and the nonhomogeneous discrete distribution examples in Chapter 13, for example, were modeled with a state dimension $k = 1$, using a first order trend model.

Each density function is approximated by a piecewise linear (first order spline) function. The densities are specified by the number of segments, location of nodes, and the value at each node. It is assumed that the density is negligible outside the two outermost nodes. Specifically, we use the following notation: One step ahead prediction density is $p(x_n|Y_{n-1})$, the number of segments is k_p, the nodes are xp_{n_i} $(i = 0,\ldots,k_p)$, and the values at the nodes are p_{n_i} $(i = 0,\ldots,k_p)$. For the filter density, $p(x_n|Y_n)$, the number of segments is k_f, the nodes are xf_{n_i} $(i = 0,\ldots,k_f)$, and the values at the nodes are f_{n_i} $(i = 0,\ldots,k_f)$. For the smoothing density, $p(x_n|Y_N)$, the number of segments is k_s, the nodes are xs_{n_i} $(i = 0,\ldots,k_s)$, and the values at the nodes are s_{n_i} $(i = 0, \ldots, k_s)$. (See Kitagawa 1987 for more details.)

The values at the nodes naturally define the entire function under the piecewise linearity assumption on the function. We denote these functions by $p_n(x)$, $f_n(x)$,

$s_n(x)$ and $q(x)$, respectively. Using this notation the filtering and smoothing algorithms are realized as follows.

One step ahead prediction:

For each $x = xp_i$ $(i = 0,\ldots,k_p)$, from (6.5) and $w_n = G^{-1}(x_n - Fx_{n-1})$, $p_n(x)$ is obtained by

$$\begin{aligned} p_n(x) &= \int_{-\infty}^{\infty} p(z|Y_{n-1})p(x|z)dz \\ &= \int_{-\infty}^{\infty} f_{n-1}(z)q(G^{-1}(x - Fz))dz \\ &= \sum_{i=1}^{m} \int_{z_{i-1}}^{z_i} f_{n-1}(z)q(G^{-1}(x - Fz))dz, \end{aligned} \tag{6.10}$$

where $\{z_0, z_1, \ldots, z_m\}$ is the ordered set which is obtained from the union of two sets $\{xf_{n_0}, \ldots, xf_{n_k}\}$ and $\{G^{-1}(x - Fxq_{n_{kq}}), \ldots, G^{-1}(x - Fxq_{n_0})\}$. Since $f_n(z)$ and $q(G^{-1}(x - Fz))$ both are linear functions on the interval $[z_{j-1}, z_j]$ each integral in the last line of (6.10) is approximated by

$$(f_{j-1}q_{j-1} + f_jq_j)(z_j - z_{j-1})/2. \tag{6.11}$$

Here $f_j = f_{n-1}(z_j)$ and $q_j = q(G^{-1}(x - Fz_j))$.

Filtering :

For each $x = xf_i$ $(i = 0,1,\ldots,k_f)$, $f_n(x)$ is obtained by

$$f_n(x) = p_n(x)r(y - Hx)/C. \tag{6.12}$$

Here y is the given observation at that stage and C is the normalizing constant given by

$$\begin{aligned} C &= \int_{-\infty}^{\infty} p_n(x)r(y - Hx)dx \\ &= \sum_{i=1}^{m} \int_{z_{i-1}}^{z_i} p_n(x)r(y - Hx)dx \qquad (6.13) \\ &= \sum_{i=1}^{m} \{p_n(xf_i)r(y - Hxf_i) + p_n(xf_{i-1})r(y - Hxf_{i-1})\}(xf_i - xf_{i-1})/2. \end{aligned}$$

Smoothing:

For each $x = xs_i$ $(i = 0,1, \ldots,k_s)$, $s_n(x)$ is obtained by

$$\begin{aligned} s_n(x) &= f_n(x)\int_{-\infty}^{\infty} s_{n+1}(z)q(G^{-1}(x - Fz))/f_{n+1}(z)dz \\ &= f_n(x)\sum_{i=1}^{k_s} \int_{z_{i-1}}^{z_i} s_{n+1}(z)q(G^{-1}(x - Fz))/f_{n+1}(z)dz, \end{aligned} \tag{6.14}$$

where $\{z_0, \ldots, z_i\}$ is the ordered set that is obtained from the union of $\{xs_{n+1,0}, \ldots, xs_{n+1,ks}\}$, $\{xf_{n+1,0}, \ldots, xf_{n+1,kf}\}$ and $\{G^{-1}(x - Fxq_0), \ldots, G^{-1}(x - Fxq_{kq})\}$. The integral in the summation is approximated by

$$(a_{j-1}b_{j-1}c_{j-1}^{-1} + a_j b_j c_j^{-1})(z_j - z_{j-1})/2. \tag{6.15}$$

Here $a_j = s_{n+1}(z_j), b_j = q(G^{-1}(x - Fz_j)), c_j = f_{n+1}(z_j)$.

The first example of the application of this method is in Chapter 8 in the modeling of trends with abrupt discontinuities. Other applications include the estimation of nonstationary variance in Chapter 10 and the modeling of inhomogeneous discrete processes in Chapter 13.

6.4 The Gaussian Sum-Two Filter Formula Approximation

For higher order state space models such as that for the seasonal adjustment of monthly data, (Chapter 9), the application of the direct numerical method of Section 6.3, is impractical because of the huge amount of computation involved in numerical integration. In this section we show a Gaussian sum-two filter smoothing method which permits the non-Gaussian and general state space modeling of higher order state space models. The Gaussian sum approximation method is introduced first and that is followed by brief descriptions of the two-filter smoothing method and a method for the control of the state dimension. Additional details on the Gaussian sum and two-filter smoothing methods are in Kitagawa (1989) and Kitagawa (1994), respectively. Applications of the Gaussian sum-two filter smoothing method to modeling seasonal time series and time series with outliers appear in Chapter 9. An application to nonlinear state space modeling is in Chapter 15.

6.4.1 The Gaussian Sum Approximation

One practical way to mitigate the computational burden in non-Gaussian filtering, has been via the use of a Gaussian-sum filter (Sorenson and Alspach 1971, Alspach and Sorenson 1972, Harrison and Stevens 1976 and Anderson and Moore 1979). In Kitagawa (1989), it was shown that such a Gaussian-sum filter can be derived from the non-Gaussian filter by using Gaussian mixture approximations to the related densities. Specifically, the following approximations were used:

$$\begin{aligned} p(w_n) &= \sum_{i=1}^{K_w} \alpha_i \varphi_i(w_n), & p(\varepsilon_n) &= \sum_{j=1}^{K_\varepsilon} \beta_j \varphi_j(\varepsilon_n) \\ p(x_n|Y_{n-1}) &= \sum_{k=1}^{L_n} \gamma_{kn} \varphi_k(x_n|Y_{n-1}), & p(x_n|Y_n) &= \sum_{\ell=1}^{M_n} \delta_{\ell n} \varphi_\ell(x_n|Y_n). \end{aligned} \tag{6.16}$$

Here φ_i denotes a properly defined Gaussian density. Substituting into the equations for one-step-ahead prediction (6.5), and filtering (6.6), immediately yields the following algorithm for the Gaussian-sum filtering:

One step ahead prediction:

$$\begin{aligned} p(x_n|Y_{n-1}) &= \sum_{i=1}^{K_w}\sum_{\ell=1}^{M_{n-1}} \gamma_{i\ell,n}\varphi_{i\ell}(x_n|Y_{n-1}). \\ &\equiv \sum_{k=1}^{L_n} \gamma_{k,n}\varphi_j(x_n|Y_{n-1}). \end{aligned} \tag{6.17}$$

where $\equiv$ means renumbering the double summation by a single summation, and therefore $L_n = K_w M_{n-1}$. Here $\gamma_{i\ell,n} = \alpha_i \delta_{\ell,n-1}$ and $\varphi_{i\ell}(x_n|Y_{n-1})$ is the one-step-ahead predictor of x_n obtained under the assumptions that the filter of x_{n-1} is $\varphi_\ell(x_{n-1}|Y_{n-1}) \sim N(x^\ell_{n-1|n-1}, V^\ell_{n-1|n-1})$ and that $p(w_n) = \varphi_i(w_n) \sim N(0, Q_i)$. Therefore, $\varphi_{i\ell}(x_n|Y_{n-1})$ is also Gaussian and its mean and covariance can be obtained by the ordinary Kalman predictor:

$$\begin{aligned} x^{i\ell}_{n|n-1} &= F_n x^\ell_{n-1|n-1} \\ V^{i\ell}_{n|n-1} &= F_n V^\ell_{n-1|n-1} F^T + G_n Q_i G_n^T. \end{aligned} \tag{6.18}$$

Filtering:

$$\begin{aligned} p(x_n|Y_n) &\propto \sum_{j=1}^{K_\varepsilon}\sum_{k=1}^{L_n} \delta_{jk,n}\varphi_{jk}(x_n|Y_n) \\ &\equiv \sum_{\ell=1}^{M_n} \delta_{\ell n}\varphi_\ell(x_n|Y_n). \end{aligned} \tag{6.19}$$

Here, $\delta_{jk,n} = \beta_j \gamma_{kn}\varphi_{jk}(y_n|Y_{n-1})$, $M_n = K_\varepsilon L_n$ and $\varphi_{jk}(x_n|Y_n) \sim N(x^{jk}_{n|n}, V^{jk}_{n|n})$ is the filter of x_n obtained under the assumption that the one-step-ahead predictor of x_n is $N(x^k_{n|n-1}, V^k_{n|n-1})$ and $p(\varepsilon_n) = \varphi_j(\varepsilon_n) \sim N(0, \sigma_j^2)$. Therefore, it can be obtained by the following Kalman filter:

$$\begin{aligned} K^{jk}_n &= V^k_{n|n-1}H_n^T(H_n V^k_{n|n-1}H_n^T + \sigma_j^2)^{-1} \\ x^{jk}_{n|n} &= x^k_{n|n-1} + K^{jk}_n(y_n - H_n x^k_{n|n-1}) \\ V^{jk}_{n|n} &= (I - K^{jk}_n H_n)V^k_{n|n-1}. \end{aligned} \tag{6.20}$$

6.4.2 The Two-filter Formula and Gaussian Sum Smoothing

As just shown, a Gaussian-sum filter can be easily obtained from the non-Gaussian filter. Our concern is with its application to higher order state space modeling. However, the presence of division by non-Gaussian densities in (6.7), is an obstacle that must be overcome before attempting to derive a Gaussian-sum smoother. That division by non-Gausian densities, can ruin the potential usefulness of the Gaussian-sum approximation. Therefore a formula that does not explicitly contain the division by a non-Gaussian density is needed. The details of a derivation of such a formula appear in Kitagawa (1994).

From Kitagawa (1994), the smoothed density $p(x_n|Y_N)$ can be expressed by

$$p(x_n|Y_N) = p(x_n|Y_{n-1})p(Y^n|x_n)p(Y^n|Y_{n-1})^{-1}. \tag{6.21}$$

Here $Y^n \equiv \{y_n, \ldots, y_N\}$ is the information from the present and future observations. From (6.21), obtaining the smoothed density requires $p(Y^n|x_n)$. This term is evaluated by the indicated backward filtering operation.

Initialization

$$p(Y^N|x_N) = p(y_N|x_N). \tag{6.22}$$

Backward Filtering

$$\begin{aligned} p(Y^{n+1}|x_n) &= \int_{-\infty}^{\infty} p(Y^{n+1}|x_{n+1})p(x_{n+1}|x_n)dx_{n+1} && (6.23)\\ p(Y^n|x_n) &= p(Y^{n+1}|x_n)p(y_n|x_n). && (6.24)\end{aligned}$$

We note that the quantity $p(Y_n|x_n)$ in (6.24) is not a conditional state density. Rather, it is an entity, similar to the likelihood. With care, it can be formally evaluated, (Kitagawa 1990), by the backward filtering algorithm. Some details of the Gaussian-sum smoother are shown next.

Since the Gaussian-sum version of the filter was given in (6.16)–(6.20), it suffices to show the implementation of smoothing (6.24).

Assume that $p(x_n|Y_{n-1})$ and $p(Y^n|x_n)$ are expressed by

$$\begin{aligned} p(x_n|Y_{n-1}) &= \sum_{k=1}^{L_n} \gamma_{kn}\varphi_k(x_n|Y_{n-1})\\ p(Y^n|x_n) &= \sum_{\ell=1}^{M'_n} \delta'_{\ell n}\varphi_\ell(Y^n|x_n), \end{aligned} \tag{6.25}$$

where $\varphi_k(x_n|Y_{n-1}) \equiv N(x^k_{n|n-1}, V^k_{n|n-1})$ and $\varphi_\ell(Y_n|x_n) \equiv N(z^\ell_{n|n}, U^\ell_{n|n})$. Then by analogy to the derivation of the Gaussian-sum filter (6.19) for the linear state space model in (6.2), the Gaussian-sum smoother is obtained by

$$\begin{aligned} p(x_n|Y_N) &\propto p(Y^n|x_n)p(x_n|Y_{n-1})\\ &= \sum_{\ell=1}^{M'_n}\sum_{k=1}^{L_n} \delta'_{\ell n}\gamma_{kn}\varphi_\ell(Y^n|x_n)\varphi_k(x_n|Y^{n-1})\\ &\equiv \sum_{\ell=1}^{M'_n}\sum_{k=1}^{L_n} \delta'_{\ell n}\gamma_{kn}\varphi_{\ell k}(x_n|Y_N). \end{aligned} \tag{6.26}$$

Here $\varphi_{\ell k}(x_n|Y_N)$ is the Gaussian density whose mean and the covariance are obtained by replacing H_n by I and y_n by $z^\ell_{n|n}$,

$$\begin{aligned} J^{\ell k}_n &= V^k_{n|n-1}(V^k_{n|n-1} + U^\ell_{n|n})^{-1}\\ x^{\ell k}_{n|N} &= x^k_{n|n-1} + J^{\ell k}_n(z^\ell_{n|n} - x^k_{n|n-1})\\ V^{\ell k}_{n|N} &= (I - J^{\ell k}_n)V^k_{n|n-1}. \end{aligned} \tag{6.27}$$

6.4.3 Remarks on the Gaussian Mixture Approximation

Some remarks on the Gaussian mixture approximation are in order. In the Gaussian mixture approximation, each Gaussian component can be evaluated by the Kalman filter. This is the most appealing feature of this method. However, this method has a severe drawback. Namely, the number of necessary Gaussian components increases to infinity very rapidly as the time proceeds. If the the number of Gaussian components used for the noise densities are K_w and K_ε, respectively, then the number of necessary Gaussian components for the state density at the n-th step is $(K_w \cdot K_\varepsilon)^n$ times that of the initial density. In this section, we show a method to circumvent this difficulty.

Reduction of the Number of Gaussian Components

To avoid the explosion of the number of Gaussian components, we re-approximate the densities by a reduced number of Gaussian components at each time step, (Harrison and Stevens 1976). The reapproximation is motivated by the observation that a relatively small number of Gaussian densities can reasonably approximate a large class of distributions and by the anticipation that the complexity of the density will not increase very significantly with the evolution of the time step.

A practical way is to approximate the prediction and filtering densities by a fixed number of components at each step of the recursion such as:

$$
\begin{aligned}
p(x_n|Y_{n-1}) &= \sum_{k=1}^{m_{pn}} \gamma_{kn}\varphi_k(x_n|Y_{n-1}) \longrightarrow \sum_{k=1}^{m} \gamma'_{kn}\varphi'_k(x_n|Y_{n-1}) \\
p(x_n|Y_n) &= \sum_{\ell=1}^{m_{fn}} \delta_{\ell n}\varphi_\ell(x_n|Y_n) \longrightarrow \sum_{\ell=1}^{m} \delta'_{\ell n}\varphi'_\ell(x_n|Y_n).
\end{aligned}
\tag{6.28}
$$

In principle, this can be realized by finding the minimizer of a criterion for the dissimilarity of the true and approximated densities. We exploit the Kullback-Leibler information number, (Section 2.1.1),

$$
I(p;q) = \int \log \frac{p(x)}{q(x)} p(x)dx, \tag{6.29}
$$

with $p(x)$ and $q(x)$ the true and the approximated densities, respectively. Generally this evaluation involves computationally costly numerical integration and nonlinear optimization with respect to many parameters and is not practical especially for higher order models. A seemingly appealing method of adopting Gaussian terms with largest m probabilities γ_{kn} or δ_{ln} was found to be inefficient. This is mainly due to the fact that even if the weight of a Gaussian component is very small at a certain time point, it may become large at the next time step. Typically that can be seen for a Gaussian component with large variance when an outlier is present or a parameter jump occurs. Ignoring such a component will be disastrous. From our experience, a practical way to reduce the number of Gaussian components is to measure the similarity of two Gaussian components of filtered densities by

$$
\begin{aligned}
D(\varphi_k, \varphi_l) &= -2\delta_k\delta_\ell\{I(\varphi_k;\varphi_l) + I(\varphi_l;\varphi_k)\} \qquad (6.30)\\
&= \delta_k\delta_\ell\{V_k^{-1}V_l + V_l^{-1}V_k + (\mu_k - \mu_l)^t(V_k^{-1} + V_l^{-1})(\mu_k - \mu_l)\}
\end{aligned}
$$

and pool the pair of Gaussian densities, φ_k and φ_l, in a way that minimizes $D(\varphi_k, \varphi_l)$. Here μ_k and V_k are the mean and the covariance of the Gaussian component φ_k. The pooling of two Gaussian densities, φ_k and φ_l, can be realized by the following calculations:

$$\begin{aligned} \mu_{kl} &= \frac{\delta_{kn}\mu_k + \delta_{ln}\mu_l}{\delta_k + \delta_l} \\ V_{kl} &= \frac{\delta_k\{V_k + (\mu_{kl} - \mu_k)(\mu_{kl} - \mu_k)^t\} \ + \ \delta_l\{V_l + (\mu_{kl} - \mu_l)(\mu_{kl} - \mu_l)^t\}}{\delta_k + \delta_l}. \end{aligned} \tag{6.31}$$

By repeating this procedure, we can get an approximation to the conditional state density with a fixed number of Gaussian components.

An illustrative example

Here, we show how the procedure works in reducing the number of Gaussian components. A worked example is illustrated in Figure 6.1. The true density $g(y)$ was arbitrarily defined using 16 Gaussian components. In Figure 6.1, the fine curves show the true density $g(y)$. The bold curves show the Gaussian-sum approximations with m components, obtained by successively applying the indicated procedure. It can be seen that fairly reasonable approximations are obtained for $m \geq 3$.

In the second column of Table 6.1 the Kullback-Leibler information numbers of the Gaussian-sum approximations with m components obtained by the indicated method are shown. The third column of Table 6.1 shows the K-L information numbers of the best Gaussian-sum approximations. They are obtained by using the computationally costly numerical integration and numerical optimization in $(3m-1)$-dimensional parameter space. It is quite time consuming and is impractical to use in filtering and smoothing algorithms. The table shows that the increase of the K-L information number with the decreasing number of Gaussian components induced by the

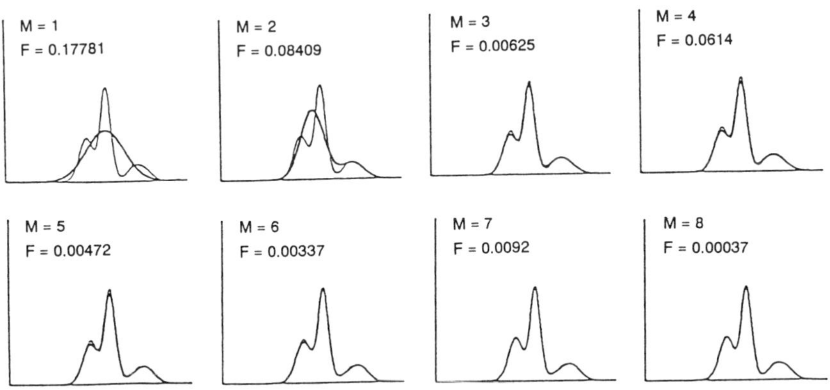

Figure 6.1. Gaussian-sum approximations with reduced number of Gaussian components. Bold curves show the approximated density and the fine curves show the true density.

Table 6.1 Kullback-Leibler numbers of the Gaussian-sum approximation with smaller number of components. The second column shows the ones for the approximation by the proposed method. The third column shows the ones for the best approximations.

m	approximation	best	m	approximation	best
15	0.00000	0.00000	7	0.00092	0.00012
14	0.00000	0.00000	6	0.00337	0.00012
13	0.00011	0.00000	5	0.00472	0.00013
12	0.00012	0.00000	4	0.00614	0.00013
11	0.00013	0.00000	3	0.00625	0.00258
10	0.00013	0.00000	2	0.08409	0.06967
9	0.00013	0.00000	1	0.17781	0.17781
8	0.00037	0.00001			

approximations can be considerable. The example also reveals the merit of using the Gaussian mixture as compared with the single approximating Gaussian density.

Other examples

The validity of our implementation of the Gaussian-sum smoother is illustrated in Chapter 8 by comparing its performance with the estimates obtained by the "exact" non-Gaussian numerical integration methods in the problem of modeling a time series with an abruptly changing mean value. The more important comparison of the performance of the relatively large state dimensional Gaussian and Gaussian mixture-two filter formula on seasonal models is shown in Chapter 9. In addition in Chapter 9, we also illustrate the value of the Gaussian mixture-two filter formula in modeling time series with outliers.

6.5 A Monte Carlo Filtering and Smoothing Method

A Monte Carlo method for non-Gaussian nonlinear filtering and smoothing is shown next. In this method, each conditional distribution is expressed by many of its realizations, and the movement of each "particle", (sample from a distribution), is simulated by using the assumed model.

6.5.1 Introduction

The use of Monte Carlo-Gibbs sampling methods for statistical data analysis received a considerable impetus from Gelfand and Smith (1990), Gelfand *et al.* (1990) and many subsequent papers. More recently Monte Carlo methods have been applied to time series analysis, and to non-Gaussian and nonlinear modeling as well, (Carlin *et al.* 1992, Carter and Kohn 1994 and Tanizaki (1993) for example), however they have primarily been used to achieve numerical integration. Here, we show a direct Monte

Carlo method for state space prediction, filtering and smoothing. The algorithm is based on the approximation of each of the prediction, filtering and smoothing conditional probability density functions by many of their realizations. It can be applied to a wide class of nonlinear non-Gaussian higher dimensional state space models if the dimensions of the system noise and the observational noise are low. Examples of the application of the Monte Carlo method include the modeling of time series with abrupt trend discontinuities, (Chapter 8), time series with abrupt trend *and* seasonal discontinuities, (Chapter 9), and nonlinear time series, (Chapter 15). Additional examples and many technical details that are not shown here appear in Kitagawa (1993, 1996). An interesting variation on the resampling scheme in the filtering procedure is in Higuchi (1995).

6.5.2 Non-Gaussian Nonlinear State Space Model and Filtering

As in Section 6.2, consider the non-Gaussian nonlinear state space model,

$$\begin{aligned} x_n &= f(x_{n-1}, w_n)' &\quad (6.32)\\ y_n &= h(x_n, \varepsilon_n), &\quad (6.33) \end{aligned}$$

with x_n a k-dimensional state vector, w_n and ε_n ℓ-dimensional and 1-dimensional white noise sequences with densities $q(w)$ and $r(\varepsilon)$, respectively. f and h are possibly nonlinear functions $\mathrm{R}^k \times \mathrm{R}^\ell \to \mathrm{R}^k$ and $\mathrm{R}^k \times \mathrm{R} \to \mathrm{R}$, respectively. We also assume that given x, $y = h(x, \varepsilon)$ has an inverse function g such that $\varepsilon = g(y, x)$. The initial state vector x_0 is distributed according to the density $p(x_0)$.

Our objective is to achieve state estimation and in particular to evaluate the conditional density $p(x_n|Y_t)$, where Y_t is the set of observations $\{y_1, \cdots, y_t\}$. In the following presentation of the algorithms for prediction, filtering and smoothing algorithms, each conditional density is expressed by samples from its realizations.

One Step Ahead Prediction:

In this step, we obtain the conditional probability density function of the one step ahead predictor $p(x_n|Y_{n-1})$, given $p(x_{n-1}|Y_{n-1})$, the conditional density function of the filtered state, x_{n-1}, and the system noise density, $q(w)$. Let $\{s_{n-1}^{(1)}, \cdots, s_{n-1}^{(m)}\}$ and $\{w_n^{(1)}, \cdots, w_n^{(m)}\}$ be independent realizations of $p(x_{n-1}|Y_{n-1})$ and $q(w)$, respectively. That is, for $i = 1, \ldots, m$

$$\begin{aligned} s_{n-1}^{(i)} &\sim p(x_{n-1}|Y_{n-1})\\ w_n^{(i)} &\sim q(w). &\quad (6.34) \end{aligned}$$

m is the number of "particles" used for the approximation. (The effect of the selection of m on the accuracy of approximation will be discussed later.) Using these realizations, define an i-th sample at time n, $t_n^{(i)}$ by

$$t_n^{(i)} = f(s_{n-1}^{(i)}, w_n^{(i)}). \quad (6.35)$$

Note that w_n is independent of both x_{n-1} and Y_{n-1} and $x_n = f(x_{n-1}, w_n)$. Therefore, the individual elements of $t_n^{(1)}$, $\cdots$, $t_n^{(m)}$ are distributed as $p(x_n|Y_{n-1})$. That is, $\{t_n^{(1)}, \cdots, t_n^{(m)}\}$ can be considered as realizations of the one step ahead predictor, $p(x_n|Y_{n-1})$.

Filtering:

Given the current observation y_n and $t_n^{(j)} \sim p(x_n|Y_{n-1})$, compute the

$$\alpha_n^{(j)} = p(y_n|t_n^{(j)}) = r(g(y_n, t_n^{(j)})) \left| \frac{\partial g}{\partial y} \right| \tag{6.36}$$

for $j = 1, \ldots, m$. Note that g is the inverse function of h, and r is the density of the observation noise w. Next, obtain $s_n^{(1)}, \cdots, s_n^{(m)}$, the filtered observation at the n-th time step, by the resampling of $t_n^{(1)}, \cdots, t_n^{(m)}$ with the probabilities proportional to $\alpha_n^{(1)}, \cdots, \alpha_n^{(m)}$, respectively. Namely, define $s_n^{(j)}$ by

$$s_n^{(j)} = \begin{cases} t_n^{(1)} & \text{with probability } \alpha_n^{(1)}/(\alpha_n^{(1)} + \cdots + \alpha_n^{(m)}) \\ \vdots & \vdots \\ t_n^{(m)} & \text{with probability } \alpha_n^{(m)}/(\alpha_n^{(1)} + \cdots + \alpha_n^{(m)}). \end{cases} \tag{6.37}$$

Then the $s_n^{(1)}, \cdots, s_n^{(m)}$ can be considered as the realizations of the filter, $p(x_n|Y_n)$.

This can be verified as follows. When m independent realizations $t_n^{(1)}, \cdots, t_n^{(m)}$ from $p(x_n|Y_{n-1})$ are given, the distribution function of $p(x_n|Y_{n-1})$ can be approximated by the empirical distribution function

$$P_n(x) = \frac{1}{m} \sum_{i=1}^{m} I(x, t_n^{(i)}), \tag{6.38}$$

where $I(x,a)$ is the indicator function defined by $I(x,a) = 0$ if $x < a$ and $I(x,a) = 1$ otherwise. This means that $p(x_n|Y_{n-1})$ is approximated by the probability function

$$\Pr(x_n = t_n^{(j)}|Y_{n-1}) = \frac{1}{m} \qquad \text{for } j = 1, \cdots, m. \tag{6.39}$$

Then, given the observation, y_n, the posterior probability be seen to be given by

$$\Pr(x_n = t_n^{(j)}|Y_n) = \frac{\alpha_n^{(j)} \cdot \frac{1}{m}}{\sum_{i=1}^{m} \alpha_n^{(i)} \cdot \frac{1}{m}} = \frac{\alpha_n^{(j)}}{\sum_{i=1}^{m} \alpha_n^{(i)}}. \tag{6.40}$$

This means that the probability distribution function associated with $\Pr(x_n = t_n^{(j)}|Y_n)$ can be expressed by a step function

$$\frac{1}{\sum_{i=1}^{m} \alpha_n^{(i)}} \sum_{j=1}^{m} \alpha_n^{(j)} I(x, t_n^{(j)}) \tag{6.41}$$

which has jumps only at $t_n^{(1)}$, $\cdots$, $t_n^{(m)}$ with step sizes proportional to $\alpha_n^{(1)}$, $\cdots$, $\alpha_n^{(m)}$.

For the next prediction step, it is necessary to represent this distribution function by an empirical distribution of the form

$$\frac{1}{m}\sum_{i=1}^{m} I(x, s_n^{(i)}). \tag{6.42}$$

This can be done by generating m independent realizations $\{s_n^{(1)}, \cdots, s_n^{(m)}\}$ by the resampling of $\{t_n^{(1)}, \cdots, t_n^{(m)}\}$ with probabilities

$$\Pr(s_n^{(i)} = t_n^{(j)}|Y_n) = \frac{\alpha_n^{(j)}}{\alpha_n^{(1)} + \cdots + \alpha_n^{(m)}} \qquad \text{for } j = 1, \ldots, m. \tag{6.43}$$

An Algorithm for Filtering:

The following is a summary of the prediction and filtering algorithm:

1. *Determine m, the number of realizations to be used for the approximation of each distribution.*
2. *Generate a k-dimensional random number $s_0^{(j)} \sim p_0(x)$ for $j = 1, \ldots, m$.*
3. *Repeat the following steps for $n = 1, \ldots, N$.*
   (a) *Generate an ℓ-dimensional random number $w_n^{(j)} \sim q(w)$ for $j = 1, \ldots, m$.*
   (b) *Compute $t_n^{(j)} = f(s_{n-1}^{(j)}, w_n^{(j)})$ for $j = 1, \ldots, m$.*
   (c) *Compute $\alpha_n^{(j)} = r(g(y_n, t_n^{(j)}))\,|\partial g/\partial y|$ for $j = 1, \ldots, m$.*
   (d) *Generate $s_n^{(j)} \sim (\sum_{i=1}^{m} \alpha_n^{(i)})^{-1} \sum_{i=1}^{m} \alpha_n^{(i)} I(x, t_n^{(i)})$ for $j = 1, \ldots, m$ by the resampling of $t_n^{(1)}, \ldots, t_n^{(m)}$.*

Likelihood of the Model:

Given the observations $y_1, \ldots, y_N$, the likelihood of the parameter θ of the model is obtained by

$$L(\theta) = p(y_1, \ldots, y_N|\theta) = \prod_{n=1}^{N} p(y_n|y_1, \ldots, y_{n-1}, \theta) = \prod_{n=1}^{N} p(y_n|Y_{n-1}). \tag{6.44}$$

Therefore, by using the approximation

$$\begin{aligned} p(y_n|Y_{n-1}) &= \int p(y_n|x_n)p(x_n|Y_{n-1})dx_n \\ &\doteq \frac{1}{m}\sum_{j=1}^{m} p(y_n|t_n^{(j)}) \\ &= \frac{1}{m}\sum_{j=1}^{m} \alpha_n^{(j)}, \end{aligned} \tag{6.45}$$

the log-likelihood can be approximated by

$$\begin{aligned} \ell(\theta) &= \sum_{n=1}^{N} \log p(y_n|Y_{n-1}) \\ &\doteq \sum_{n=1}^{N} \log\Big(\sum_{j=1}^{m} \alpha_n^{(j)}\Big) - N\log m. \end{aligned} \tag{6.46}$$

An Illustrative Example

Consider a one dimensional state space model

$$x_n = x_{n-1} + w_n \tag{6.47}$$

$$y_n = x_n + \varepsilon_n, \tag{6.48}$$

where w_n and ε_n are white noises distributed as the Cauchy distribution, Cauchy(0,0.01), with density $q(w) = 0.1\pi^{-1}(w^2 + 0.01)^{-1}$, and the standard normal distribution, $N(0,1)$, respectively. To be specific, we further assume that $p(x_{n-1}|Y_{n-1})$ is also the standard normal distribution, $N(0,1)$, and consider the predictive and the filter distributions, $p(x_n|Y_{n-1})$ and $p(x_n|Y_n)$, respectively.

The panels (a) and (c) in Figure 6.2 show the density functions of $p(x_{n-1}|Y_{n-1})$ and $q(w_n)$ and 100 realizations. In (b) and (d), the bold curves show the empirical distribution functions obtained by 100 realizations shown in (a) and (c), respectively. The fine curves, which were obtained by numerical intergration, show the true distribution functions. Panel (e) shows 100 realizations of $p(x_n|Y_{n-1})$ directly computed by (6.47). On the other hand, the curve in (e) shows the "exact" density function obtained by the numerical convolution of the densities shown in (a) and (c). Panel (f) shows the corresponding distribution function and the empirical distribution function, respectively. Panel (g) shows the "exact" filtered density $p(x_n|Y_n)$. The 100 realizations in (g) are the same as the one in (e). However, the probability of the j-th each realization is proportional to $\alpha_n^{(j)}$, and the corresponding distribution function shown in (h) is quite different from the one in (f) which has equal probability. Panel (i) shows 100 realizations obtained by the resampling of the data shown in (g) according to the probability given by (6.37). It can be seen that the empirical distribution function shown in (j) resembles the one in (i) and is close to the "exact" one shown by the fine curve.

Some details of the resampling scheme are mentioned here. The basic algorithm is:

1. Rearrange $t_n^{(1)}, \cdots, t_n^{(m)}$ in order of magnitude and the results are denoted by $t_{(n)}^{(1)}, \cdots,$ $t_{(n)}^{(m)}$.
2. For $j = 1, \ldots, m$, repeat the following steps (a)–(c).
   (a) Generate a uniform random number $u_n^{(j)} \in U[0,1]$.

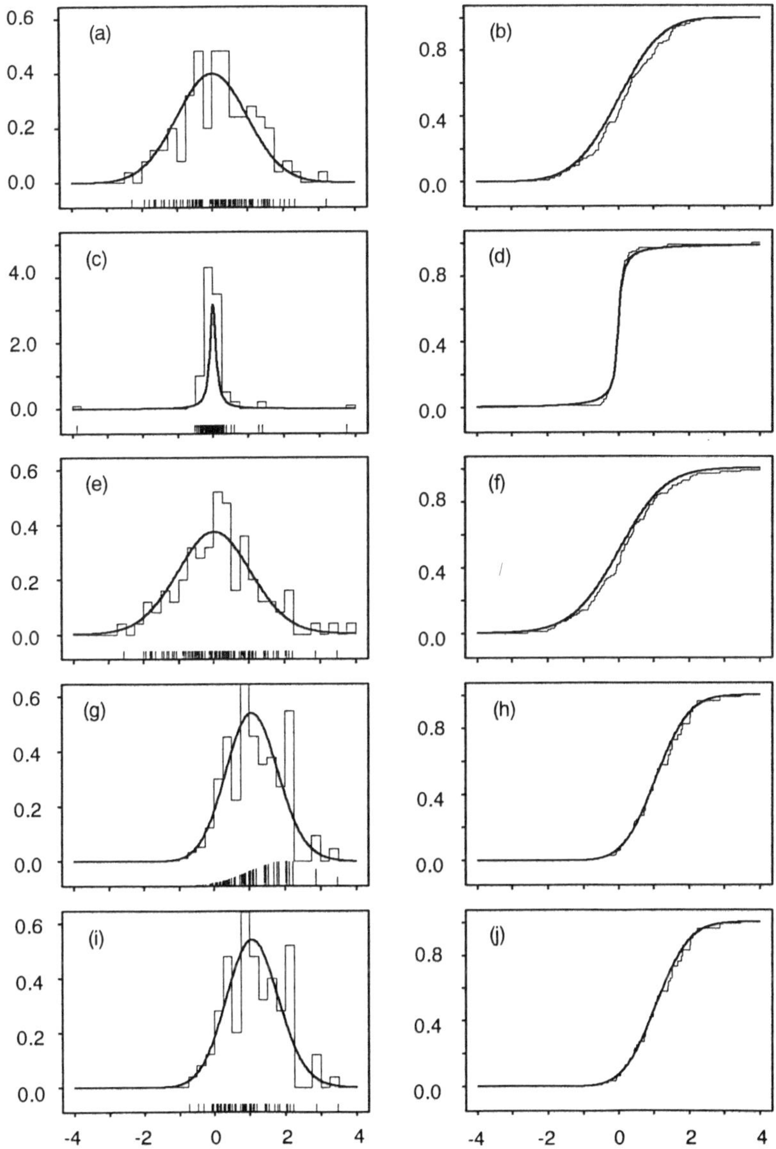

Figure 6.2. Density functions, 100 realizations, distribution functions and empirical distribution functions. (a) and (b): $p(x_{n-1}|Y_{n-1})$. (c) and (d): $q(w_n)$. (e) and (f): $p(x_n|Y_{n-1})$. (g) and (h): $p(x_n|Y_n)$. (i) and (j): $p(x_n|Y_n)$ after resampling.

(b) Find i such that $C\sum_{\ell=1}^{i-1}\alpha_{(n)}^{(\ell)} < u_n^{(j)} \le C\sum_{\ell=1}^{i}\alpha_{(n)}^{(\ell)}$ where $C=\sum_{\ell=1}^{m}\alpha_{(n)}^{(\ell)}$ and $\alpha_{(n)}^{(\ell)}$ is the posterior weight corresponding to $t_{(n)}^{(\ell)}$.

(c) Define $s_n^{(j)} = t_n^{(i)}$.

Some delicate computational considerations and several modifications of the basic algorithm are shown in Kitagawa (1996).

6.5.3 Smoothing

In principle, the algorithm for smoothing can be obtained by a simple generalization of the filter algorithm. Here too, for the careful treatment required to avoid numerical difficulties, see Kitagawa (1996). In the following subsections, we show two alternative smoothing algorithms.

Smoothing by Storing the State Vector

Here, let the joint distribution of $x_1, \ldots, x_n$, be represented by,

$$p(t_1^{(j)},\cdots,t_n^{(j)}|Y_\ell) = \Pr(x_1 = t_{1|\ell}^{(j)},\cdots,x_n = t_{n|\ell}^{(j)}|Y_\ell). \tag{6.49}$$

Assume that $p(s_1^{(j)},\cdots,s_{n-1}^{(j)}|Y_{n-1}) = 1/m$ and $w_n^{(j)} \sim q(w)$, and define $(t_1^{(j)},\cdots,t_n^{(j)})$ by

$$t_{i|\ell}^{(j)} = \begin{cases} s_i^{(j)} & \text{for } i = 1,\ldots,n-1 \\ f(s_{n-1}^{(j)}, w_n^{(j)}) & \text{for } i = n. \end{cases} \tag{6.50}$$

Then $(t_{1|\ell}^{(j)},\cdots,t_{n|\ell}^{(j)})$ can be considered as realizations from $p(t_1^{(j)},\cdots,t_n^{(j)}|Y_\ell)$.

Next, given the observation y_n, $p(t_1^{(j)},\cdots,t_n^{(j)}|Y_{n-1})$ is updated as follows:

$$p(t_1^{(j)},\cdots,t_n^{(j)}|Y_n) = \frac{p(y_n|t_n^{(j)})p(t_1^{(j)},\cdots,t_n^{(j)}|Y_{n-1})}{p(y_n|Y_{n-1})}. \tag{6.51}$$

This indicates that realization of the fixed interval smoother, $p(x_1,\cdots,x_n|Y_n)$ can be obtained by storing and resampling m sets of realizations $(t_1^{(j)},\cdots,t_n^{(j)})$, $j = 1,\ldots,m$, with the same probability as for the filtering case.

In principle, this algorithm realizes the fixed interval smoothing for the nonlinear non-Gaussian state space model. However in practice, since the number of realizations is finite, the repetition of the resampling will gradually decrease the number of different realizations and will deteriorate the shape of the distribution. Therefore, it is recommended to stop the smoothing algorithm after repeating the resampling fewer times (at the largest, less than 5 per cent of m is recommended).

The outcome is equivalent to applying an L-lag fixed lag smoother rather than the fixed interval smoother. The increase of lag, L, will improve the accuracy of the

$p(x_n|Y_{n+L})$ as an approximation to $p(x_n|Y_N)$, while it will decrease the accuracy of $\{s_n^{(1)}, \cdots, s_n^{(m)}\}$ as representatives of $p(x_n|Y_{n+L})$. Since $p(x_n|Y_{n+L})$ usually converges quickly to $p(x_n|Y_N)$, it is recommended to take L not so large.

Smoothing by the Two-filter Formula

Another way of achieving fixed interval smoothing, is to apply the two-filter formula used in Kitagawa (1994). Define $Y^n = \{y_n, \cdots, y_N\}$, then we have $Y_N = Y_{n-1} \cup Y^n$. Therefore, the smoother is divided as follows:

$$\begin{aligned} p(x_n|Y_N) &= p(x_n|Y_{n-1}, Y^n) \\ &\propto p(x_n, Y^n|Y_{n-1}) \\ &= p(Y^n|x_n)p(x_n|Y_{n-1}). \end{aligned} \tag{6.52}$$

The quantity $p(Y^n|x_n)$ can be evaluated by the indicated backward filtering.

$$\begin{aligned} p(Y^N|x_N) &= p(y_N|x_N) \\ p(Y^{n+1}|x_n) &= \int p(Y^{n+1}|x_{n+1})p(x_{n+1}|x_n)dx_{n+1} \\ p(Y^n|x_n) &= p(y_n|x_n)p(Y^{n+1}|x_n). \end{aligned} \tag{6.53}$$

Details for the realization of the two-filter formula (6.52) and the backward filtering (6.53) appear in Kitagawa (1994). Examples of the application of the Monte Carlo filtering method are in Chapters 8, 9 and 15. Additional examples are shown in Kitagawa (1996).

6.6 A Derivation of the Kalman filter

In this section, the Kalman filter formulas are derived from the general state space model.

6.6.1 Preparations

Let V, U and R be $m \times m$, $m \times m$ and $\ell \times \ell$ symmetric matrices, respectively, and H be an arbitrary $\ell \times m$ matrix. Then, assuming the indicated inverses exist, we have the following lemmas:

Lemma 1.

$$\text{(i)} \quad (V^{-1} + H^T R^{-1} H)^{-1} = V - VH^T(HVH^T + R)^{-1}HV \tag{6.54}$$

$$\text{(ii)} \quad (V^{-1} + H^T R^{-1} H)^{-1} H^T R^{-1} = VH^T(HVH^T + R)^{-1} \tag{6.55}$$

$$\text{(iii)} \quad (V^{-1} - U^{-1})^{-1} = V(U - V)^{-1}U \tag{6.56}$$

$$\text{(iv)} \quad U + (V^{-1} - U^{-1})^{-1} = U(U - V)^{-1}U \tag{6.57}$$

Proof. (i) is the well-known matrix inversion lemma which can be easily proved by multiplying the inverse of the left term to the right term. (ii) can be proved by using (i) as follows

$$\begin{aligned}
&(V^{-1}+H^TR^{-1}H)^{-1}H^TR^{-1}\\
&= \{V-VH^T(HVH^T+R)^{-1}HV\}H^TR^{-1}\\
&= VH^TR^{-1}-VH^T(HVH^T+R)^{-1}HVH^TR^{-1}\\
&= VH^T(HVH^T+R)^{-1}(HVH^T+R)R^{-1}-VH^T(HVH^T+R)^{-1}HVH^TR^{-1}\\
&= VH^T(HVH^T+R)^{-1} \qquad (6.58)
\end{aligned}$$

(iii) and (iv) are immediate. ▮

Lemma 2. If $p(\theta) \sim N(\bar{\theta}, V)$ and $p(x|\theta) \sim N(F\theta, R)$, then $p(x) \sim N(F\bar{\theta}, FVF^T + R)$.

Proof. From the assumption

$$p(\theta) = \left(\frac{1}{\sqrt{2\pi}}\right)^m |V|^{-\frac{1}{2}} \exp\left\{-\frac{1}{2}(\theta-\bar{\theta})^TV^{-1}(\theta-\bar{\theta})\right\} \qquad (6.59)$$

$$p(x|\theta) = \left(\frac{1}{\sqrt{2\pi}}\right)^\ell |R|^{-\frac{1}{2}} \exp\left\{-\frac{1}{2}(x-F\theta)^TR^{-1}(x-F\theta)\right\}. \qquad (6.60)$$

Then $p(x) = \int p(x|\theta)p(\theta)d\theta$ and the integrand is obtained by

$$\begin{aligned}
p(x|\theta)p(\theta) &= \left(\frac{1}{\sqrt{2\pi}}\right)^{m+\ell} |V|^{-\frac{1}{2}}|R|^{-\frac{1}{2}}\\
&\times \exp\left\{-\frac{1}{2}(\theta-\bar{\theta})^TV^{-1}(\theta-\bar{\theta})-\frac{1}{2}(x-F\theta)^TR^{-1}(x-F\theta)\right\} \qquad (6.61)
\end{aligned}$$

-2 times the terms in the bracket can be expressed as

$$\begin{aligned}
&\theta^T(V^{-1}+F^TR^{-1}F)\theta-\theta^T(V^{-1}\bar{\theta}+F^TR^{-1}x)-(\bar{\theta}^TV^{-1}+x^TR^{-1}F)\theta\\
&+\bar{\theta}^TV^{-1}\bar{\theta}+x^TR^{-1}x = (\theta-\xi)^T(V^{-1}+F^TR^{-1}F)(\theta-\xi)+\Delta \qquad (6.62)
\end{aligned}$$

where

$$\xi = (V^{-1}+F^TR^{-1}F)^{-1}(V^{-1}\bar{\theta}+F^TR^{-1}x) \qquad (6.63)$$

$$\Delta = \bar{\theta}^TV^{-1}\bar{\theta}+x^TR^{-1}x-\xi^T(V^{-1}+F^TR^{-1}F)^{-1}\xi. \qquad (6.64)$$

Here, from (i) and (ii) of the Lemma 1,

$$\begin{aligned}
\Delta &= x^T\{R^{-1}-R^{-1}F(V^{-1}+F^TR^{-1}F)^{-1}F^TR^{-1}\}x\\
&- x^TR^{-1}F(V^{-1}+F^TR^{-1}F)^{-1}V^{-1}\bar{\theta}\\
&- \bar{\theta}^TV^{-1}(V^{-1}+F^TR^{-1}F)^{-1}F^TR^{-1}x\\
&+ \bar{\theta}^TV^{-1}\bar{\theta} - \bar{\theta}^TV^{-1}(V^{-1}+F^TR^{-1}F)^{-1}V^{-1}\bar{\theta}\\
&= (x-F\bar{\theta})^T(FVF^T+R)^{-1}(x-F\bar{\theta}). \qquad (6.65)
\end{aligned}$$

Therefore, we have

$$\begin{aligned} p(x) &= \int p(x|\theta)p(\theta)d\theta \\ &= \left(\frac{1}{\sqrt{2\pi}}\right)^m |V|^{-\frac{1}{2}}|R|^{-\frac{1}{2}}|V^{-1}+F^TR^{-1}F|^{-\frac{1}{2}} \\ &\quad \times \exp\left\{-\frac{1}{2}(x-F\bar{\theta})^T(FVF^T+R)^{-1}(x-F\bar{\theta})\right\}. \end{aligned} \tag{6.66}$$

This shows that $p(x) \sim N(F\bar{\theta}, FVF^T + R)$ and completes the proof. ∎

Corollary. $|V||R||V^{-1} + F^TR^{-1}F| = |FVF^T + R|.$ (6.67)

Lemma 3. If $p(\theta) \sim N(\bar{\theta}, V)$ and $p(y|\theta) \sim N(H\theta, R)$, then

$$p(\theta|y) \sim N(\zeta, U), \tag{6.68}$$

where again we have the standard normal theory result,

$$\begin{aligned} \zeta &= \bar{\theta} + VH^T(HVH^T+R)^{-1}(y-H\bar{\theta}) \\ U &= V - VH^T(HVH^T+R)^{-1}HV. \end{aligned} \tag{6.69}$$

Proof.

$$\begin{aligned} p(\theta|y) &\propto p(y|\theta)p(\theta) \\ &= \left(\frac{1}{\sqrt{2\pi}}\right)^{\ell+m} |R|^{-\frac{1}{2}}|V|^{-\frac{1}{2}} \\ &\quad \times \exp\left\{-\frac{1}{2}(y-H\theta)^TR^{-1}(y-H\theta) - \frac{1}{2}(\theta-\bar{\theta})^TV^{-1}(\theta-\bar{\theta})\right\}. \end{aligned} \tag{6.70}$$

Here -2 times the terms in the bracket can be expressed as

$$\begin{aligned} &(y-H\theta)^TR^{-1}(y-H\theta) + (\theta-\bar{\theta})^TV^{-1}(\theta-\bar{\theta}) \\ &= \theta^T(H^TR^{-1}H+V^{-1})\theta \; - \; \theta^T(H^TR^{-1}y+V^{-1}\bar{\theta}) \\ &\quad - (y^TR^{-1}H+\bar{\theta}V^{-1})\theta \; + \; y^TR^{-1}y \; + \; \bar{\theta}^TV^{-1}\bar{\theta} \\ &= (\theta-\zeta)^T(H^TR^{-1}H+V^{-1})(\theta-\zeta) \\ &\quad + (y-H\bar{\theta})^T(V+HRH^T)^{-1}(y-H\bar{\theta}) \end{aligned} \tag{6.71}$$

where

$$\zeta = (H^TR^{-1}H+V^{-1})^{-1}(H^TR^{-1}y+V^{-1}\bar{\theta}). \tag{6.72}$$

From (i) and (ii) of the Lemma 1, since

$$\begin{aligned} (H^TR^{-1}H+V^{-1})^{-1}H^TR^{-1} &= VH^T(HVH^T+R)^{-1} \\ (H^TR^{-1}H+V^{-1})^{-1}V^{-1} &= \{V - VH^T(HVH^T+R)^{-1}HV\}V^{-1} \\ &= I - VH^T(HVH^T+R)^{-1}H, \end{aligned}$$

(6.70) can be expressed as

$$\zeta = \bar{\theta} + VH^T(HVH^T + R)^{-1}(y - H\bar{\theta}). \tag{6.73}$$

∎

6.6.2 Derivation of the Filter and Smoother

In this section, we shall show that the Kalman filter and fixed interval smoother, (shown in Chapter 5), can be easily derived from the non-Gaussian filter and the non-Gaussian smoother described in Section 6.2.

A Derivation of the Kalman Filter

In Lemma 2, if we put $x = x_n$, $\theta = x_{n-1}$, $\bar{\theta} = x_{n-1|n-1}$, $V = V_{n-1|n-1}$, $F = F_n$ and $R = G_n Q_n G_n^T$, then we have

$$p(x_n|Y_{n-1}) \sim N(x_{n|n-1}, V_{n|n-1}) \tag{6.74}$$

where

$$\begin{aligned} x_{n|n-1} &= F_n x_{n-1|n-1} \\ V_{n|n-1} &= F_n V_{n-1|n-1} F_n^T + G_n Q_n G_n^T. \end{aligned}$$

Next, by putting $\theta = x_n$, $\bar{\theta} = x_{n|n-1}$, $V = V_{n|n-1}$, $H = H_n$, and $R = R_n$, Lemma 3 yields $p(\theta|Y_n) \sim N(x_{n|n}, V_{n|n})$ where

$$\begin{aligned} K_n &= V_{n|n-1} H_n^T (H_n V_{n|n-1} H_n^T + R_n)^{-1} \\ x_{n|n} &= x_{n|n-1} + K_n(y - H_n x_{n|n-1}) \\ V_{n|n} &= V_{n|n-1} - K_n H V_{n|n-1}. \end{aligned} \tag{6.75}$$

∎

A Derivation of the Fixed Interval Smoother

From (6.7) we have that

$$p(x_n|Y_N) = \int p(x_{n+1}|Y_N) \frac{p(x_n|Y_n)p(x_{n+1}|x_n)}{p(x_{n+1}|Y_n)} dx_{n+1} \tag{6.76}$$

Here, by putting, $\theta = x_n$, $\bar{\theta} = x_{n|n}$, $V = V_{n|n}$, $y = x_{n+1}$, $H = F_{n+1}$, $R = G_{n+1}Q_{n+1}G_{n+1}^t$, Lemma 3 yields

$$\frac{p(x_n|Y_n)p(x_{n+1}|x_n)}{p(x_{n+1}|Y_n)} \sim N(x_{n|N}^*, V_{n|N}^*), \tag{6.77}$$

where

$$\begin{aligned} x_{n|N}^* &= x_{n|n} + V_{n|n} F_{n+1}^T (F_{n+1} V_{n|n} F_{n+1}^T + G_{n+1} Q_{n+1} G_{n+1}^T)^{-1} (x_{n+1} - F_{n+1} x_{n|n}) \\ &= x_{n|n} + V_{n|n} F_{n+1}^T V_{n+1|n}^{-1} (x_{n+1} - F_{n+1} x_{n|n}) \\ V_{n|N}^* &= V_{n|n} - V_{n|n} F_{n+1}^T (F_{n+1} V_{n|n} F_{n+1}^T + G_{n+1} Q_{n+1} G_{n+1}^T)^{-1} F_{n+1} V_{n|n} \\ &= V_{n|n} - V_{n|n} F_{n+1}^T V_{n+1|n}^{-1} F_{n+1} V_{n|n}. \end{aligned}$$

Then, by equating $\theta = x_{n+1}$, $\bar{\theta} = x_{n+1|N}$, $V = V_{n+1|N}$, $F = F_{n+1}$, $R = G_{n+1}Q_{n+1}G_{n+1}^T$, $x = x_n$ and by applying Lemma 2, we obtain

$$p(x_n|Y_N) \sim N(x_{n|N}, V_{n|N}), \tag{6.78}$$

where

$$\begin{aligned} x_{n|N} &= x_{n|n} + V_{n|n}F_{n+1}^T V_{n+1|n}^{-1}(x_{n+1|N} - x_{n+1|n}) \\ V_{n|N} &= V_{n|n}F_{n+1}^T V_{n+1|n}^{-1} V_{n+1|N} V_{n+1|n}^{-1} F_{n+1} V_{n|n} + V_{n|n} - V_{n|n}F_{n+1}^T V_{n+1|n}^{-1} F_{n+1} V_{n|n} \\ &= V_{n|n} + V_{n|n}F_{n+1}^T V_{n+1|n}^{-1}(V_{n+1|N} - V_{n+1|n})V_{n+1|n}^{-1} F_{n+1} V_{n|n}. \end{aligned}$$

∎

7

Applications of Linear Gaussian State Space Modeling

Some applications of the state space models that were described in Chapter 5 are presented in this chapter. In particular, the modeling of the famous Canadian lynx data by an AR state space model, the modeling of irregularly spaced data and an example of the decomposition of an observed time series into a signal, background noise and observation noise are shown.

First, for illustrative purposes, the lynx data is modeled by a standard state space model. AR models up to order 15 are computed and the AIC best AR model is selected. Two different features of state space analysis, prediction and the filling in of missing data are also illustrated in the context of state space AR modeling of the lynx data. In addition, some Kullback-Leibler (KL), computational results are shown. We compute the KL number between models of different AR orders and the minimum AIC AR model. Underfitted models are known to be biased, while overfitted models tend to have excessive variance. The KL numbers give a sense of magnitude of the discrepancy between the true and alternative models orders for both underfitted and overfitted models. In a second KL number computational example, the KL number is computed for situations in which a model is fitted to limited data. In one version, the available data is a foreshortened version of the lynx data. In an alternative version, the same length of data is available, however it is randomly selected from the original lynx data. The results are instructive. They illustrate the importance of the state space modeling property that easily permits the modeling of data with missing values.

The second problem topic treats the modeling of irregularly spaced data. This work was stimulated by Jones (1980). It is a slightly revised version of that same topic treated in Kitagawa (1984).

The third topic treats the decomposition of an observed signal into several components by a state space time domain analysis that exploits the AR state space model. The data analyzed is a microearthquake signal embedded in background noise. The problem is a version of the signal extraction problem, the objective being to extract the signal from the non-white noise background.

7.1 AR Time Series Modeling

Here we use the state space model shown in Chapter 5 for fitting autoregressive models to assumed stationary time series data. For illustrative purposes, the Canadian Lynx, $N = 114$ data of observations from 1821-1934, (from Elton and Nicholson 1942), was modeled. A maximum likelihood procedure that uses a Davidon-Fletcher-Powell type numerical optimization routine, (Section 2.3), realized the computations. Models of increasing AR order $m = 1, \ldots, 15$ were fitted using the state space model. The initial parameter estimates required as inputs to the state space model, were computed by a Householder transformation least squares fit, (Section 2.2), to the lynx data. That latter procedure is illustrated by the program UNIMAR, (univariate autoregressive modeling), in the TIMSAC-78 package, Akaike *et al.* (1978). Shown below in Table 7.1 for the successive values of the fitted model order, are the values of the estimated innovations variance $\hat{\sigma}^2$, the log-likelihood and the AIC.

Table 7.1: State space-maximum likelihood estimates of the innovations variance, the log-likelihood and the AIC of the autoregressive models.

model order	innovations variance	log likelihood	AIC
1	0.11523	−39.090	82.179
2	0.05111	6.462	−6.923
3	0.05038	7.265	−6.530
4	0.04828	9.618	−9.237
5	0.04737	10.659	−9.318
6	0.04713	10.935	−7.870
7	0.04455	13.976	−11.952
8	0.04377	14.926	−11.851
9	0.04319	15.634	−11.268
10	0.04113	18.215	−14.430
11	0.03618	24.911	−25.822
12	0.03551	25.857	−25.714
13	0.03539	26.017	−24.035
14	0.03539	26.027	−22.053
15	0.03535	26.080	−20.160

In our experience, the results in Table 7.1 are typical of such analyses. The innovations variance and the log-likelihood are respectively non-increasing and non-decreasing functions of model order. Model order 11 is the AIC best model order. That conclusion is in agreement with the least squares estimation results shown in TIMSAC-78 for the UNIMAR program. The rate of decrease of the innovations variance and corresponding decreasing rate of increase of log-likelihood is apparent for models whose order exceeds the AIC best AR model order. That rate of change is more than offset by the increase in the value of the AIC as additional parameters are added to the model.

For completeness, in Table 7.2 we show the coefficients of the AR order 11 models estimated by both the Householder transformation least squares and the state space-

Table 7.2: Estimated AR(11) model parameters, Householder least squares and state space-maximum likelihood methods.

parameter order	least squares method	maximum likelihood
1	1.18132	1.16633
2	−0.55157	−0.54188
3	0.23137	0.26229
4	−0.17804	−0.30520
5	0.01987	0.15187
6	−0.06257	−0.14598
7	0.02856	0.05803
8	−0.05071	−0.03096
9	0.19989	0.13782
10	0.16182	0.19952
11	−0.33909	−0.33758

maximum likelihood computational methods. The values of the estimated innovations variance for the least squares and state space maximum likelihood modeling methods are respectively 0.03318 and 0.003605. The least squares model was fitted with a lag of 20. That means that in the least squares procedure, the first 20 observations are treated as initial conditions and the model is actually fitted to the data from observation 21 to 114.

The differences in the AR coefficients values between the maximum likelihood and least squares methods have no particular significance for this example. The state space maximum likelihood method is preferred to the least squares method of AR modeling in the situation in which the data length is short and it is desirable not to have to exclude any portion of the data for use as initial values as is required in the Householder transformation least squares modeling. Also because of its greater numerical stability, the maximum likelihood method should be used in preference to the least squares method when the zeros of the characteristic equation of the AR model are close to the unit circle. Finally, an examination of estimates of the power spectral density obtained by both methods computed from the formula (4.11), (Section 4.2), showed no significant differences and is not shown here.

Prediction

The performance of the state space model in prediction is also illustrated for this data set. The first 100 data points of the lynx data were used to compute the AIC's of successively increasing model orders by the state space-maximum likelihood method. AR model order 11 was again selected as the AIC best model of that data. The state space model was then run once more with AR model order 11 and the missing data points 101-114 were predicted. (This is referred to as increasing horizon prediction.)

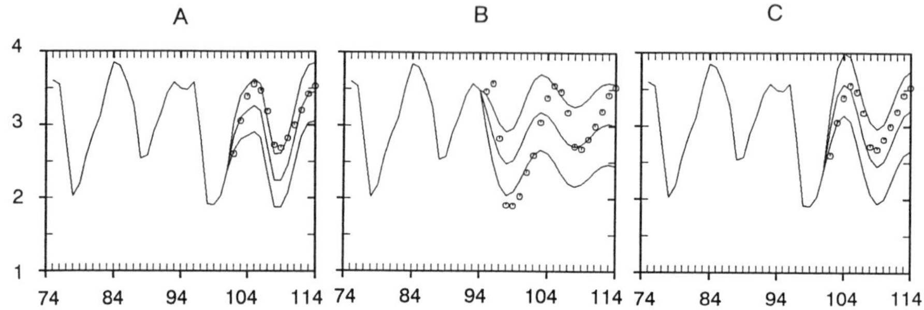

Figure 7.1 A: Predictions AR(11) model starting from $n = 101$. B: Predictions AR(5) model starting from $n = 101$. C: Predictions AR(5) model starting from $n = 90$.

In Figure 7.1A, B, C those results as shown compared with the prediction of an AR(5) model of the data starting from $n = 101$ and an AR(5) model starting from $n = 90$. In Figure 7.1 the original data for the first n modeled data points is shown by a line, the values of the original data for successive data points are shown by circles. The other lines in that figure corresponding to the predicted data points are the predicted value and the predicted value plus and minus one sigma. Some interesting observations can be made: The best prediction results are from the AR(5) model, not from the AIC best prediction model, (Shibata 1981, notwithstanding) and the best prediction results, (Figure 7.1C), may be dependent upon the starting point of the prediction. More importantly, it is clear that the best increasing horizon prediction model may not be identical to the best one-step-ahead prediction model, which is the AIC best model.

Missing Observations

Here, again using state space AR modeling of the Canadian lynx data, we illustrate the performance of state space modeling in the presence of missing data. As pointed out in Section 5.4, to fill in missing data from a regularly spaced observations, it is merely necessary to skip the filter step in the Kalman filter recursions. Some illustrations of the filling in of missing data are shown in Figure 7.2.

7.2 Kullback-Leibler Computations

In this section the Kullback-Leibler number between the assumed true distribution associated with the minimum AIC, (MAICE), model, AR(11), of the lynx data and two other sets of distributions of interest are computed. In one situation we compute the KL number between the assumed true distribution and fitted AR models of

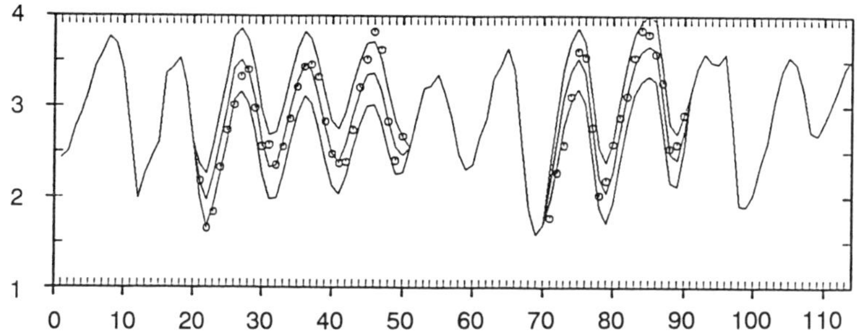

Figure 7.2: Examples of the filling in of missing data.

different orders. The intentions are two-fold. One is to demonstrate that such computations are feasible and the second is to get a quantitative feeling for a measure of the KL number discrepancy in AR model orders. In the second case, again under the assumption that the AR(11) model is "true", the alternative models considered here are the models computed from the lynx data under two different reduced data situations, (i) that the model is estimated from an incomplete or short segment of the lynx data and (ii) that the model is estimated from a randomly sampled segment of the lynx data of the same sample size as in (i). The models in (ii) are estimated by the missing observations procedure described in Section 5.4.

If the distributions in question are known, the Kullback-Leibler number may in general be computed by Monte Carlo methods. In the special case of computing KL numbers between different AR models we use (4.27), (Section 4.2). That formula yields the entropy or the negative of the KL number in terms of the covariance function of the true model and the coefficients of the fitted alternative models. For convenience that formula is reproduced here,

$$\frac{1}{2}\left[1 - \log\left(\frac{\hat{\sigma}^2}{\sigma^2}\right) - \left\{\gamma_0 - 2\sum_{m=1}^{M} \hat{a}_m \gamma_m + \sum_{m=1}^{M}\sum_{l=1}^{M} \hat{a}_m \hat{a}_l \gamma_{m-l}\right\}\right]. \tag{7.1}$$

In (7.1), σ^2 and $\hat{\sigma}^2$ are respectively the innovations variance of the true model and the innovations variance of the fitted model. γ_m is the autocovariance function determined by the true model.

The results of the computations are in Tables 7.3 and 7.4. Table 7.3 shows the KL numbers computed between the true AR(11) model and AR(k) model orders $k = 1, \ldots, 18$. The three columns of numbers in Table 7.4 are left to right respectively the data length $\tilde{N}$ with which the KL numbers were computed when (i) the model is estimated from a short segment of lynx data with length $\tilde{N}$ and (ii) that the lynx data model is estimated from a randomly sampled segment of length $\tilde{N}$ from the lynx data.

As expected, in Table 7.3, when the fitted model orders are less than the "true" model order, the KL numbers are a decreasing function of k until the "true" model

Table 7.3: AR model order and KL number from the MAICE AR model.

m	KL number	m	KL number	m	KL number
1	0.5100	7	0.0982	13	0.0103
2	0.1598	8	0.0899	14	0.0104
3	0.1544	9	0.0857	15	0.0107
4	0.1342	10	0.0608	16	0.0181
5	0.1254	11	0.0000	17	0.0182
6	0.1230	12	0.0093	18	0.0249

Table 7.4: KL numbers between true model and models of systematic and random samples of length $\tilde{N}$.

$\tilde{N}$	systematic	random
104	0.0024	0.0053
94	0.0047	0.0080
84	0.0064	0.0109
74	0.0393	0.0232
64	0.0989	0.0684
54	0.0823	0.0512

order AR(11). Also as expected, when the model order k increases to be larger than the "true" model order, the KL numbers are an increasing function of k. We note however that systematically, the difference in KL numbers around the true model order is an asymmetric function of the difference between true and fitted model orders. The rate of increase of KL number with model orders increasing beyond the true mode order is considerably slower than the rate of increase of KL numbers with model orders decreasing form the true model order. This result was also expected. Underfitted models are not rich enough to explain the data and contribute bias errors while overfitted models contribute only variance or sampling errors. While this fact is well known, (Sakamoto *et al.*, 1986), we don't know if it has been exhibited before.

The results in Table 7.4 show that systematically as the available data for modeling decreases, the KL measure of discrepancy between true and fitted model increases. The perhaps unanticipated result is that if the shorter length of available data is only a consequence of randomly missed data the KL number measure of true and fitted model discrepancy is substantially less than if the data is systematically missed by data length shortening. Also we note that the computed KL numbers indicate that for up to the 35% randomly missed data case, the KL number discrepancy is not as severe as is the case of fitting an AR(10) model instead of the true AR(11) model to the data. These results clearly illustrate the importance of the state space modeling property that easily permits the modeling of data with missing values.

7.3 Smoothing Unequally Spaced Data

Here we show an application of the continuous time-discrete time state space method of modeling irregularly spaced data that was described in Section 5.3. The method is applied to a simulated data set.

Consider an example of the $N = 100$ unequally spaced observations generated by

$$\begin{aligned} v_n &\sim U(-6,6) \\ f(v_n) &= 2.5(v_n^5 - 10v_n^3 + 15v_n)\exp\left\{-\frac{t_n^2}{2}\right\} \\ y_n &= f(v_n) + \varepsilon_n \end{aligned} \tag{7.2}$$

where $\varepsilon_n \sim N(0, 0, 09)$ and U signifies the uniform distribution. The y_n sequence was modeled by the state space model for unequally spaced data described in Chapter 5. Table 7.5 shows the values of τ^2, -2 log-likelihood and AIC computed for continuous time AR differential equation models of order k. Graphical results are shown in Figure 7.3. The estimated curve obtained by the first order model is too wiggly. The curve obtained by the fourth order model is too sensitive at the end of the data. The curves obtained by the second or the third order models look quite reasonable. The order selected by the AIC criterion is $k = 2$.

Table 7.5: Values of τ^2 and -2 log-likelihood for different values of differential equation order k.

differential order	τ^2	log-likelihood
1	1.15	65.478
2	13.2	63.636
3	109	70.863
4	914	79.545

7.4 A Signal Extraction Problem

This is one of the time series modeling applications in which an observed time series is decomposed into several components. In this particular case, the problem of extracting a microearthquake signal from background noise in the presence of added observation noise, with both the relatively small amplitude microearthquake and background noise is modeled by independent AR processes.

Earthquake ground movement is conventionally recorded by a seismometer. The earth's surface is actually in continuous movement due to a variety of natural forces such as the continuation or ringing effects of past earthquakes, and by wave, wind and tide effects for seismometers located near ocean shores and by a variety of human induced sources. We consider the sum of these background effects to be a stationary process. If the amplitude of the earthquake signal is very small, it will be quite

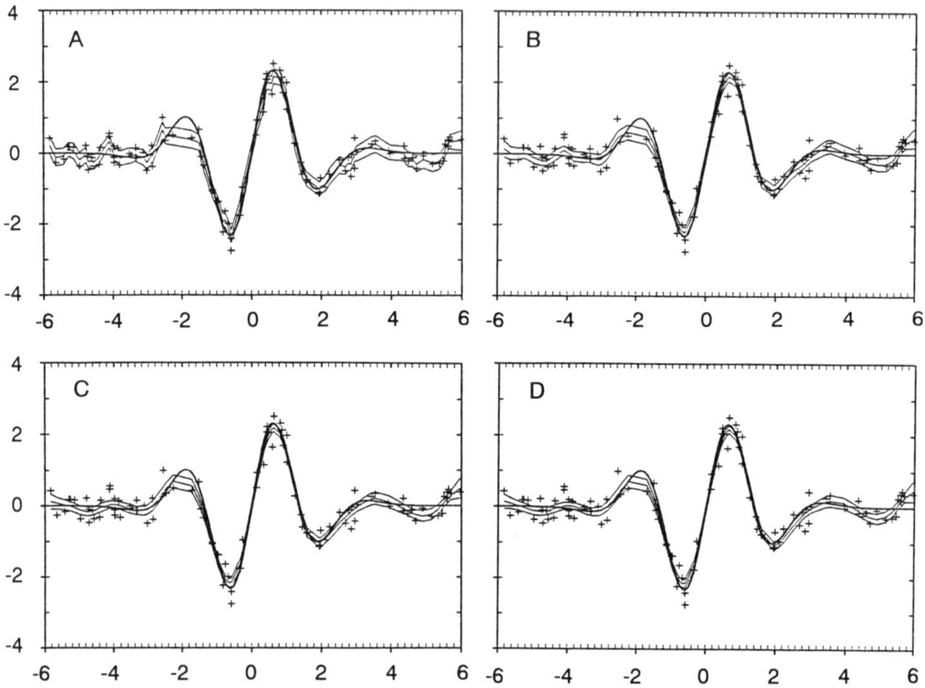

Figure 7.3: Smoothing unequally spaced data, for model orders 1,. . . ,4.

difficult to distinguish it from the background noise. A second source of difficulty in modeling this earthquake plus background noise data is that if we consider the microearthquake to be represented by an AR model it will have to be one with a time varying innovations variance. A conventional approach to modeling nonstationary covariance time series is to segment the series and and to model each segment separately. If the segment duration is short enough to follow the change of variance it might not be large enough to yield a statistically reliable estimate. Here, a local likelihood approach is taken to model the changing variance. That approach eliminates the messy problem of choosing segment lengths. An alternative non-Gaussian method is provided in Chapter 10.

The Earthquake Model

The model we consider for the decomposition of the microearthquake in background noise signal is

$$y_n = r_n + s_n + \varepsilon_n \tag{7.3}$$

where r_n represents the background noise process and s_n represents the signal process. Also, for convenience in state space modeling, ε_n is introduced as an assumed zero-mean normally distributed white observation noise, $\varepsilon_n \sim N(0, \sigma^2)$ and r_n and s_n are each (independent) AR processes. That is,

$$\begin{aligned} r_n &= \sum_{i=1}^{m} a_i r_{n-i} + u_{r,n}, \\ s_n &= \sum_{i=1}^{l} b_i s_{n-i} + u_{s,n}. \end{aligned} \tag{7.4}$$

Here, $u_{r,n}$ and $u_{s,n}$ are normally distributed white noise sequences with $u_{r,n} \sim N(0, \tau_1^2)$ and $u_{s,n} \sim N(0, \tau_2^2)$. The models in (7.3) and (7.4) can be combined and represented in the usual state space model form

$$\begin{aligned} x_n &= F x_{n-1} + G w_n \\ y_n &= H x_n + \varepsilon_n \end{aligned} \tag{7.5}$$

where x_n is an $(m + \ell)$-dimensional state vector, the system or process noise w_n is a two-dimensional zero mean normally distributed white noise vector that is independent of the ε_n process and y_n is the observation at time n. The matrices F, G and the process noise and observation noise processes are defined by

$$\begin{aligned} F &= \left[\begin{array}{cccc|cccc} a_1 & a_2 & \cdots & a_m & & & & \\ 1 & & & & & & & \\ & \ddots & & & & & & \\ & & 1 & 0 & & & & \\ \hline & & & & b_1 & b_2 & \cdots & b_\ell \\ & & & & 1 & & & \\ & & & & & \ddots & & \\ & & & & & & 1 & 0 \end{array}\right], \quad G = \left[\begin{array}{cc} 1 & 0 \\ 0 & 0 \\ \vdots & \vdots \\ 0 & 0 \\ \hline 0 & 1 \\ 0 & 0 \\ \vdots & \vdots \\ 0 & 0 \end{array}\right], \\ H &= [\, 1\ 0\ \cdots\ 0 \mid 1\ 0\ \cdots\ 0\,], \\ & \left[\begin{array}{c} w_n \\ \varepsilon_n \end{array}\right] \sim N\left(\left[\begin{array}{c} 0 \\ 0 \\ 0 \end{array}\right], \left[\begin{array}{ccc} \tau_1^2 & 0 & 0 \\ 0 & \tau_2^2 & 0 \\ 0 & 0 & \sigma^2 \end{array}\right] \right). \end{aligned} \tag{7.6}$$

In fact, using the state space method of fitting an AR model to data described in Section 7.1, the parameters of the background signal, were estimated independently of the microearthquake from a segment of data preceding the arrival of the microtremor signal. Those parameters are m, the background signal AR order and $a_1, \ldots, a_m$, the background AR process coefficients, the innovations variance τ_2^2 and observation noise variance σ^2, respectively. Then the model in (7.6) is fitted using the Kalman filter algorithm described in Section 5.2 with the background AR process parameters known and only the microtremor AR model parameters $\ell, b_1, \ldots, b_\ell$ and τ_2^2 need to be estimated.

7.4.1 Estimation of the Time Varying Variance

The variance of the second autoregressive signal is related to the magnitude of the earthquake signal and is time varying. That variance, τ_2^2, is zero before the earthquake signal arrives, becomes positive and then decreases back to zero as the tail of the earthquake signal (coda) dies out. This parameter has the role of a signal to noise ratio, we estimate it by a local likelihood method. From Section 5.2, the log-likelihood of the state space model is

$$\ell(\theta) = \log f(y_1, \ldots, y_N) = -\frac{1}{2}\left\{N \log 2\pi + \sum_{n=1}^{N} \log v_n + \sum_{n=1}^{N} \frac{e_n^2}{v_n}\right\}, \tag{7.7}$$

where v_n the observation noise variance at time n, and e_n the residual at time n, are as defined in Section 5.2. From this expression the quantity

$$\ell_n = \frac{1}{2}\left\{\log 2\pi v_n + \frac{e_n^2}{v_n}\right\} \tag{7.8}$$

has a natural interpretation as the log-likelihood of the model at time n. On the assumption that the true model changes quite slowly and smoothly, then correspondingly the log-likelihood should also change quite smoothly. For the smoothing of the log-likelihood we use the procedure used for the smoothing of a chi-square variate that appears later in Chapter 10. For completeness in this section, that procedure is briefly described here. Define, $t_m = \log z_m + \gamma, m = 1, \ldots, n/2$ where $z_m = (\ell_{2m-1} + \ell_{2m})/2$ and $\gamma \simeq 0.57721$ is the Euler constant, (Davis and Jones 1968). Thus t_m is almost normally distributed with moments $E[t_m] \simeq \log c_m, Var(t_m) \simeq \pi^2/6$ where c_m is the average of the true expected log-likelihood at time $2m-1$ and $2m$. Upon approximating the distribution of t_m by a normal distribution and using the smoothness priors trend model, $t_m = 2t_{m-1} - t_m + v_m, v_m \sim N(0, \rho^2)$ we obtain the smoothed value $\hat{t}_m$ of t_m. The local log-likelihood of that model is then obtained by $\hat{\ell}_{2m} = \hat{\ell}_{2m-1} = \exp\{\hat{t}_m\}$.

For simplicity here, τ_2^2, the value of the innovations variance of the earthquake signal at any one instant in time, is only allowed to take on discrete values. (A more sophisticated procedure which permit a continuous estimate of time series variance is in Chapter 10.) Then, a summary of the procedure used here for the estimation of the changing variance based on the local likelihood method is as follows:

1. For each possible value of $\tau_2^2 = c_0 2^{-k}, k = 1, \ldots, k_{max}$.
   (a) Compute $\ell_{k,n},\ n = 1, \ldots, N$.
   (b) Obtain the smoothed value $\hat{\ell}_{k,n}, n = 1, \ldots, N$.
2. For each time instant $n,\ n = 1, \ldots, N$
   (a) Find k^* for which $\ell_{k,n},\ k = 1, \ldots, k_{max}$ is maximum.
   (b) Set $\tau_{2,n}^2 = c_0 2^{k^*}$.

7.4.2 Separating a Micro Earthquake From Noisy Data

Figure 7.4 shows the seismogram record of the North-South component of a foreshock of the 1982 Urakawa-Oki earthquake observed at the Erimo station of the Research Center for Earthquake Prediction, Hokkaido University, Japan at 11:32 AM, March 21 1982, (Suzuki *et al.* 1986, Takanami 1991). (This earthquake record and many others are available to the public through the ISMLIB in www home page `http://www.ism.ac.jp`) The magnitude was $M = 2.30$, and the epicenter was about 50 km away from the location of the recording seismometers which in turn are located 0.5 km from the shoreline of the Pacific. The observed signal is miniscule relative to the background noise. The original record of 83.85 seconds duration was regularly sampled at 0.01084 second intervals (92.3 samples/second), and subsequently 4 consecutive observations were averaged to produce the individual observations in Figure 7.4A. As a result of this sampling scheme, the sampling error variance of this record is 1/48.

The background noise was modeled using the first 500 observations of the data shown in Figure 7.4A by the state space AR modeling procedure described earlier in this chapter, Section 7.1. An AR(7) model was the best AIC fit model. Subsequently using the fitted AR model parameters, $\hat{a}_1, \ldots, \hat{a}_7, \hat{\tau}_1^2, \hat{\sigma}^2$, the AR(7) + AR($\ell$) + *white noise* model was fitted to the data from $n = 725$ to $n = 1450$ where the earthquake signal apparently exists. The second AR model, the one for the earthquake signal is estimated by the Kalman filter algorithm for computing the likelihoods and the MAICE procedure for selecting the AIC best AR earthquake signal model. An AR(7) model was the AIC best model for the earthquake data.

Figures 7.4B through 7.4E respectively illustrate the derived background noise, the extracted microearthquake signal, the observation noise and the estimated time changing variance. It is clear from Figures 7.4A, 7.4B and 7.4C that the earthquake and background noise levels are of comparable magnitude, and that the earthquake process variance $\tau_{2,n}^2$ is never very large. That variance was only permitted to take on discrete values. It is also clear that the estimation of the time varying variance by the local likelihood method appears to work well.

In effect, from a frequency domain interpretation, in each time interval, we have decomposed the spectrum of the times series into the three components

$$S_{f,n} = \tau_1^2 S_{1,f} + \tau_{2,n}^2 S_{2,f} + \sigma^2. \tag{7.9}$$

The inherent complexity of this problem is also apparent from Figure 7.4F in which the stationary spectrums of the background and the earthquake processes are shown. There is considerable overlap in the two spectra. The two-step state space time domain analysis has been successful here. It is unlikely that a frequency domain approach would work as well. Additional more detailed examples of the extraction of microtremors from background noise appear in Kitagawa and Takanami (1985).

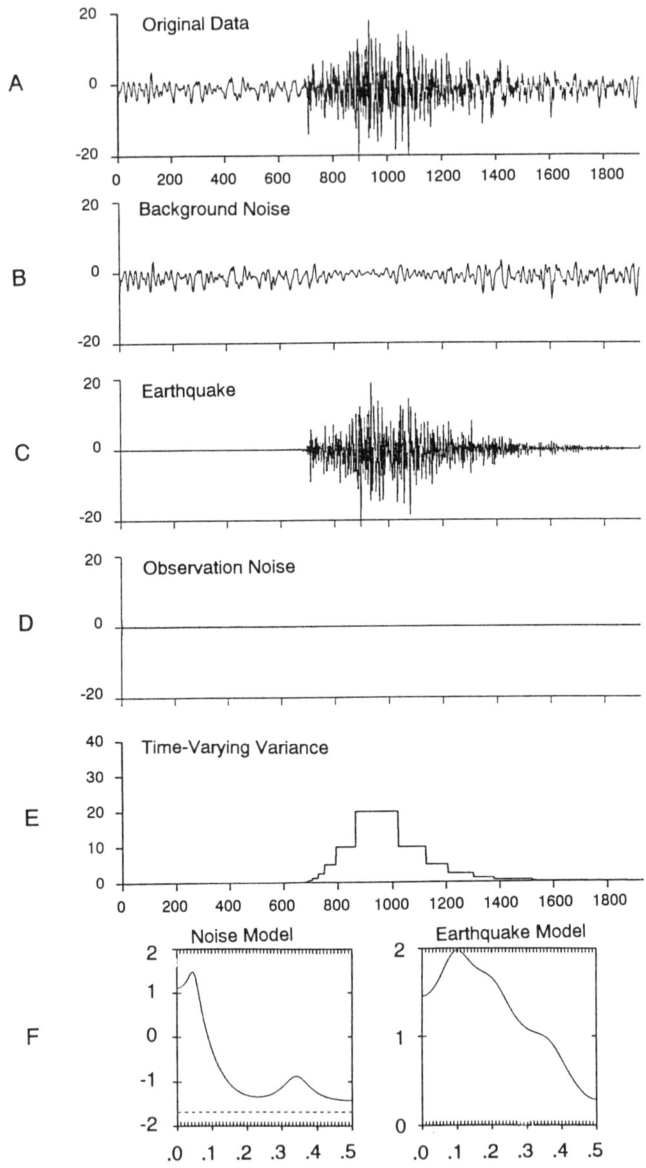

Figure 7.4 A: Seismogram record, Urakawa-Oki earthquake, March 21, 1982. B: Estimated background signal. C: Extracted microearthquake signal. D: Estimated observation noise. E: Estimated time varying signal variance. F: Power spectra, background and earthquake signals.

7.4.3 A Second Example

Here we illustrate a related but slightly more complex signal extraction problem. In this case, the earthquake signal is stronger and we attempt to extract both the seismic P-wave and S-wave as well as the background noise from the observed data. (Both the faster primary or P-wave and secondary or S-wave are elastic waves that propagate from within the earth. The alternating compression and dilation P-wave motion is like that of a sound wave. The up-and-down and side-to-side S-wave motion, perpendicular to the direction of wave travel is the wave motion that is so damaging to structures, Bolt 1993.) The analysis in this case requires an independent processing of three AR processes, one each for the background signal, the P-wave and the S-wave. The results are shown in Figure 7.5. In that illustration we see the original data, the extracted background noise, the P-wave and the S-wave, as well as the the estimated time changing innovations variances for the P-wave and S-wave. The power spectral densities of the background signal and the P-wave and S-wave are also shown. The considerable overlap in the P-wave and S-wave spectral densities are an indication of the inherent difficulty in this problem. Very likely, time domain methods, such as the ones shown here, are the most effective way of treating the overlapping spectrum signal decomposition problem.

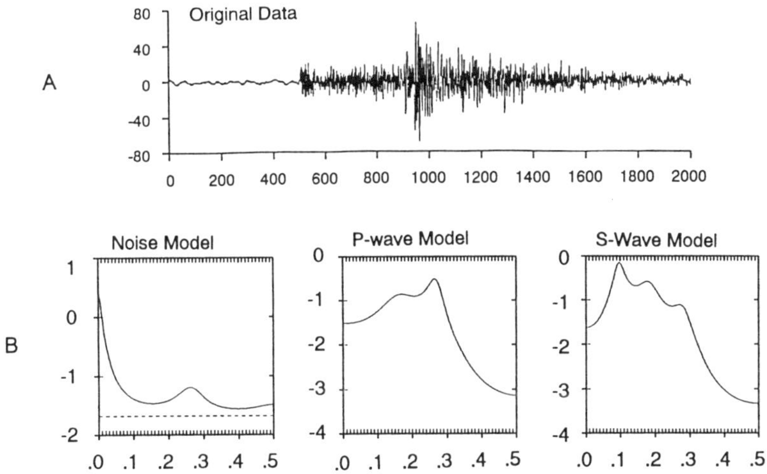

Figure 7.5 A: Seismogram record. B: Power spectra, backgound signal and P-wave and S-wave.

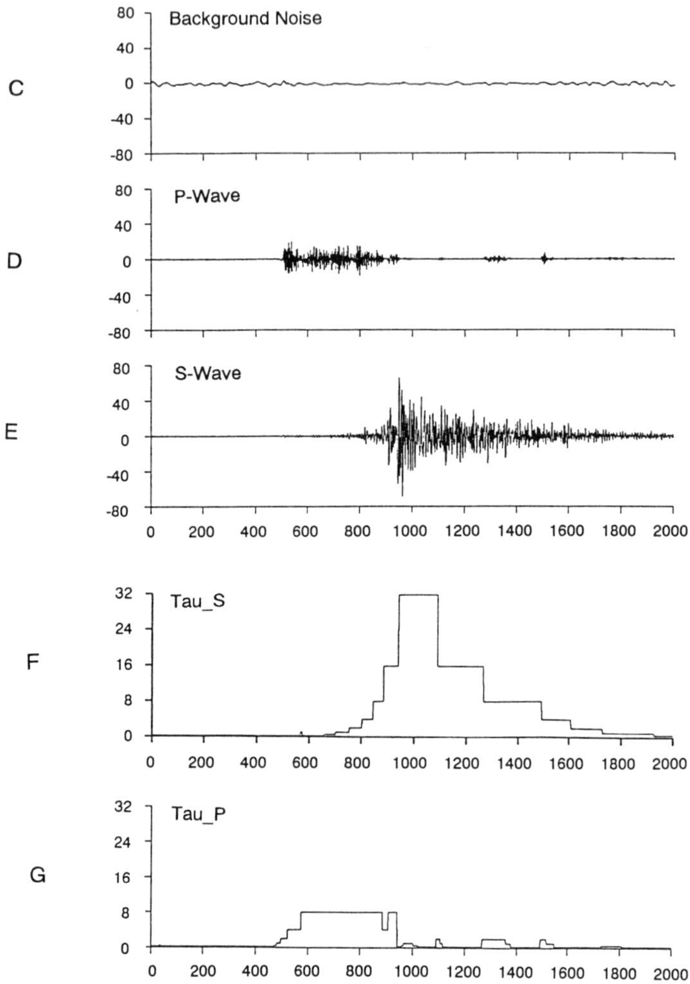

Figure 7.5 C: Estimated background noise. D: Extracted P-wave signal. E: Extracted S-wave signal. F: Time varying innovations variance, P-wave, G: Time varying innovations variance, S-wave.

8

Modeling Trends

In this chapter we consider the modeling of nonstationary mean time series, or *trends*, by state space methods. Initially we consider a Gaussian disturbances state space smoothness priors modeling that is appropriate when the trend is a smooth function. An example of the analysis of simulated data illustrates the method of analysis. Secondly we model simulated time series data with local trend and globally stochastic components. A generalization of the latter procedure is shown in the analysis of a collection of 22 years of daily maximum temperatures recorded in Tokyo. Evidence is shown to justify modeling this data with a common trend plus an individual annual AR processes model. The analysis of the Tiao and Tsay (1985, 1989) flour price data, another example of the analysis of multiple time series with common trend and individual AR process is also shown. On the basis of these examples, the common trend individual AR process modeling is offered as a candidate parsimonious model of multiple nonstationary mean time series.

Subsequently the trend model is generalized by considering the modeling of trends with abrupt discontinuities. In this case the system noise and the observation noise, (individually and together), are permitted to be non-Gaussian. The results of three different non-Gaussian state space analysis methods; numerical integration, Gaussian sum approximations and the Monte Carlo path tracing method, (each method was introduced in Chapter 6), are shown and compared in the analysis of a simulated data example.

The concept of nonstationary mean time series includes time series with seasonal components as well as local polynomial trend and potentially globally stochastic AR components. This topic is of considerable practical importance in the analysis of econometric, meteorological and other physical data. This topic is treated separately in Chapter 9. The treatment of outliers in the modeling of real physical data with trends, another important topic, is also treated by general state space methods under the somewhat more general setting of time series with trends and seasonalities in Chapter 9.

8.1 State Space Trend Models

Here we return to the original problem posed by Whittaker, that of modeling time series with nonstationary mean values or trends. The term trend usually refers to relatively long-run movements in the series. In contrast with the smoothness priors-least squares computational method of modeling trends in Chapter 4 here, nonstationary mean time series are modeled by state space models. The observation model of such a time series is given by

$$y_n = t_n + \varepsilon_n, \quad n = 1, \ldots, N \tag{8.1}$$

where $t_n, n = 1, \ldots, N$ represents the nonstationary mean value function of the time series and is the function to be estimated and $\varepsilon_n, n = 1, \ldots, N$ represents an added i.i.d. noise component sequence. In Chapter 4, the trends were assumed to be continuous functions and the added noise was assumed to be i.i.d. Gaussian distributed. In contrast with those assumptions in this chapter we permit the trend to be discontinuous. Also, the additive noise model is generalized to permit i.i.d. noise that is non-Gaussian and is defined by a parametric distribution or even more generally not defined by a parametric distribution.

As before, we consider trend models to be of the form $\nabla^k t_n = w_n$ with w_n assumed to be a normally distributed zero-mean sequence with unknown variance τ^2. For example for $k = 1, 2, 3$, those models are:

$$\begin{aligned} t_n &= t_{n-1} + w_n, \\ t_n &= 2t_{n-1} - t_{n-2} + w_n, \\ t_n &= 3t_{n-1} - 3t_{n-2} + t_{n-3} + w_n. \end{aligned} \tag{8.2}$$

The general state space model is

$$\begin{aligned} x_n &= F_n x_{n-1} + G_n w_n \\ y_n &= H_n x_n + \varepsilon_n, \end{aligned} \tag{8.3}$$

where in general the joint distribution of the system noise and observation noise is

$$\begin{bmatrix} w_n \\ \varepsilon_n \end{bmatrix} \sim P\left(\begin{bmatrix} 0 \\ 0 \end{bmatrix}, \begin{bmatrix} Q_n & 0 \\ 0 & R_n \end{bmatrix} \right). \tag{8.4}$$

In (8.4), the notation P, denotes an arbitrary zero mean and uncorrelated joint distribution for w_n, ε_n with potentially time dependent variances.

In the general state space trend model, define the state vector x_n and the matrices F_n, G_n, H_n by

$$x_n = \begin{bmatrix} t_n \\ t_{n-1} \\ \vdots \\ t_{n-p+1} \end{bmatrix}, \quad F_n = \begin{bmatrix} c_1 & \cdots & c_{p-1} & c_p \\ 1 & & & 0 \\ & \ddots & & \vdots \\ & & 1 & 0 \end{bmatrix}, \quad G_n = \begin{bmatrix} 1 \\ 0 \\ \vdots \\ 0 \end{bmatrix}, \quad H_n^T = \begin{bmatrix} 1 \\ 0 \\ \vdots \\ 0 \end{bmatrix} \tag{8.5}$$

where the c_i's are binomial coefficients, $c_i = (-1)^{i-1}\binom{p}{i}$, which satisfy $c_1 + c_2 + \cdots + c_p = 1$. (The sequence of p-th order binomial coefficients is in concordance with the sequence of coefficients in the p-th order differences.) In this model the first state component, t_n, can be considered as the trend of the nonstationary mean time series y_n. With $p = 1$ and $c_1 = 1$, t_n obeys a random walk model $t_n = t_{n-1} + w_n$. This model will be satisfactory if the trend is approximately constant in short time spans. In this case we call t_n a locally constant trend. For $p = 2$, $c_1 = 2$ and $c_2 = -1$, we get a locally linear trend which characterizes a trend component with a locally fixed slope. Similarly with $p = 3$, $c_1 = 3, c_2 = -3$ and $c_3 = 1$ we get a locally quadratic trend which is suited to a smoothly changing trend component.

8.2 State Space Estimation of Smooth Trend

Several different examples that involve smooth trend estimation by state space smoothness priors methods and Gaussian disturbances are shown in this section. Here we consider the $N = 200$ simulated data smooth trend example that was treated by least squares methods in Chapter 4. Results similar to those obtained in Chapter 4 are obtained here. In a next level of complexity, the analysis of a simulated data set consisting of a smooth trend plus a globally stochastic AR component observed in the presence of additive Gaussian noise is shown. A third example is a more complex version of the last example in which a set of the 22 years of daily records of the maximum daily temperature in Tokyo from 1971 to 1992 is analyzed by a common trend plus 22 individual annual AR components model. A similar model analysis is made of the Tiao and Tsay (1985, 1989) trivariate flour price data.

We do note here that the k-th order state space trend models have transition matrices systems with zeros on the unit circle and that from Chan *et al.* (1984), the Kalman filter computations are stable in this situation. That result is also relevant for the Gaussian modeling of time series with both smooth trend and smooth seasonal components (Chapter 9).

8.2.1 Estimation of a Smooth Trend

Here we consider the same $N = 200$ simulated data asymmetrically truncated Gaussian shaped smooth trend function contaminated by additive observation Gaussian noise that was treated in Chapter 4. The problem is: Given the noisy observations $y_n, n = 1, \ldots, N$, estimate the unknown smooth function that is in the noise, i.e. specify $\hat{t}_{n|N}, n = 1, \ldots, N$.

In detail, if we consider a locally constant trend model with $p = 1$ the resulting state space model is

$$\begin{aligned} x_n &= x_{n-1} + w_n \\ y_n &= x_n + \varepsilon_n, \end{aligned} \tag{8.6}$$

where the state x_n is defined by $x_n = t_n$.

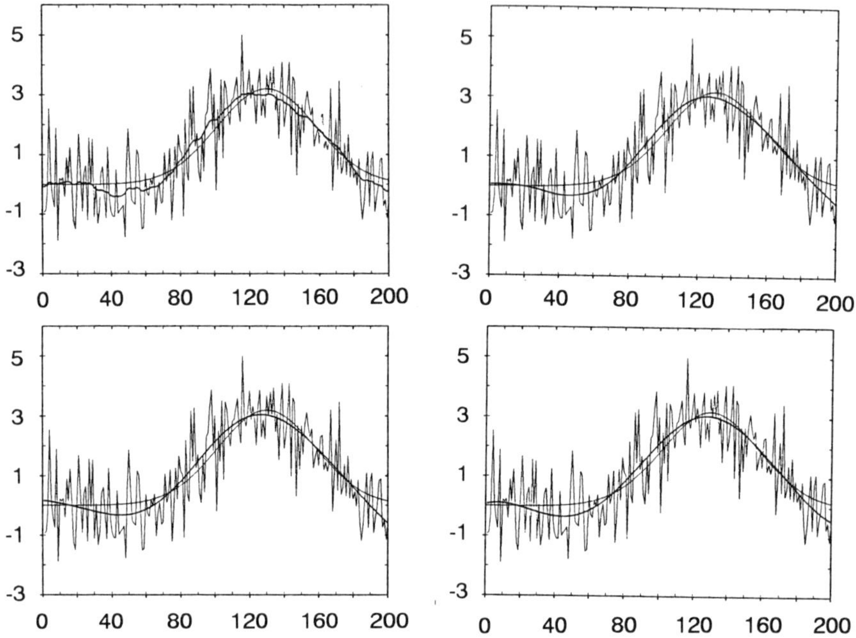

Figure 8.1: Optimal trend modeling, trend model orders 1,. . . ,4.

If we consider a locally linear trend model with $p = 2$, the state is $x_n = (t_n, t_{n-1})^T$ and the corresponding state space model is

$$\begin{aligned} x_n &= \begin{bmatrix} 2 & -1 \\ 1 & 0 \end{bmatrix} x_{n-1} + \begin{bmatrix} 1 \\ 0 \end{bmatrix} w_n, \\ y_n &= \begin{bmatrix} 1 & 0 \end{bmatrix} x_n + \varepsilon_n. \end{aligned} \tag{8.7}$$

Here in our analyses of both the trend order model $p = 1$ and $p = 2$ cases, the state or system noise vector and observation noise w_n and ε_n are assumed to be Gaussian distributed, independent of each other and i.i.d. with zero mean and diagonal covariance matrix. That is,

$$\begin{bmatrix} w_n \\ \varepsilon_n \end{bmatrix} \sim N\left(\begin{bmatrix} 0 \\ 0 \end{bmatrix}, \begin{bmatrix} \tau^2 & 0 \\ 0 & \sigma^2 \end{bmatrix} \right). \tag{8.8}$$

Equation (8.5) is the general state space model that was used for analysis. Trend model orders $k = 1, \ldots, 4$ were considered. In each case, the single hyperparameter τ^2 was estimated by a simple discrete search algorithm, (Section 2.3). The maximum likelihood estimate of σ^2 was obtained free of computational cost within the Kalman filter algorithm. Numerical results for those models which include the values of the hyperparameter, the estimated observation variance, -2 log-likelihood, (LL), and the AIC are shown in Table 8.1. Illustrations of the hyperparameter best computational

Table 8.1: Smooth trend model order and summary statistics.

k	τ^2	σ^2	LL	AIC
1	0.334×10^{-1}	0.871	-288.18	582.36
2	0.790×10^{-4}	0.875	-283.77	575.54
3	0.180×10^{-6}	0.873	-285.25	580.51
4	0.383×10^{-9}	0.870	-287.23	586.47

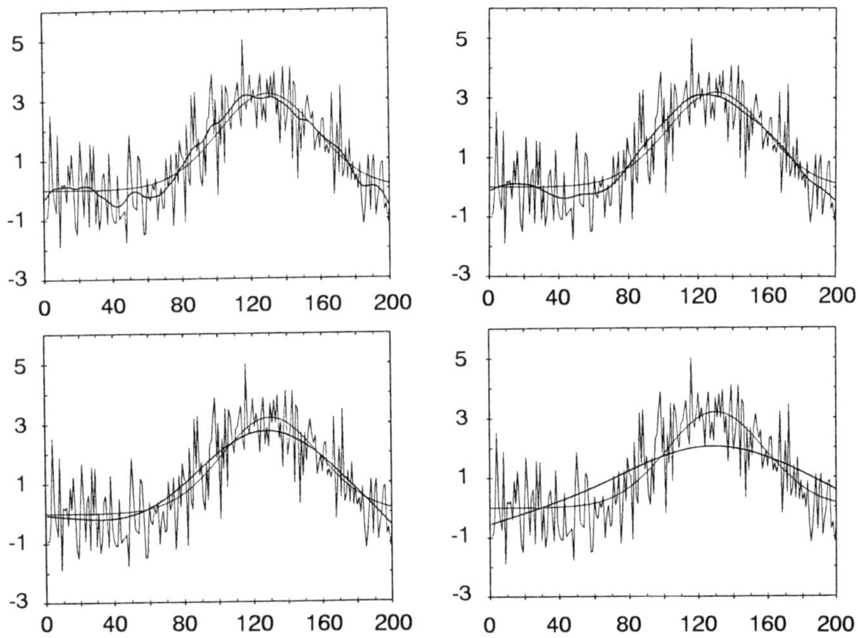

Figure 8.2: Trend model order 2, for different hyperparameter values.

results for the different trend model orders are shown in Figure 8.1. The $k = 2$ order trend model is the AIC best model, the $k = 1$ model is too wiggly and it is interesting to note that the visual appearance of the $k = 2, 3, 4$ results are indistinguishable.

Then, similar to the illustration in Section 4.1, graphical results for the fixed trend model order $k = 2$ for the values of the hyperparameter $\tau^2 = (8.0 \times 10^{-8}, 8.0 \times 10^{-7}, 8.0 \times 10^{-5}, 8.0 \times 10^{-3})$, are shown in Figure 8.2. The critical role of the hyperparameter is clearly exhibited in this example. The estimated trend is the most irregular for the smallest value of τ^2 and becomes increasingly smooth with increasing values of τ^2.

8.2.2 Smooth Trend Plus Autoregressive Model

Here we consider the modeling of an artificially generated data set that consists of the sum of a smooth mean value function, a second order AR process and a small amount of a zero mean, constant variance normally distributed observation noise. The model for generating this data set is

$$y_n \;=\; t_n \;+\; v_n \;+\; \varepsilon_n. \tag{8.9}$$

The trend process t_n was generated by

$$t_n \;=\; \frac{8}{2\pi}\exp\left\{-\frac{(n-335)^2}{20000}\right\}, \qquad n=1,\ldots,500. \tag{8.10}$$

The autoregressive process, v_n was generated by

$$v_n \;=\; 1.394v_{n-1} - 0.752v_{n-2} + u_n. \tag{8.11}$$

Both the observation noise and the AR process noise are taken to be Gaussian distributed, with $\varepsilon_n \sim N(0, 0.02^2)$, and $u_n \sim N(0, 0.02)$, $n = 1, \ldots, N$

A state space model for modeling this trend plus AR component set is

$$F_n = \left[\begin{array}{cccc|cccc} c_1 & \cdots & c_{k-1} & c_k & & & & \\ 1 & & & & & & & \\ & \ddots & & & & & & \\ & & 1 & 0 & & & & \\ \hline & & & & a_1 & \cdots & a_{p-1} & a_p \\ & & & & 1 & & & \\ & & & & & \ddots & & \\ & & & & & & 1 & 0 \end{array}\right], \quad G_n = \left[\begin{array}{cc} 1 & 0 \\ 0 & 0 \\ \vdots & \vdots \\ 0 & 0 \\ \hline 0 & 1 \\ 0 & 0 \\ \vdots & \vdots \\ 0 & 0 \end{array}\right], \tag{8.12}$$

$$H_n \;=\; [\,1\;0\;\cdots\;0\;|\;1\;0\;\cdots\;0\,]$$

$$x_n \;=\; [t_n, \ldots, t_{n-k+1}, y_n, \ldots, y_{n-p+1}]^T.$$

In (8.12), k is the difference order of the trend model, the c_i, $i = 1, \ldots, k$ are the trend model parameters and p is the order of the AR model with a_i, $i = 1, \ldots, p$ the AR model parameters. The dimensions of the state vector x_n, the state transition matrix F, the input matrix G and the output matrix H are respectively $(k+p) \times 1$, $(k+p) \times (k+p)$, $(k+p) \times 2$ and $1 \times (k+p)$.

Table 8.2 shows the summary statistics, including the log-likelihood, number of fitted parameters and the AIC's of models with combinations of trend orders $k = 1, \ldots, 4$ and AR model orders, $p = 1, \ldots, 4$ fitted to this artificially generated data. The model with local linear trend ($k = 2$) and second order AR process ($p = 2$) attains the minimum of the AIC.

Viewing downward, Figure 8.3 respectively shows the AIC best fitted trend superimposed on the observed data, the estimated noise component, the fitted AR component, and the fitted trend plus fitted AR model. In this situation, the observation noise was intentionally minuscule and as a result, the original data and best fitted trend plus best fitted AR model are visually indistinguishable.

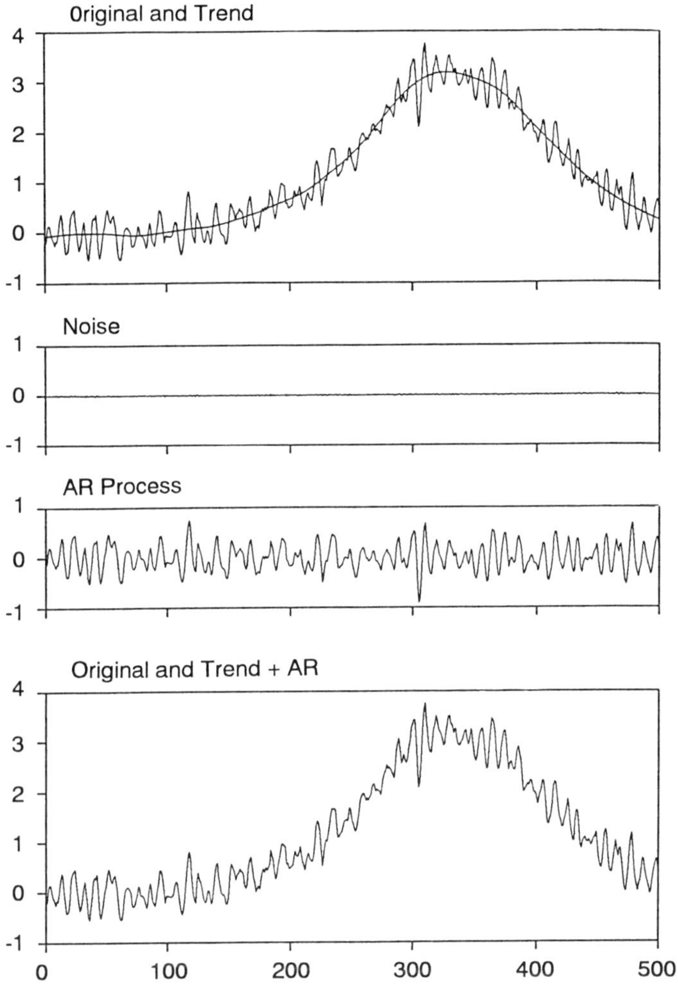

Figure 8.3 A: Trend plus AR model data and superimposed AIC best trend. B: The estimated noise component. C: The fitted AR component. D: The fitted trend plus fitted AR component.

Table 8.2: Summary statistics of Trend plus AR model

Trend order	AR order	AIC	$\hat{\tau}_1^2$	$\hat{\tau}_2^2$	$\hat{\sigma}^2$
	0	-323.67			
	1	-328.42	2.22×10^4	5.36×10^4	3.80×10^{-7}
1	2	-669.19	3.56×10^1	6.43×10^1	1.11×10^{-4}
	3	-698.52	1.82×10^3	3.17×10^3	2.26×10^{-6}
	4	-699.67	6.21×10^2	8.48×10^2	7.22×10^{-6}
	0	150.87	5.15×10^{-5}		6.97×10^{-2}
	1	-337.76	1.02×10^1	7.69×10^4	3.57×10^{-7}
2	2	-727.39	5.96×10^{-3}	1.79×10^1	5.63×10^{-4}
	3	-725.52	6.00×10^{-3}	1.79×10^1	5.63×10^{-4}
	4	-725.21	6.02×10^{-3}	1.79×10^1	5.57×10^{-4}
	0	152.69	3.51×10^{-8}		7.03×10^{-2}
	1	-334.58	2.61×10^{-4}	3.35×10^3	8.16×10^{-6}
3	2	-725.71	4.61×10^{-6}	1.93×10^1	5.30×10^{-4}
	3	-724.29	2.82×10^{-5}	1.33×10^2	8.79×10^{-5}
	4	-723.03	4.61×10^{-6}	1.95×10^1	5.25×10^{-4}

8.3 Multiple Time Series Modeling: The Common Trend Plus Individual Component AR Model

In this section two examples of the analyses of multiple time series with common trends and individual AR processes are shown. The first is the analysis of the maximum daily temperatures recorded in Tokyo in the 22 year interval from 1971 through 1992. The second is the analysis of Tiao and Tsay (1989) flour price data. We suggest the common trend model as a candidate parsimonious model for the modeling of nonstationary multiple time series.

8.3.1 Maximum Daily Temperatures 1971-1992

A generalization of the trend plus AR process model in the analysis of a set of real data, the record of the maximum daily temperature in degrees centigrade from January 1, 1971 to December 31 1992, is considered next. Some examples of that individual annual data, and corresponding superimposed AIC best fitted trend, (typically trend order $k = 1$), are illustrated in Figure 8.4. Viewed downward, Figure 8.5 shows the 22 superimposed individual annual estimated trends, a scatter diagram of the maximum daily temperatures for each year versus day number ($1 \leq n \leq 365$), and the overall common fitted trend, (from a model to be described below). The evidence in Figure 8.5 suggests that the 22 years of daily maximum temperature data can be well modeled as a common trend plus individual annual AR component model.

An observation model for the set of 22 years of data with common trend and

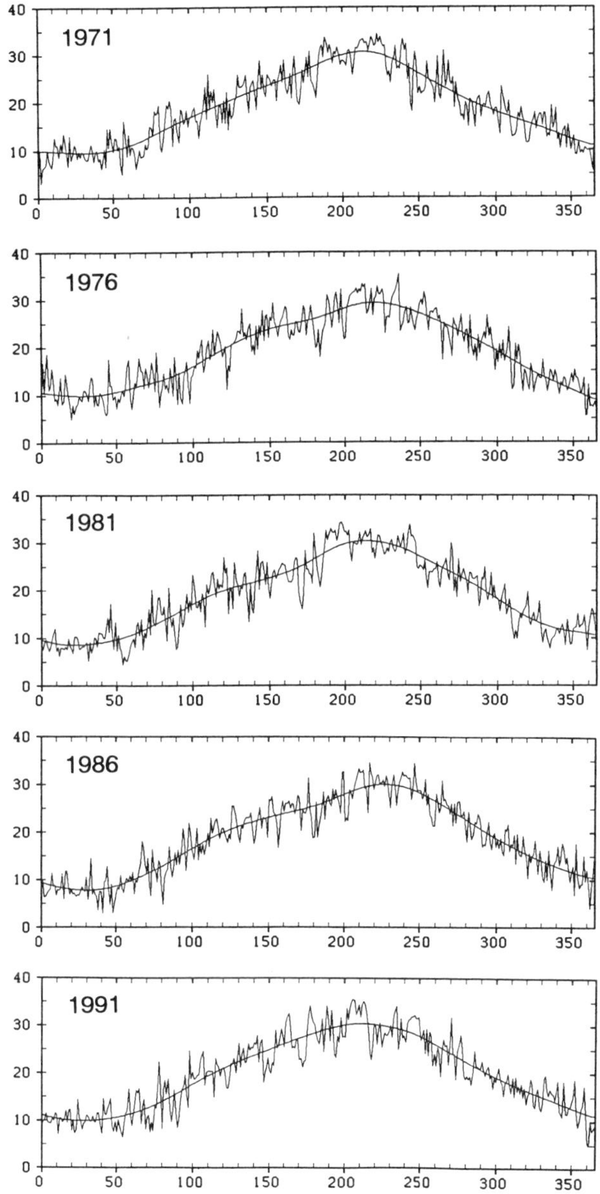

Figure 8.4: Samples of maximum daily temperatures and superimposed fitted trends.

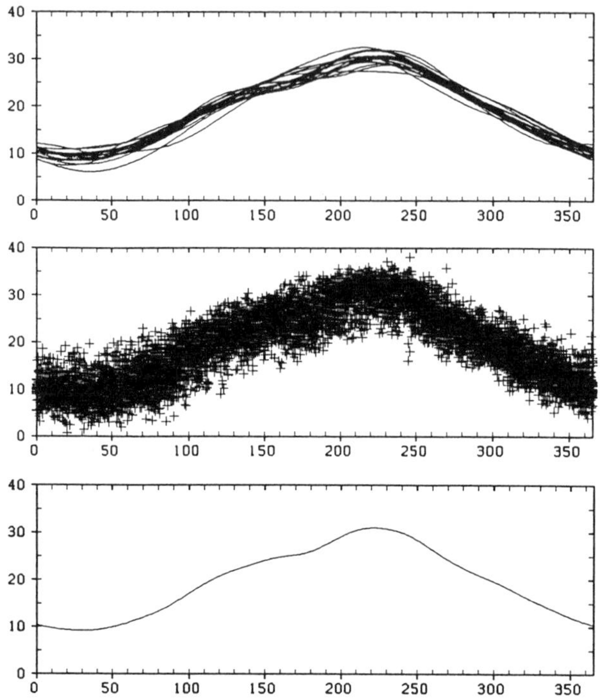

Figure 8.5: Superimposed annual trends, scatter diagram of all the data, and the fitted common trend.

individual AR components taking the form

$$y_{i,n} \;=\; t_n \;+\; v_{i,n} \;+\; \varepsilon_{i,n}, \quad i = 1, \ldots, 22 \tag{8.13}$$

where t_n, $v_{i,n}$, $\varepsilon_{i,n}$ are respectively the common trend, the individual annual AR processes and the unobserved added noise process for each year.

An alternative approach to modeling this data, is represented in the observation model for the set of 22 years of data with common trend and individual AR components taking the form

$$y_n \;=\; t_n \;+\; v_n \;+\; \varepsilon_n, \tag{8.14}$$

where t_n is the common trend that satisfies the seasonal condition, $t_n = t_{n-365}$ and v_n, ε_n are respectively the AR process and the unobserved added noise process. In this model t_n is treated as an annual "seasonal" component with period length 365 and we might apply the seasonal adjustment model treated in Chapter 9. Seasonal adjustment of daily data would require a state dimension larger than 365, and while it is feasible, it is more convenient to use the multivariate model form described above.

The common trend component, individual annual component AR processes and noise processes for the first described modeling approach are respectively given by

$$\begin{aligned} \nabla^k t_n &= \eta_n, \quad t_n = t_{n-365}, \\ v_{i,n} &= \sum_{j=1}^{p} \alpha_{i,j} v_{i,n-j} + \xi_n^i, \quad v_{i-1,n} = v_{i,n-365} \\ \varepsilon_{i,n} &\sim N(0, \sigma_i^2), \quad i = 1, \ldots, 22. \end{aligned} \tag{8.15}$$

In (8.15), the trend system noise and the AR components innovations noise, $\eta_n \sim N(0, \sigma_n^2)$, $\xi_n^i \sim N(0, \tau_i^2)$ are all assumed independent and independent of each other. A state space model with common trend order $k = 1$ and AR order 2 (for each individual time series) for this data set is given by

$$\begin{bmatrix} t_n \\ p_n^1 \\ p_{n-1}^1 \\ \vdots \\ p_n^\ell \\ p_{n-1}^\ell \end{bmatrix} = \left[\begin{array}{c|cc|c|cc} 1 & & & & & \\ \hline & a_1 & a_2 & & & \\ & 1 & 0 & & & \\ \hline & & & \ddots & & \\ \hline & & & & a_1 & a_2 \\ & & & & 1 & 0 \end{array}\right] \begin{bmatrix} t_{n-1} \\ p_{n-1}^1 \\ p_{n-2}^1 \\ \vdots \\ p_{n-1}^\ell \\ p_{n-2}^\ell \end{bmatrix} + \left[\begin{array}{c|c|c|c} 1 & & & \\ \hline & 1 & & \\ & 0 & & \\ \hline & & \ddots & \\ \hline & & & 1 \\ & & & 0 \end{array}\right] \begin{bmatrix} v_n \\ u_n^1 \\ \vdots \\ u_n^\ell \end{bmatrix},$$

$$\begin{bmatrix} y_n^1 \\ \vdots \\ y_n^\ell \end{bmatrix} = \left[\begin{array}{c|cc|c|cc} 1 & 1 & 0 & & & \\ \hline \vdots & & & \ddots & & \\ \hline 1 & & & & 1 & 0 \end{array}\right] \begin{bmatrix} t_n \\ p_n^1 \\ p_{n-1}^1 \\ \vdots \\ p_n^\ell \\ p_{n-1}^\ell \end{bmatrix} + \begin{bmatrix} w_n^1 \\ \vdots \\ w_n^\ell \end{bmatrix}. \tag{8.16}$$

The dimensions of the state vector x_n, the state transition matrix F, the input matrix G and the output matrix H with common trend model order $k = 1$, and the number of years $\ell = 22$ and common AR model order p are respectively $(k + \ell \cdot p) \times 1$, $(k + \ell \cdot p) \times (k + \ell \cdot p)$, $(k + \ell \cdot p) \times (1 + \ell)$ and $l \times (k + \ell \cdot p)$. The cyclic conditions $t_n = t_{n-365}$ and $v_{i-1,n} = v_{i,n-365}$ are not explicitly expressed in the model. To assure that these conditions hold, we augment the data in the following way:

$$\left.\begin{array}{ccc|cccccc|ccc} y_{\ell,366-L} & \cdots & y_{\ell,365} & y_{1,1} & y_{1,2} & \cdots & y_{1,364} & y_{1,365} & & y_{2,1} & \cdots & y_{2,L} \\ y_{1,366-L} & \cdots & y_{1,365} & y_{2,1} & y_{2,2} & \cdots & y_{2,364} & y_{2,365} & & y_{3,1} & \cdots & y_{3,L} \\ \vdots & & \vdots & \vdots & \vdots & & \vdots & \vdots & & \vdots & & \vdots \\ y_{\ell-1,366-L} & \cdots & y_{\ell-1,365} & y_{\ell,1} & y_{\ell,2} & \cdots & y_{\ell,364} & y_{\ell,365} & & y_{1,1} & \cdots & y_{1,L} \end{array}\right. \tag{8.17}$$

The likelihood is evaluated only for the middle part of the data. L can be set, for example to 365. In this example, the value $L = 50$ was sufficient. Inclusion of $y_{\ell,366-L}, \cdots, y_{\ell,365}$ to the left of $y_{1,1}$ and of $y_{1,1}, \cdots, y_{1,L}$ to the right of $y_{\ell,365}$ is technically questionable. To be more rigorous, that data can be considered as missing values. Doing this will cause a very minor violation of the cyclic condition. The left hand side of Figure 8.6 is a duplication of Figure 8.4 showing several samples of the original annual maximum daily temperature data and individual fitted annual

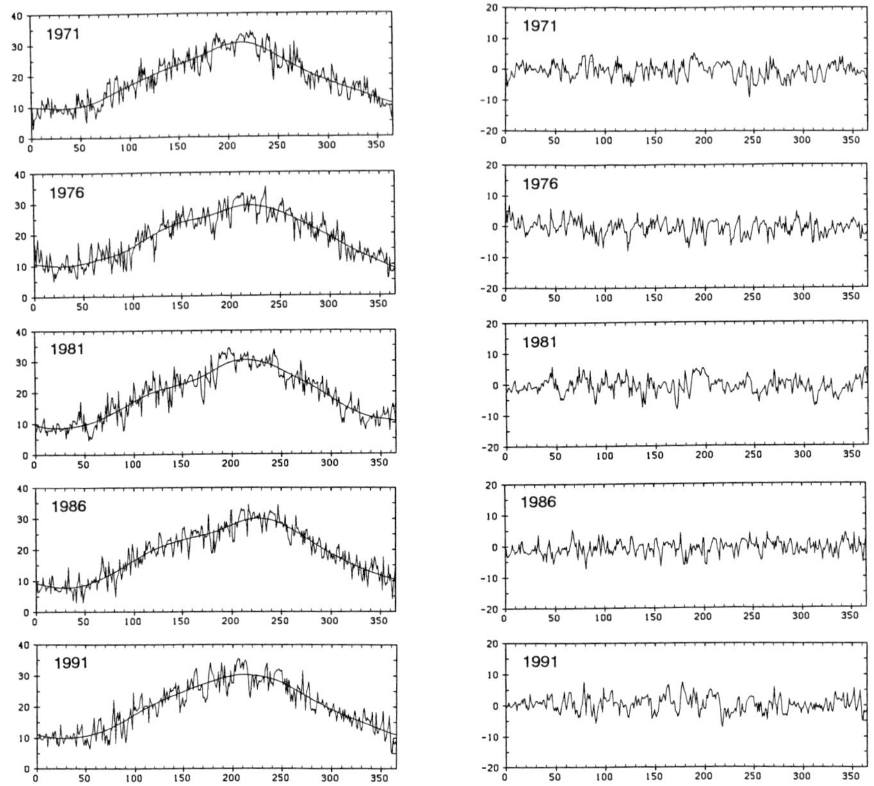

Figure 8.6: Left hand side, a duplication of Figure 8.4A and right hand side, the corresponding fitted AR components.

trends. The right hand side of Figure 8.6 shows the individual fitted AR processes for the corresponding samples of yearly data. The AR processes show a good deal of similarity to the wiggling lines about each fitted trend but do not duplicate those wiggles exactly because of the fact that they have been extracted from the annual data with a best fit annual AR model with a common trend.

8.3.2 Tiao and Tsay Flour Price Data

The trivariate Tiao-Tsay flour price data is the logarithm of indices of monthly flour process in three cities Buffalo, Minneapolis, and Kansas City, over the period from August 1972 to November 1980. Results of the analysis of the data are illustrated in Figure 8.7. The data is characterized by similar appearing nonstationary mean time series. Respectively Tiao and Tsay (1985, 1989) consider transformations of the original series, canonical correlation analyses, AR(2) and ARMA(1,1) models. The fitted AR(2) model was a full model with 18 AR parameters and 5 innovation

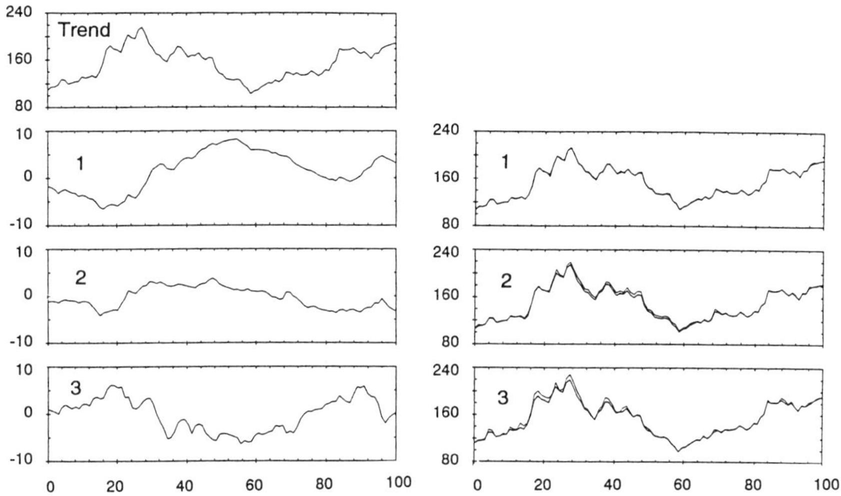

Figure 8.7: (Viewing downward),Tiao and Tsay flour price data common trend, individual AR components, common trend plus added individual fitted AR components and superimposed original data.

variance parameters. The objective of the 1989 paper is to achieve parsimonious multivariate time series modeling. The ARMA(1,1) model conceptually can have as many as 18 AR-MA parameters and 6 innovation matrix parameters. The Tiao-Tsay model actually fitted to the flour price data consists of 9 AR-MA parameters and 5 innovations variance parameters. In addition, the Tiao-Tsay model also requires computation, via a fairly complex analysis procedure, of a 3×3 data transformation matrix.

Graphical results of our analysis are shown in Figure 8.7. In Figure 8.7, viewing downward are shown the common trend, the individual fitted AR components and the common trend plus added fitted AR component superimposed on the original data for each time series. The best fitted common trend model appears to fit the data well.

Table 8.3 shows the tabulated AIC's of different trend order and different AR order models common trend individual AR processes model fitted to that data.

Our conclusion is that a common trend model of order 1 and individual AR_1 process models is the AIC best model. This procedure requires only 1 hyperparameter for the common trend, two offset or intercept parameters to "line-up" the trends and two parameters (an AR coefficient and an innovations variance), for each residual time series for a total of 9 parameters to model the flour price data. In this case then the common trend individual AR process model is considerably more parsimonious than the Tiao-Tsay modeling of the flour price data. We suggest this model as a candidate parsimonious model for the modeling of multiple time series.

Table 8.3: AIC's of Trend plus AR models

AR order	Trend order $m = 1$	Trend order $m = 2$
1	−1629.57	−1650.06
2	−1629.48	−1650.04
3	−1629.26	−1649.58

8.4 Modeling Trends with Discontinuities

Here we consider the modeling of a synthesized data set time series with a trend which has abrupt discontinuities. To achieve reliable modeling of such time series it is necessary to use the more general state space model with either non-Gaussian state and/or non-Gaussian observation noise processes. The performance of several different means of achieving the trend estimation are shown. In order, we examine the model performance of the Gaussian disturbances smooth trend model, and the numerical integration, Gaussian sum and Monte Carlo methods of non-Gaussian state space modeling that were described in Chapter 6.

We considered the data generated from the following model,

$$\begin{aligned} Y_n &\sim N(t_n, 1) \\ t_n &= \begin{cases} 0 & n = 1, \ldots, 100 \\ 1.5 & n = 101, \ldots, 200 \\ -1 & n = 201, \ldots, 300 \\ 0 & n = 301, \ldots, 400 \end{cases} \end{aligned} \tag{8.18}$$

In the top left insert in Figure 8.8, the theoretical constant mean trend values in successive intervals, are superimposed on the observed data. The problem is to estimate the abruptly changing mean value function t_n.

8.4.1 Pearson Family, Gaussian Mixture and Monte Carlo Filter Estimation of an Abruptly Changing Trend

For this data we used the model

$$\begin{aligned} \nabla^k t_n &= w_n \\ y_n &= t_n + \varepsilon_n. \end{aligned} \tag{8.19}$$

As before, ∇ is the difference operator defined by $\nabla t_n = t_n - t_{n-1}$ and w_n and ε_n are white noise sequences that are not necessarily normally distributed. For simplicity in the analysis, we assume that the difference order k is one. In this case, the state space model takes on a particularly simple form with the state, transition, input and observation matrices respectively given by $x_n = t_n$, $F = G = H = 1$. As a consequence of earlier experience, the following system and observation noise model

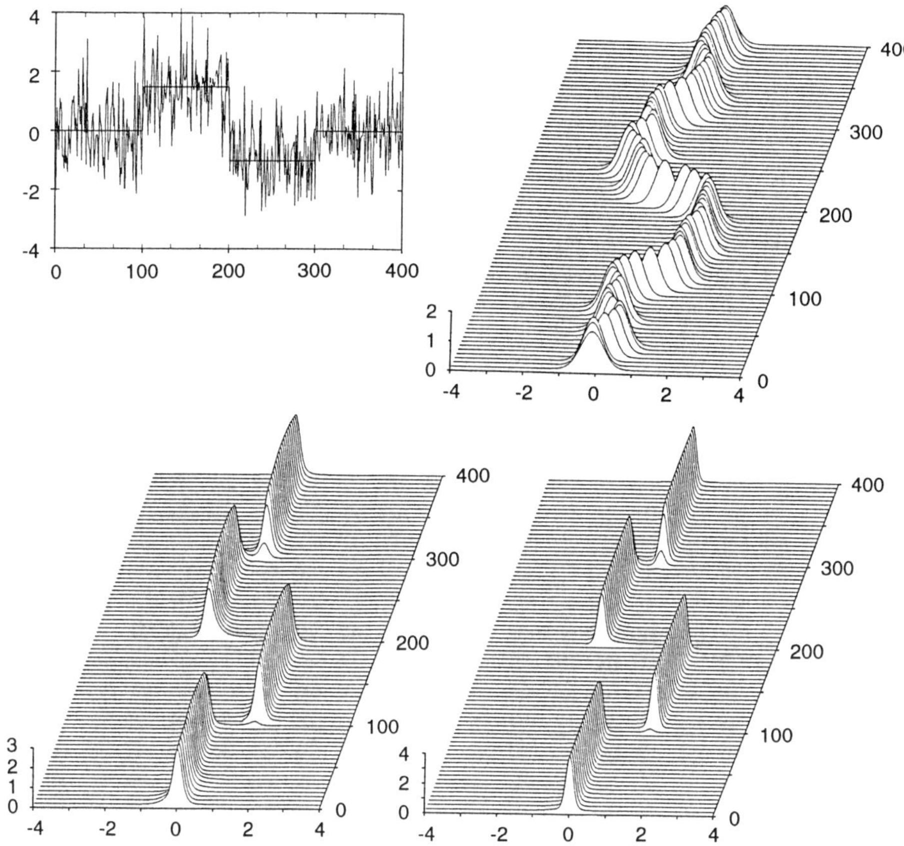

Figure 8.8: Nonstationary mean data with an abruptly changing theoretical mean value superimposed and Marginal posterior densities, for the best Gaussian , Pearson and sum of Gaussian models.

classes were considered:

$$\begin{aligned} &\text{Model (a):} \quad \varepsilon_n \sim N(0,1), \quad w_n \sim \alpha N(0,\tau^2) + (1-\alpha)N(0,\tau_x^2) \\ &\text{Model (b):} \quad \varepsilon_n \sim N(0,1), \quad w_n \sim Q(b,\tau^2). \end{aligned} \tag{8.20}$$

Model(a) denotes a mixture of Gaussian noises. In (8.20), where it appears, $N(0,r)$ denotes a Gaussian distribution with mean 0 and variance r. In Model(b), $Q(b,\tau^2)$ denotes the distribution of the Pearson system with density $q(x;b,\tau^2) = C(\tau^2+x^2)^{-b}$ with $\frac{1}{2} < b \leq \infty$ and $C = \tau^{2b-1}\Gamma(b)/\Gamma\left(b-\frac{1}{2}\right)\Gamma\left(\frac{1}{2}\right)$, (Johnson and Kotz 1970). This broad family of distributions includes the Cauchy distribution ($b = 1$) and the

Gaussian distribution ($b = \infty$). These models were fitted by the numerical integration method.

In the analysis of this data, in Model(a), τ_x^2 was arbitrarily set to 4.0, approximately the sample variance of the simulated data. The maximum likelihood estimate of τ^2 for the Gaussian model, Model(a), with $\alpha = 1.0$ or equivalently Model(b) with $b = \infty$, was $\hat{\tau}^2 = 0.0197$. The AIC of the model was 1189.44. For the mixture of Gaussian system noises model, $\hat{\alpha} = 0.99$, $\hat{\tau}^2 = 0.000004$ and AIC$= 1179.59$. We tried six Pearson family models: $b = 0.6, 0.75, 1.0, 1.5, 3.0$ and ∞. $b = 0.60$ is the AIC best Pearson family model with $\hat{\tau}^2 = 0.90 \times 10^{-10}$ and AIC$= 1180.98$. The AIC best model is the mixture of Gaussian noises model.

The marginal posterior density $p(x_n|Y_N)$ versus time n is also shown in Figure 8.8 for the Gaussian model, the best Pearson system and the mixture of Gaussians system noise model, respectively. For the Gaussian model, the densities obtained are identically shaped except for the ends of the time interval, where the densities become slightly broader. For the best Pearson system model, the shape of the posterior density varies with time. When the mean value shifts, the density becomes heavy tailed on one side. The Gaussian mixtures model also exhibits the latter behavior.

Figure 8.9 respectively shows the mean (bold) and $\pm$ 1, 2, 3 σ intervals of $p(x_n|Y_N)$ versus n for the Gaussian model and the median (bold) and corresponding $0.13, 2.27, 15.87, 84.13, 97.73$ and 99.87 percentage points of the Pearson system model, and the Gaussian mixtures model. Results of the Monte Carlo filter method for a Gaussian system and observation noise model and a Cauchy system noise and Gaussian observation noise model are in the third row of Figure 8.9. For the Gaussian distributions model by the Monte Carlo method, the likelihood best system noise and observation noise respectively were, $w_n \sim N(0, 1.22 \times 10^{-2})$ and $\varepsilon_n \sim N(0, 1.043)$. On the other hand, the estimates based on a Monte Carlo computation non-Gaussian model were

$$w_n \sim \text{Cauchy}(0, 3.48 \times 10^{-5}), \quad \varepsilon_n \sim N(0, 1.022). \tag{8.21}$$

In Figure 8.9, for the Gaussian model, the estimated mean value function is a wiggly curve and does not reflect the sudden change of the mean value. The estimated median of the Pearson system and the Gaussian mixtures models do capture the sudden change of the signal mean value. The multimodal or skewed distribution and jumps of the mean value are typical of the phenomenon that are seen in non-Gaussian modeling. As expected, because numerical integration is conceptually the most accurate way to compute in the non-Gaussian case, the Monte Carlo results will only approximate the results obtained by numerical integration. On the other hand, the Monte Carlo results are sufficiently similar to the results of the numerical integration approach that they are a vindication of that approach.

We have observed that in the presence of time series with slowly varying trends, computations with Cauchy and Gaussian process noise are visually almost indistinguishable. So despite the fact that the computational burden with Cauchy process noise and numerical integration is greater than Kalman filter computations with Gaussian noise, in the presence of a trend with unknown properties, we systematically start our data analysis with non-Gaussian model with Cauchy noise.

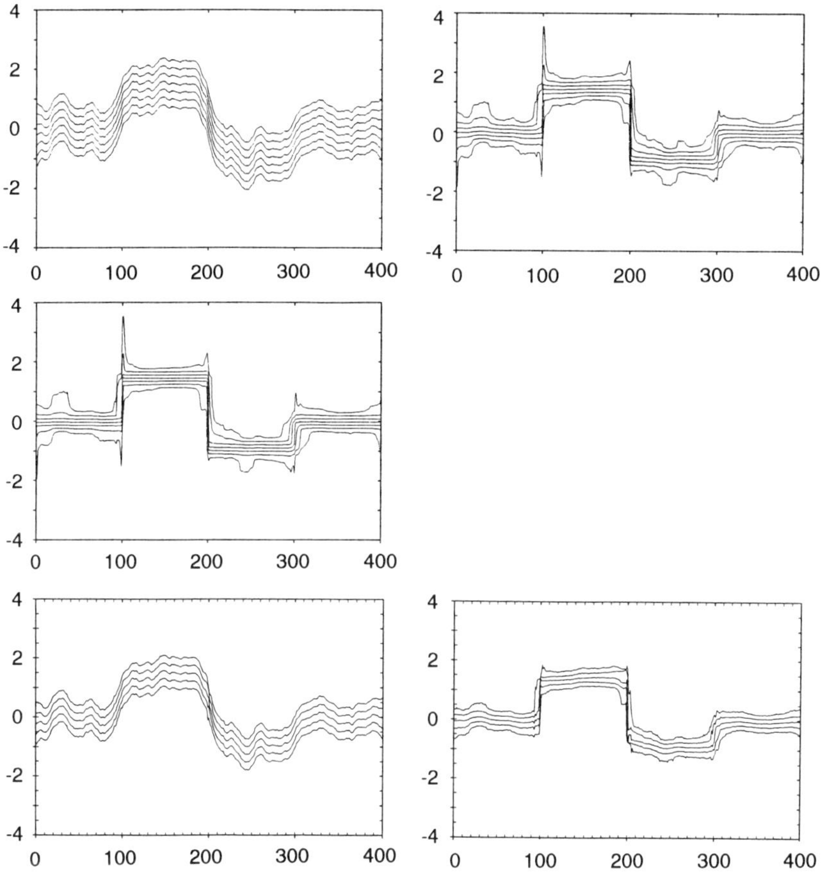

Figure 8.9: Mean and $\pm 1, 2, 3\sigma$ intervals for the best Gaussian model, and median and corresponding probability intervals for the best Gaussian mixtures and Monte Carlo models.

9

Seasonal Adjustment

Time series with seasonal components is an important generalization of nonstationary mean time series. Such time series occur for example in meteorological, oceanographic and econometric studies. In this section we consider the modeling of nonstationary mean time series with trend and seasonal and perhaps other components, (such as globally stochastic AR and trading day components), by state space methods. As in the modeling of the simpler nonstationary mean trend alone time series, first we consider the modeling of time series with smooth trend and smooth seasonal components. Such time series can be well modeled by Gaussian process noise and Gaussian observation noise using the Kalman filter/smoother algorithms. As in the trend alone modeling, when we consider that there might be outliers or abrupt changes in either the trend or seasonal components, it will be necessary to model such time series by the more general non-Gaussian state space modeling methods. In the more general non-Gaussian state space trend plus seasonal modeling, state estimation is achieved here in two different way, using the Gaussian sum and Monte Carlo filter methods.

9.1 Introduction

The econometric applications of seasonal adjustment time series has provoked an extensive literature and a variety of approaches. Essentially four classes of approaches to seasonal decomposition of time series can be distinguished: Methods based on moving averages, (e.g., X-11, Shiskin *et al.* 1976, SABL, Cleveland *et al.*, 1982, and STL, Cleveland *et al.* 1990); Methods based on multiple regression and, methods based on time series ARIMA modeling(e.g., Box *et al.*, 1978). The fourth method, introduced in Akaike (1980b), BAYSEA for "Bayesian Seasonal Adjustment", makes Gaussian assumptions about the underlying probability model, admits computationally straightforward penalized leasts squares methods, and allows for the incorporation of likelihood and Bayesian inferential tools as a means of model selection and for providing predictive intervals. Kitagawa (1981), Gersch and Kitagawa (1983a) and Kitagawa and Gersch (1984) extended the ideas in Akaike (1980b) to state space methods and incorporated seasonal decomposition features that were not present in

Akaike (1980b). Kitagawa (1994) is a state space Gaussian mixtures-two filter formula approach to seasonal decomposition modeling. That method realizes non-Gaussian seasonal adjustment modeling of time series with abrupt trend and seasonal discontinuities and with outliers. Kitagawa (1993, 1996) introduced the Monte Carlo filter computational method that is also satisfactory for modeling time series with abruptly changing trend and seasonal components. Young (1993), is a robust seasonal decomposition method that is also capable of modeling seasonal time series with abrupt discontinuities and outliers. It is based on a double exponential error model and uses Gibbs sampling to obtain confidence bands and predictive intervals.

Zellner (1978, 1981) are collections of articles on seasonal adjustment. Ghysels (1993b) is a more recent collection of articles on seasonality and econometric models.

9.2 A State Space Seasonal Adjustment Model

Here we consider a complex nonstationary mean time series problem motivated by economic time series considerations. The economic time series nonstationary mean can be decomposed into a trend t_n, a globally stationary component v_n, a seasonal component s_n, a trading day factor d_n and an observation noise component ε_n,

$$y_n = t_n + s_n + v_n + d_n + \varepsilon_n. \tag{9.1}$$

The trading day effect model is an adjustment used because in general there are a different number of i-th days of the week ($i = 1, \dots, 7$) each month, (Cleveland and Devlin 1980, Hilmer 1982). Each of the aforementioned components in (9.1) can be modeled as a stochastically perturbed difference equation. The generic state space model for this decomposition can be expressed by

$$\begin{aligned} x_n &= Fx_{n-1} + Gw_n \\ y_n &= H_n x_n + \varepsilon_n, \end{aligned} \tag{9.2}$$

where F, G and H_n are $M \times M$, $M \times L$ and $1 \times M$ matrices, respectively. w_n and ε_n are each assumed to be zero mean independent normally distributed random variables. x_n is the state vector at time n and y_n is the observation at time n. For any particular model of the time series, the matrices F, G and H_n are known and the observations are generated recursively starting from an initial state that is assumed to be normally distributed with mean x_0 and covariance matrix V_0.

The state space model that includes the local polynomial trend, stationary AR component, trading day effects and observation error components can be written in the orthogonal decomposition form

$$\begin{aligned} x_n &= \begin{bmatrix} F_1 & 0 & 0 & 0 \\ 0 & F_2 & 0 & 0 \\ 0 & 0 & F_3 & 0 \\ 0 & 0 & 0 & F_4 \end{bmatrix} x_{n-1} + \begin{bmatrix} G_1 & 0 & 0 & 0 \\ 0 & G_2 & 0 & 0 \\ 0 & 0 & G_3 & 0 \\ 0 & 0 & 0 & G_4 \end{bmatrix} w_n, \\ y_n &= \begin{bmatrix} H_1 & H_2 & H_3 & H_4 \end{bmatrix} x_n + \varepsilon_n. \end{aligned} \tag{9.3}$$

The component models (F_j, G_j, H_j) in order, ($j = 1, \ldots, 4$) represent the trend, stationary AR, seasonal and trading day effects components, respectively. Some of the particular trend, AR, seasonal and trading day difference equation constraints that we have employed and that have representations in the (F_j, G_j, H_j) matrices in (9.3) are shown below.

As before, the trend component t_n satisfies a k-th order stochastically perturbed difference equation

$$\nabla^k t_n = w_{1,n}, \tag{9.4}$$

where $w_{1,n}$ is an i.i.d. sequence with $w_{1,n} \sim N(0, \tau_1^2)$.

The stationary AR component v_n is assumed to satisfy an AR model of order p. That is given by

$$v_n = a_1 v_{n-1} + \cdots + a_p v_{n-p} + w_{2,n}. \tag{9.5}$$

In (9.5) $w_{2,n}$ is an i.i.d. sequence with $w_{2,n} \sim N(0, \tau_2^2)$.

The seasonal component of the L period difference equation is

$$s_n = -s_{n-1} - s_{n-2} - \cdots - s_{n-L+1} + w_{3,n}. \tag{9.6}$$

In (9.6), $w_{3,n}$ is an i.i.d. sequence with $w_{3,n} \sim N(0, \tau_3^2)$.

The trading day effect model is

$$d_n = \beta_{1,n} d_{1,n} + \cdots + \beta_{6,n} d_{6,n} \tag{9.7}$$

where $\beta_{i,n}$ denotes the trading-day effect factor and $d_{i,n}$ corresponds to the number of the i-th day of the week minus the number of 7-th day of the week at time n. Implicit in (9.7) is the constraint $\sum_{i=1}^{7} \beta_{i,n} = 0$ and there is no stochastic component in (9.7).

For a general model including local polynomial trend, AR component, local seasonal component and trading day effect component, the state or system noise vector, w_n and observation noise ε_n are assumed to i.i.d. with zero mean and diagonal covariance matrix

$$\begin{bmatrix} w_n \\ \varepsilon_n \end{bmatrix} \sim N\left(\begin{bmatrix} 0 \\ 0 \\ 0 \\ 0 \end{bmatrix}, \begin{bmatrix} \tau_1^2 & 0 & 0 & 0 \\ 0 & \tau_2^2 & 0 & 0 \\ 0 & 0 & \tau_3^2 & 0 \\ 0 & 0 & 0 & \sigma^2 \end{bmatrix} \right). \tag{9.8}$$

An example of a state space model that incorporates each of the components with trend order 2, AR model order 2 and seasonal component with period $L = 4$ is,

$$\begin{bmatrix} t_n \\ t_{n-1} \\ \hline v_n \\ v_{n-1} \\ \hline s_n \\ s_{n-1} \\ s_{n-2} \\ \hline \beta_{1,n} \\ \vdots \\ \beta_{6,n} \end{bmatrix} = \left[\begin{array}{cc|cc|ccc|ccc} 2 & -1 & & & & & & & & \\ 1 & 0 & & & & & & & & \\ \hline & & a_1 & a_2 & & & & & & \\ & & 1 & 0 & & & & & & \\ \hline & & & & -1 & -1 & -1 & & & \\ & & & & 1 & 0 & 0 & & & \\ & & & & 0 & 1 & 0 & & & \\ \hline & & & & & & & 1 & & \\ & & & & & & & & \ddots & \\ & & & & & & & & & 1 \end{array}\right] \begin{bmatrix} t_{n-1} \\ t_{n-2} \\ \hline v_{n-1} \\ v_{n-2} \\ \hline s_{n-1} \\ s_{n-2} \\ s_{n-3} \\ \hline \beta_{1,n-1} \\ \vdots \\ \beta_{6,n-1} \end{bmatrix} + \begin{bmatrix} 1 & 0 & 0 \\ 0 & 0 & 0 \\ \hline 0 & 1 & 0 \\ 0 & 0 & 0 \\ \hline 0 & 0 & 1 \\ 0 & 0 & 0 \\ 0 & 0 & 0 \\ \hline 0 & 0 & 0 \\ 0 & 0 & 0 \\ 0 & 0 & 0 \end{bmatrix} \begin{bmatrix} w_{1,n} \\ w_{2,n} \\ w_{3,n} \end{bmatrix}.$$

$$y_n = [\,1\ 0 \mid 1\ 0 \mid 1\ 0\ 0 \mid d_{1,n}, \ldots, d_{6,n}] x_n + \varepsilon_n$$

Ignoring initial values, the smoothness priors problem that includes all of the components in the decomposition identified above corresponds to the maximization of

$$\exp\left\{-\frac{1}{2\sigma^2}\sum_{n=1}^{N}(y_n - t_n - s_n - d_n)^2\right\}\exp\left\{-\frac{1}{2\tau_1^2}\sum_{n=1}^{N}\left(\nabla^k t_n\right)^2\right\}$$
$$\times\exp\left\{-\frac{1}{2\tau_2^2}\sum_{n=1}^{N}\left(v_n - \sum_{i=1}^{p}a_i v_{n-i}\right)^2\right\}\exp\left\{-\frac{1}{2\tau_3^2}\sum_{n=1}^{N}\left(\sum_{i=0}^{L-1}s_{n-i}\right)^2\right\}. \quad (9.9)$$

The first term in (9.9) corresponds to the conditional data distribution. The remaining terms in (9.9), in order correspond to the priors on the trend, the globally stochastic component and the seasonal component.

The role of the hyperparameters τ_1^2 and τ_3^2 as measures of the uncertainty in the belief of the priors is clear from (9.9). Relatively large τ_1^2 (τ_3^2) imply relatively wiggly trend (seasonal) components. Relatively small τ_1^2 (τ_3^2) imply relatively smooth trend (seasonal) components. The ratios of τ_j^2/σ^2, $j = 1$ or 3, can be interpreted as signal-to-noise ratios. The value of σ^2 in (9.9) is estimated essentially free of computational cost in the Kalman filter algorithm.

9.3 Smooth Seasonal Adjustment Examples

Here we illustrate some computational and graphical results using the DECOMP program. A version of this program appeared in TIMSAC-84 (Akaike *et al.* 1985). The modeling and prediction performance of globally stochastic AR components and separately, with AR components of different order, and with and without a trading day component were fitted and the AIC best component model was determined.

Trend plus seasonal models with and without an AR component

First we examine the modeling and prediction performance of seasonal models with and without AR components. The 27 year monthly ($n = 324$), log-transformed sale of household appliance data, (U.S. Bureau of the Census), Table 9.1 shows the AR model order and corresponding values of the log-likelihood, estimated innovations variance and AICs.

Table 9.1: AR Model order and AICs

AR model order	log likelihood	variance	AIC
0	962.410	0.369×10^{-4}	3347
1	1004.602	0.277×10^{-4}	3266
2	1008.097	0.226×10^{-4}	3261
3	1008.333	0.110×10^{-4}	3263
4	1008.331	0.145×10^{-4}	3265

The AIC best model is the one with the AR model order $m = 2$. The AIC is computed from the usual formula and includes the log-Jacobian of the transformation.

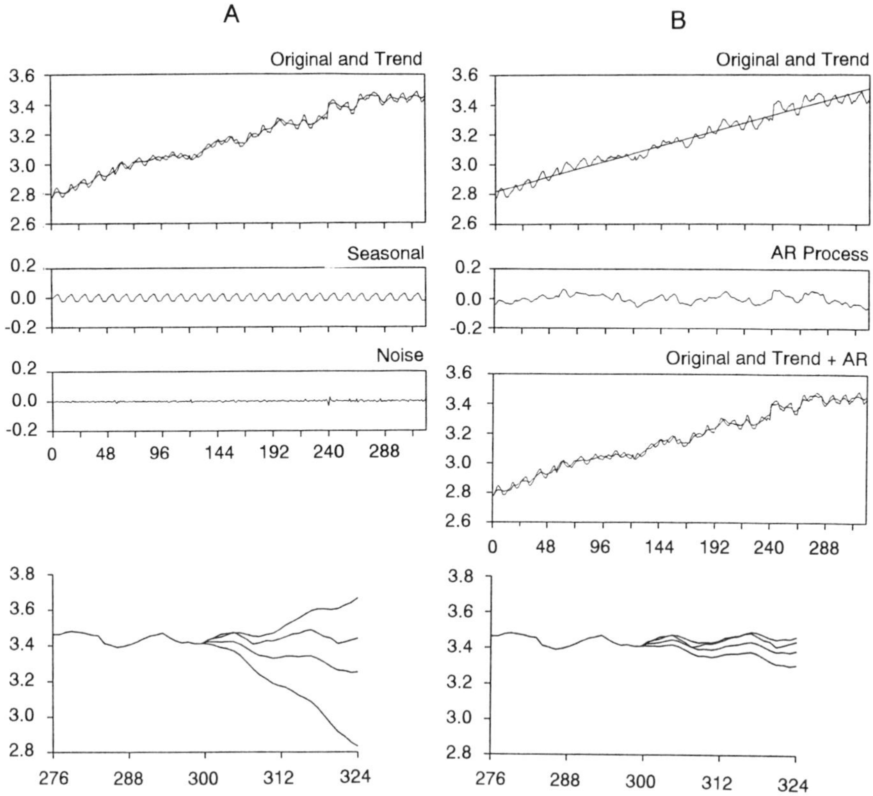

Figure 9.1: Linear Gaussian model analysis of household appliances data. A: Left hand column: Results from models without an AR component, original and trend, seasonal and noise components. True data, $n = 276, \ldots, 324$, out of sample forecast, $n = 301, \ldots, 324$, and 3σ band. B: Reft hand column: Results from models with an AR component, original and trend, AR component and original and trend plus AR components. True data, $n = 276, \ldots, 324$, out of sample forecast, $n = 301, \ldots, 324$, and 3σ band.

The number of parameters estimated is the sum of the number of hyperparameters, plus the number of AR components plus the state dimension. As expected, the largest absolute and relative changes in the log-likelihood, estimated innovations variance $\hat{\sigma}^2$ and AIC in Table 9.1 occur in going from the AR(0) model to the AR(1) model. Subsequent to identifying the AR(2) model as the single best model, we with-held the 24 observations, $n = 301, \ldots, 324$, recomputed the AR(0) and AR(2) models and predicted the with-held data.

Graphical results from these computations are in Figure 9.1. The left hand column results, in that illustration, correspond to the AR(0) model, the right hand column refer to the AR(2) model. It is noteworthy that the trend of the AR component model

Table 9.2: AR model orders and AIC's with and without trading day components

	Without TD			With TD		
AR	log-likelihood	$\hat{\sigma}^2$	AIC	log-likelihood	$\hat{\sigma}^2$	AIC
0	670.861	0.226×10^{-3}	375.20	716.265	0.137×10^{-3}	284.40
1	680.040	0.161×10^{-3}	360.85	736.817	0.583×10^{-4}	247.89
2	683.107	0.138×10^{-3}	356.71	738.003	0.427×10^{-4}	247.92
3	693.224	0.531×10^{-5}	338.48	739.738	0.338×10^{-6}	245.45
4	710.622	0.147×10^{-3}	305.68	740.535	0.440×10^{-5}	245.86

is smoother than the trend of the without-AR component model. Also, the trend plus AR component is almost indistinguishable from the trend in the without-AR component model. The seasonal component and the noise component of the AR(2) model were very similar to those of the AR(0) model and are not shown. The last row of graphs in Figure 9.1 shows the true data $n = 276, 277, \ldots, 324$, the (out-of-sample) forecast values for $n = 301, 302, \ldots, 324$, and the corresponding $\pm 1, 2, 3\sigma$ forecast bands around the forecast values. Those graphs illustrate an important property of the AIC best trend plus seasonal plus AR component model, as compared to the trend plus seasonal model. That is, the forecast bands of the AIC best, trend plus AR plus seasonal model are much tighter than those for the model without an AR component. Therefore we can be more confident about the reliability of the forecast using the trend plus AR plus seasonal model than from the trend plus seasonal model. We have seen many time series in which the trend plus AR plus seasonal model is the better AIC criterion model than the trend plus seasonal model.

Trend plus seasonal plus AR models, with and without a trading day component

In a second example, the 25 year, $n = 300$, January 1967–December 1990 monthly wholesale hardware data series, (U.S. Bureau of the Census), is analyzed. The series consists of the wholesale sales of hardware, reported in millions of dollars. The sales for each month are affected by the number of times each day of the week occurs, the trading-day effect, because buying behavior differs for each day of the week. (The sales are also affected by holidays.) Here, and generally when we model monthly seasonal time series data, we are interested in determining whether or not it is appropriate to include the trading day component and the globally stochastic AR component in the model. The AIC statistic is also used to determine those decisions. Table 9.2 shows the values of the log-likelihood, estimated innovations variance and AIC for models with different AR model orders and with and without trading day components

From the results in Table 9.2 we see that the trend plus seasonal plus AR(3) plus trading day component model, is the AIC best model. Subsequent to the examination of these results, trend plus seasonal plus AR(3) models with and without the trading day components were refitted to the $y_1, \ldots, y_{264}$ data, with the 24 data points,

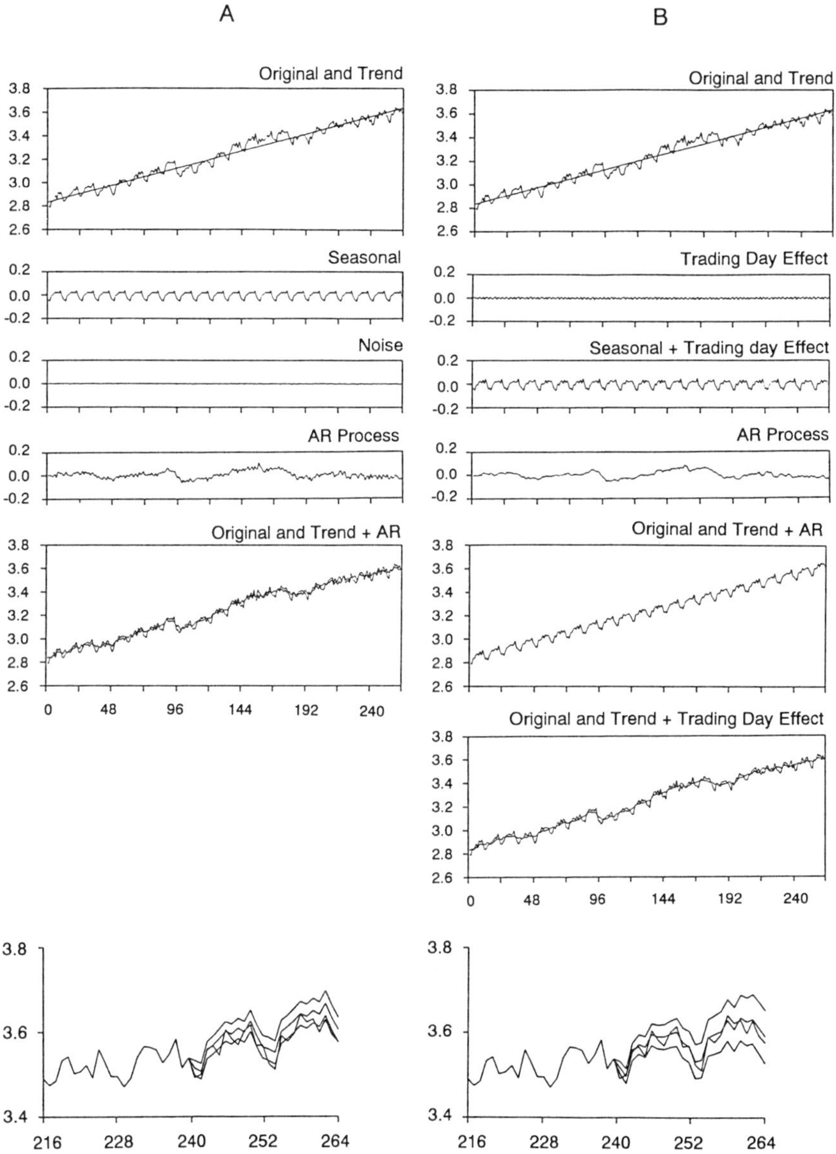

Figure 9.2: Linear Gaussian model analysis of wholesale hardware data. A: Left hand column: Results from a model without a trading day component, original and trend, seasonal, noise and AR components, original and trend plus AR components. True data, $n = 216, \ldots, 264$, out of sample forecast, $n = 241, \ldots, 264$, and 3σ band. B: Right hand column: Results from a model with a trading day component, original and trend, trading day effect, seasonal plus trading day effect, AR component, original and trend plus AR and original and trend plus trading day effect. True data, $n = 216, \ldots, 264$, out of sample forecast, $n = 241, \ldots, 264$, and 3σ band.

$y_{265}, \ldots, y_{300}$ being with-held for a study of the prediction performance of the with and without trading day component.

The graphical results that correspond to the computations of the trend plus seasonal plus AR(3) models with the trading day component are shown in the right hand side of Figure 9.2. The results for the non-trading day model are in the left column graphs of Figure 9.2. The graph of the seasonal component for the trading day component model was very similar to that for the non-trading day model and is not shown here. The seasonal component is very regular whereas the seasonal plus trading day component reveals slight irregularities. The last row of graphs in Figure 9.2 shows the true data $n = 216, 217, \ldots, 264$, the (out-of-sample) forecast values and the corresponding $\pm 1, 2, 3\sigma$ forecast bands around the forecasted values. Here we notice that the forecast bands of the non-trading day model are not as wide as those for the trading day model. We do note that the forecast values of the model with trading day component are closer to the true values and that the true values exceed the forecast intervals of the non-trading day model. Also we point out that the minimum AIC model is only optimum for one-step-ahead prediction and not for the increasing horizon prediction graphed in Figure 9.2. In Gersch and Kitagawa (1983a) prediction with a twelve-step-ahead model enhanced prediction performance at 12 steps, as compared with the one-step-ahead model's prediction performance at 12 steps. Findley (1985a) and Gersch and Sklarz (1986) considered multiple horizon models to optimize predictions at different horizons.

These results illustrate the flexibility of the decomposition of the nonstationary mean concept via smoothness priors modeling and the importance of the role of the AIC in selecting the best of alternative models.

9.4 Non-Gaussian Seasonal Adjustment

The generic state space model that we consider for non-Gaussian seasonal adjustment is the same as that in (9.1)–(9.8) with the exception that instead of Gaussian process noise and Gaussian observation noise we permit one or the other or both noise processes to be the sum of Gaussian processes. In this section we show synthetic and real data examples of non-Gaussian seasonal adjustment modeling realized by the Gaussian sum-two filter method.

An artificial abrupt trend plus seasonal component model example

For simplicity here we first consider an artificial data example without a trading day effect and without a global stochastic effect. An example of an $n = 108$ synthetic series with abrupt discontinuities in the trend and a monthly seasonal component was considered. The data was modeled by a trend order $k = 2$ and a seasonal component, with a resulting state dimension of 13. Modeling was done by the smoothness prior linear Gaussian model, and the Gaussian mixture-two filter formula method. In modeling this data by the Gaussian mixture-two filter formula method, we assumed that the observation noise was Gaussian and that the system process noise was a

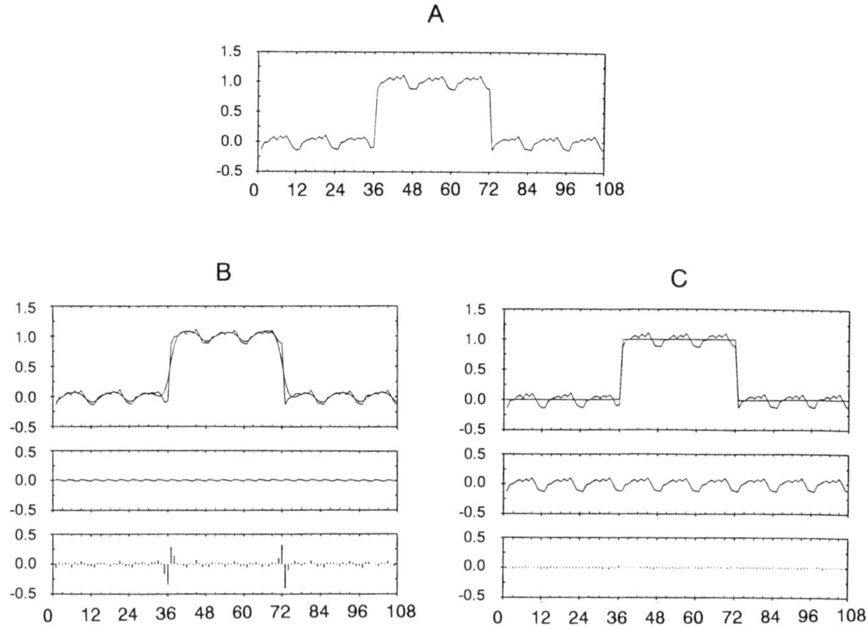

Figure 9.3: Analysis of an artificial data set with abrupt trend and seasonal components. A: Top row, the original data. B: Left hand column, Linear Gaussian model analysis, original and trend, seasonal and noise components. C: Right hand column, Gaussian mixture-two filter analysis, original and trend, seasonal and noise components.

mixture of two Gaussian distributions, namely

$$\begin{aligned} \varepsilon_n &\sim N(0,\sigma^2) \\ w_n^{(1)} &\sim \alpha N(0,\tau_1^2) + (1-\alpha)N(0,\beta_1), \\ w_n^{(2)} &\sim \alpha N(0,\tau_2^2) + (1-\alpha)N(0,\beta_2). \end{aligned} \tag{9.10}$$

The parameters estimated by a Gaussian model fit to this data are $\sigma^2 = 0.73195 \times 10^{-2}$, $\tau_1^2 = 0.4174 \times 10^{-3}$, $\tau_2^2 = 0.3126 \times 10^{-4}$. The log-likelihood of the model was 45.50. In the Gaussian mixtures modeling, the parameters β_1 and β_2 were each fixed at the value 4.0, the values of τ_1^2 and τ_2^2 were each estimated to be less than 10^{-6} and the estimate of α was 0.96281. The log-likelihood for this model was 210.336. Since the Gaussian mixtures model only had one more estimated parameter than the linear Gaussian model, the enormous difference between the likelihoods suggests that the Gaussian mixtures model is the much better model.

The results of this modeling are shown in Figure 9.3. The top row of Figure 9.3 illustrates the synthesized data. The original data plus superimposed fitted trend, seasonal component and residual of the Gaussian model results are shown in the left column. Corresponding results for the Gaussian mixture-two filter formula model are in the right column. The trend fitted by the Gaussian model smooths out the abrupt discontinuities in the trend whereas those discontinuities are captured by the Gaussian mixture model method. The residual noise in the Gaussian model has distinguishing components at the points of trend discontinuities whereas, illustrated on the same scale, the Gaussian mixtures model residuals are essentially invisible.

A real data abrupt trend seasonal model example

A real data example of the analysis of an economic time series, the quarterly series of the increase of the inventories of private companies in Japan 1965-1983 is discussed here. As in the previous example, the data was modeled by a Gaussian disturbance model and also modeled under the assumption of Gaussian observation noise and a two component Gaussian mixture process noise model. A second order trend model was used, therefore the state dimension was 5. Gaussian and Gaussian mixture modeling results by both the two filter formula and the Monte Carlo filter are shown in Figure 9.4

The original data is in the top row of Figure 9.4. The observation interval includes the time of the 1973 oil price crisis so as expected, abrupt changes can be seen in the original data. Graphical results of the estimated trend and seasonal components for the Gaussian and Gaussian mixture-two filter formula modeling methods are shown in the left and right hand columns respectively. The Gaussian model smooths out the abrupt changes in the trend. Those abrupt changes can be clearly seen in both the Gaussian mixture-two filter formula and Monte Carlo modeling methods' graphical results. Also the seasonal component in the linear Gaussian modeling is quite different than that seen in the non-Gaussian modeling method.

9.5 Modeling Outliers

The Gaussian mixture-two filter model for observational noise is very useful for handling outliers in time series. Figure 9.5A shows the results of standard seasonal adjustment of the BLSALLFOOD, (the Bureau of Labor Statistics, all employees in food industries, January 1966-December 1979, N=156), (Kitagawa and Gersch 1984). The original data, the estimated trend with $\pm 1 \sim \pm 3$ standard error intervals, the estimated seasonal component and the residuals are shown in the figure.

To check the robustness of the Gaussian mixture observational noise model, we consider an artificially contaminated series of the BLSALLFOOD data. The data shown in Figure 9.5B was obtained by replacing six randomly chosen observations, $y_{29}, y_{50}, y_{53}, y_{90}, y_{110}, y_{111}$ by the arbitrary large value, 1900. These data are considered as outliers. The result of the linear Gaussian modeling of the contaminated data is shown in Figure 9.5C, and the impact on the linear Gaussian modeling of this data as compared with the original data modeling can be seen by comparing figures 9.5A and 9.5C. The trend estimated by the standard linear Gaussian seasonal adjustment model for the contaminated data is strongly affected by the outliers and is wiggly.

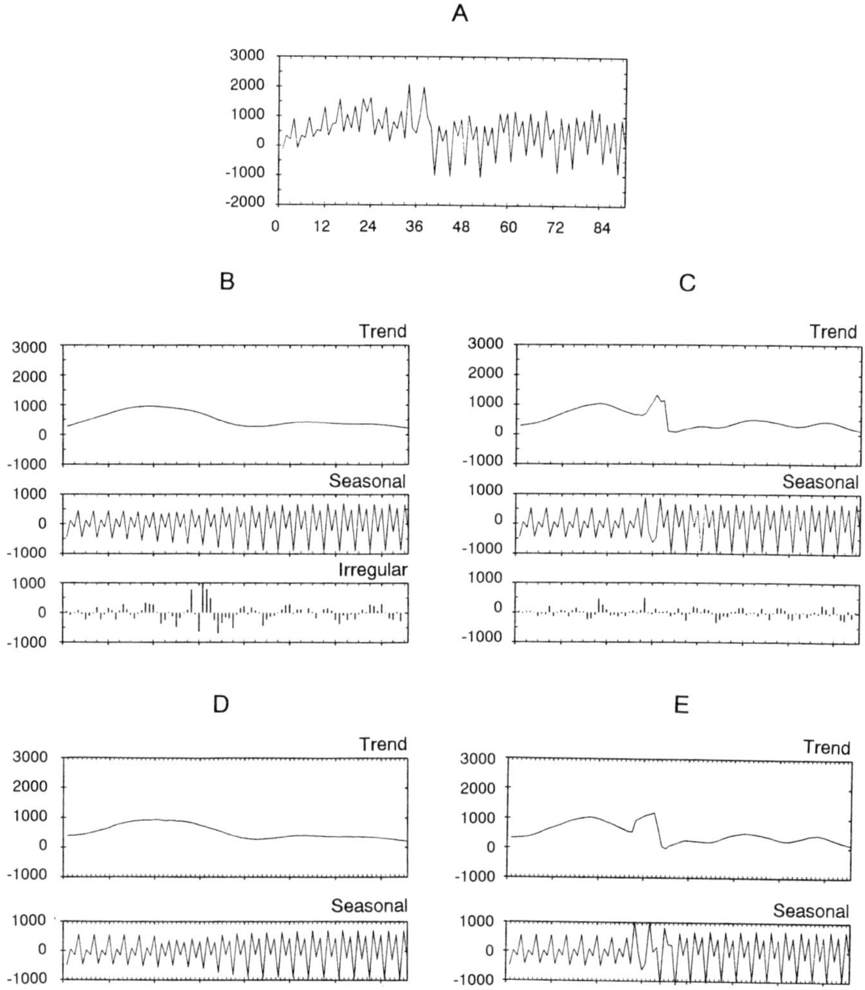

Figure 9.4: Analysis of quarterly increase of inventories series. A: Top row, the original data. B: Left hand column: Linear Gaussian model analysis, mean and $\pm 1 \sim \pm 3$ standard error intervals, seasonal and noise components. C: Right hand column: Gaussian mixture-two filter analysis, median and $\pm 1 \sim \pm 3$ equivalent standard error intervals, seasonal and noise components. D: Linear Gaussian model analysis by a Monte Carlo smoother. E: Gaussian mixture modeling by a Monte Carlo smoother.

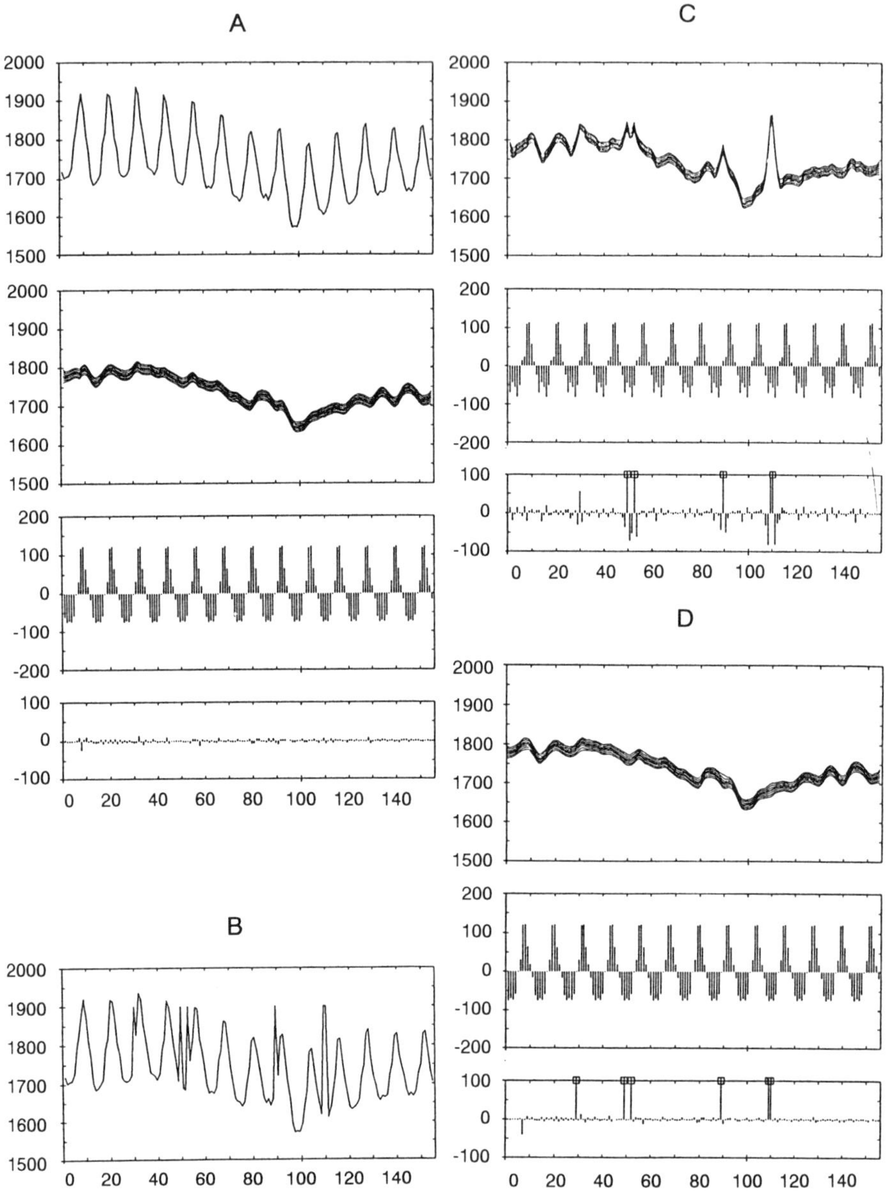

Figure 9.5: BLSALLFOOD outlier analysis example. A: Left hand column: BLSALLFOOD data, Gaussian analysis, estimated trend, seasonal and residual components. B: BLSALLFOOD data with 6 outliers C: Right hand column: (rows 1-3), Gaussian analysis of data with outliers, estimated trend, seasonal and residual components. D: Right hand column: (rows 4-6), Gaussian mixture observation noise analysis, estimated trend, seasonal and residual components.

Seasonal pattern estimated from the contaminated data is also quite different from the one shown in Figure 9.5A. The log-likelihood of the Gaussian model with the outliers is -2795.2.

Figure 9.5D shows the results obtained by assuming that the observational noise density is a mixture of two Gaussian densities

$$r(w) \sim \alpha N(0, \sigma^2) + (1-\alpha) N(\mu, \tau^2), \tag{9.11}$$

with $\alpha = 0.96$, $\sigma^2 = 30.3$, $\tau^2 = 4 \times 10^4$. The estimated trend and the seasonal components are mostly indistinguishable from the ones in Figure 9.4A. The residuals take large values corresponding to the location of the outliers. The log-likelihood of this model is -690.2 indicating that this model is significantly better than the standard Gaussian model.

Incidentally, by applying the Gaussian mixture model to the original uncontaminated data shown in Figure 9.4A, we obtain almost identical estimates of trend and seasonal components as compared with the original Gaussian modeling. The log-likelihood of the Gaussian mixture model is -650.8 whereas that of the original Gaussian model was -649.4. Therefore, it might be said that we lose very little by using the Gaussian mixture model. This example suggests that the Gaussian mixture model is robust to the presence of observation outliers and also very satisfactory when applied to data without outliers.

10

Estimation of Time Varying Variance

There are many time series whose structure involves a substantial change of variance. The decomposition of the seismic time series problem in Chapter 7 is such an example. In other practical data situations, the relatively fast wiggles of a nonstationary covariance time series appears to be modulated by a relatively slowly changing envelope function. For example, seismic measurements during an earthquake exhibit this behavior. That envelope function can be interpreted as a change of scale associated with the instantaneous innovations variance of the state space model of the time series. In this chapter the changing variance structure of the Urakawa-Oki, Hokkaido Japan March 21 1983, earthquake data, (code name MYN2F, Takanami 1991) is estimated by both Gaussian and non-Gaussian state space models. Additional applications include the change of variance modeling in the estimation of the log-periodogram of a time series and the estimation of the instantaneous variance of the collection of 21 years of daily maximum temperatures in Tokyo.

10.1 Introduction and Background

A key idea in the contemporary topic of the estimation of instantaneously changing variance has its origin in a classical topic, the estimation of the variance of a stationary process. If a stationary process has an absolutely continuous spectral distribution function with spectral density P_f, the variance of the innovations process or one step prediction variance is (Wiener 1949),

$$\sigma^2 = \exp\left\{\int_{-\frac{1}{2}}^{\frac{1}{2}} \log P_f df\right\}.$$

Estimation of the innovations variance of stationary time series using the log-periodogram was studied in Bartlett and Kendall (1946). The periodogram I_ν of a time series $y = y_1, \ldots, y_N$ is given by

$$I_v = \frac{1}{N}\left|\sum_{n=1}^{N} x_n \exp\left\{\frac{2\pi\nu(n-1)}{N}\right\}\right|^2,$$

where ν is an integer, $|\nu| \leq [N/2]$, (i.e. the integral part of $N/2$.)

It is known that the I_ν have approximately independent exponential distributions with mean values equal to the variance of the y_n process. It is also known (Davis and Jones 1968), that the mean of the log of the periodogram, is approximately the log of the true spectral density at that frequency minus Euler's constant, $\gamma \approx 0.57722$. It is also known that the variance of the log of the periodogram is approximately $\pi^2/6$.

Wahba (1980) used these results in attempting to obtain an objective optimum smoothing procedure of an estimate of the log spectral density, based on smoothing the log periodogram with a cross-validated smoothing spline.

10.2 Modeling Time-Varying Variance

First we consider a Gaussian state space modeling of changing variance. Consider a realization of white noise s_n, $n = 1, \ldots, N$, where $s_n \sim N(0, \sigma_n^2)$ with unknown time varying variance σ_n^2. Then, the stochastic process χ_m^2 defined by

$$\chi_m^2 = (s_{2m-1}^2 + s_{2m}^2)/2, \quad m = 1, \ldots, N/2 \tag{10.1}$$

constitutes an independent sequence of chi-square random variables with two degrees of freedom ($\chi_m^2 \sim \chi_2^2$). Then from Davis and Jones (1968), the transformation

$$t_m = \log \chi_m^2 + \gamma, \tag{10.2}$$

where γ=0.57722 is the Euler constant, leaves the independent random variable t_m with distribution that is almost normal and with the mean and variance

$$E[t_m] \simeq \log \sigma_{2m}^2, \quad \text{var}\,[t_m] \simeq \pi^2/6. \tag{10.3}$$

That transformation justifies our use of a smoothness priors procedure for the estimation of t_m, and hence for the estimation of the unknown variance σ_{2m}^2. Wahba (1980) used this approximate normal property of t_m for smoothing the log-periodogram.

To obtain a smooth estimate of the variance σ_n^2, consider a k-th order difference equation constraint on the log variance defined by

$$\nabla^k t_m = w_m. \tag{10.4}$$

In (10.4), $w_m \sim N(0, \tau^2)$ i.i.d. Then as before, imbed the difference equation constraint model in (10.4) into a state space form

$$\begin{aligned} x_m &= F x_{m-1} + G w_m \\ t_m &= H x_m + \varepsilon_m \end{aligned}$$
$$\begin{bmatrix} w_m \\ \varepsilon_m \end{bmatrix} \sim N\left(\begin{bmatrix} 0 \\ 0 \end{bmatrix}, \begin{bmatrix} \tau^2 & 0 \\ 0 & \sigma^2 \end{bmatrix}\right). \tag{10.5}$$

For convenience, assume here that $k = 2$. Define the state vector $x_m = (t_m, t_{m-1})^T$. Then the matrices F, G, H associated with (10.5) are

$$F = \begin{bmatrix} 2 & -1 \\ 1 & 0 \end{bmatrix}, \quad G = \begin{bmatrix} 1 \\ 0 \end{bmatrix}, \quad H^T = \begin{bmatrix} 1 \\ 0 \end{bmatrix}. \tag{10.6}$$

Applying the Kalman filter and smoothing algorithms described in Chapter 5 yields the smoothed value of $t_{n|N}$, the logarithm of the smoothed estimate of the changing variance. $\sigma^2_{2m|N} = \sigma^2_{2m-1|N} = \exp\left\{t_{m|N} - \gamma\right\}$ is then our smoothed estimate of the changing variance.

10.3 The Seismic Data

In this section, the changing variance structure of the Urakawa-Oki, Hokkaido Japan March 21 1991, earthquake data (Takanami 1991) is estimated by both Gaussian and non-Gaussian state space models.

Linear State Space Gaussian Modeling

For example, Figure 10.1A shows an $N = 1000$ seismometer earthquake signal observed in Hokkaido March 21, Japan 1982, Takanami (1991). The original data was regularly sampled at intervals of about 0.01 seconds. Four successive observations were averaged to yield samples at intervals of 0.04 seconds. The problem here is the estimation of the changing variance (envelope function) of the time series, $y_1, \ldots, y_N$. Figure 10.1B shows the sequence of observations, (10.1), χ^2_m, $m = 1, \ldots, N/2$. Apparently there are two abrupt changes in that sequence. As expected, they correspond to the arrivals of the P-wave and the S-wave. Since the density of the observational noise is assumed known in this case, the model has only one parameter, τ^2, the variance of the system noise. The maximum likelihood estimate of the parameter was $\hat{\tau}^2$=0.04318 and the AIC=1861.64.

The estimated mean value by this model is wiggly (Figure 10.1C). If we use a smaller value of τ^2, we can get a smoother curve but that curve cannot follow the sudden change of the variance. It is possible to alleviate the difficulty by allowing several outliers in the system noise, but that requires the positioning of outliers and thus is applicable only when we know the position of the outliers beforehand (e.g. the exact arrival times of P-wave and S-wave), or when an objective and effective outlier detection procedure is available.

Linear State Space Non-Gaussian Modeling

Alternatively, if we consider modeling the transformed series by a smoothness priors non-Gaussian model, we may define the system and observation noise distributions respectively by

$$\begin{aligned} q(x) &= \tau\{\pi(\tau^2 + x^2)\}^{-1}, \\ r(x) &= \exp\{x - e^x\}. \end{aligned} \tag{10.7}$$

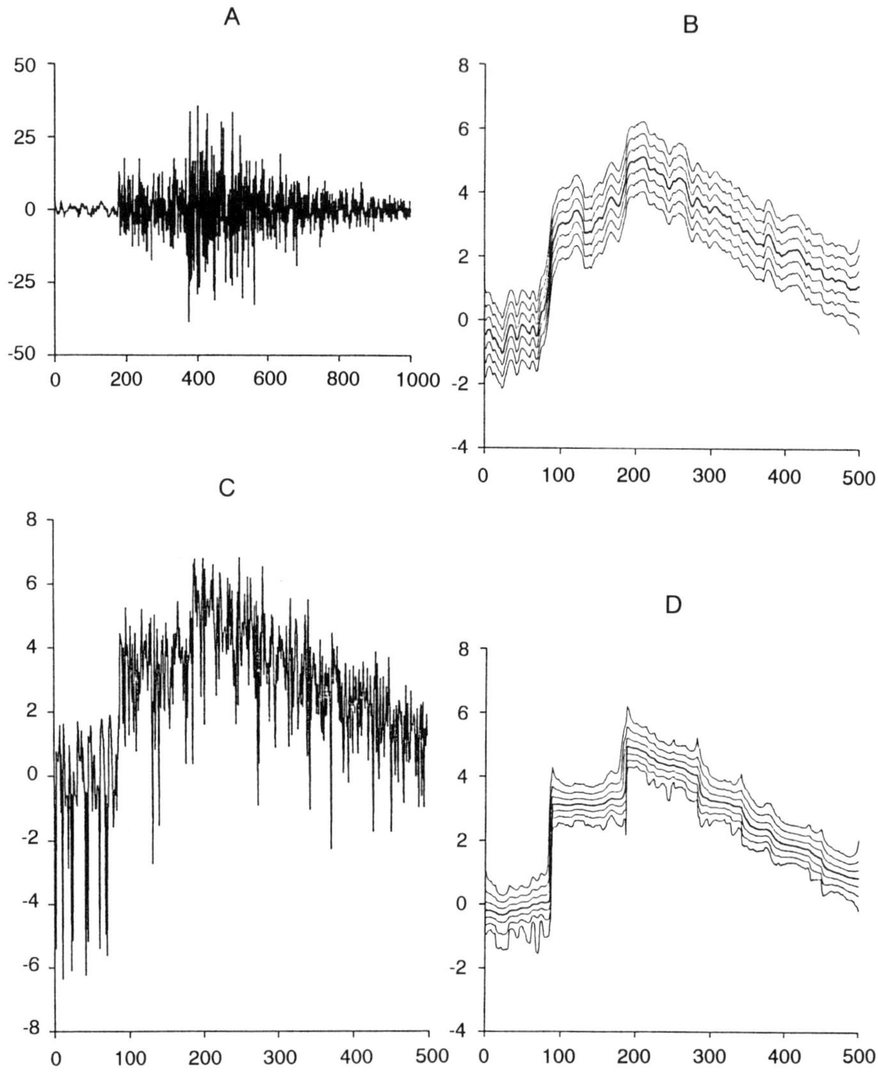

Figure 10.1: A: Seismometer recording of an earthquake observed in Hokkaido Japan 1983. B. The transformed series χ^2_m. C: Changing log-variance estimated by the best Gaussian model, mean and $\pm 1, 2, 3\sigma$ intervals. D: Changing log-variance estimated by a non-Gaussian model, median and 0.13,2.27,15.87,84.13,97.73 and 99.87 percentage intervals.

Since χ^2_m is exponentially distributed, the distribution of the transformed data t_m, $r(x)$, is double exponential, (Johnson and Kotz 1970).

These distributions are used in the state space model defined in (10.5). The numerical integration procedure described in Chapter 6 is employed to compute the likelihood of the unknown parameter τ^2. Application of the non-Gaussian model yields the maximum likelihood estimate of τ^2 to be 0.000112. The value of the AIC, (AIC =1718.15), indicates that in this case the non-Gaussian model is significantly better than the Gaussian model.

Graphical results for the analysis of the changing variance problem are in Figure 10.1. Figures 10.1A, 10.1B are respectively the original seismic data and the data transformed in accordance with (10.1). Figures 10.1C and 10.1D are respectively the Gaussian and non-Gaussian estimates of the log-periodogram. The estimated curves shown in Figure 10.1D capture the abrupt changes in the sequence of transformed variables. Elsewhere they are smoother and have less variability than the corresponding Gaussian modeling. This figure clearly shows that the non-Gaussian models with heavy-tailed noise distributions have the ability to estimate both smooth changes and jumps of parameters simultaneously. That was not possible with the simple Gaussian model. The robustness of the non-Gaussian model to large deviations is apparent in this example. Due to the nature of the sampling distribution of χ^2_m, there are frequent large negative deviations. Those deviations act as outliers for the Gaussian model. As a result, the estimated curves are wiggly. For the non-Gaussian model however, they are no longer outliers and do not severely affect the estimates. In Figure 10.D, we see that only the lowermost 0.13 per cent curve is affected by these large deviations.

It is interesting to note that if we use non-Gaussian modeling, the data can be modeled directly without requiring the use of the transformations in (10.1) to (10.3). If as before, $s_n \sim N(0, \sigma_n^2)$, we can immediately use $y = \log s^2$ and the distribution of y is the double exponential $y \sim \frac{1}{2\pi}\exp\{\frac{y}{2} - \frac{1}{2}e^y\}$.

10.4 Smoothing the Periodogram

The problem of estimating the spectral density for a stationary time series is of fundamental interest in statistics and in engineering. Windowed periodogram (also windowed log-periodogram) and AR model methods being the most popular. Window estimates are obtained by smoothing the periodogram using window functions, the degree of smoothing being controlled by the window width. Recent approaches to the window width selection problem include the application of cross-validation, (Hurvich and Beltrão 1990), and the bootstrap, (Franke and Härdle 1992, Politis and Romano 1992). Wahba (1980) and Kooperberg *et al.* (1995), are nonparametric smoothing spline regression smoothings of the log-periodogram. Here we consider an example of smoothness priors smoothing of the log-periodogram using both the Gaussian and non-Gaussian modeling methods.

Since the periodogram is distributed as an exponential distribution, as a natural application of this method, we can also smooth the log-periodogram. Wahba (1980) approximated the density $r(w)$ by a Gaussian distribution with the same first and second moments as those of $r(w)$. But here by our method, the series can be smoothed

Table 10.1 Summary Statistics. Smoothing the Log-Periodogram

trend order	observation noise	system noise	τ^2	log likelihood	AIC
1	D. Exp.	Gauss	0.110	−437.47	876.94
	D. Exp.	Cauchy	0.0013	−431.13	864.26
	Gauss	Gauss	0.081	−455.21	912.42
	Gauss	Cauchy	0.0014	−452.62	907.24
2	D. Exp.	Gauss	0.18×10^{-2}	−448.40	898.80
	D. Exp.	Cauchy	0.13×10^{-4}	−439.48	880.96
	Gauss	Gauss	0.43×10^{-3}	−461.32	924.64
	Gauss	Cauchy	0.88×10^{-5}	−460.10	922.20

without invoking a Gaussian approximation.

Table 10.1 shows the summary statistics for the cases of smoothing the log-periodogram, of a particular $N = 500$ sample, of the roll motion of a ship at sea. Gaussian and non-Gaussian Cauchy system noise and Gaussian and double exponential observation noise models were considered for trend order models $k = 1$ and $k = 2$. The AIC best model, was the first order trend model ($k = 1$) with Cauchy system noise and double exponential observation noise.

Figure 10.2 illustrates the results obtained. Figure 10.2A shows the log-periodogram of the data set. In the left-hand column, Figures 10.2B–E show the smoothed log-periodogram for trend model order $k = 1$ and the combinations of system and observation noise in the order shown in Table 10.1. In the right-hand column, Figures 10.2F–I show the corresponding smoothed periodograms for the case of trend order model $k = 2$

The results in Table 10.1 indicate that the trend models for $k = 1$ are systematically better than those for trend models $k = 2$. The single AIC best model is that for which $k = 1$ with double exponential observation noise and Cauchy system noise. The Gaussian observation noise models are systematically the poorest models. We also note that the smoothed periodogram results for the $k = 1$ trend model order captures the discontinuities that are apparent in the data better and are systematically better AIC models than their smoother $k = 2$ trend order model counterparts.

10.5 The Maximum Daily Temperature Data

In this case, consider each of the consecutive years of the detrended maximum daily temperature to be periodic with period 365. (This data set was discussed in Chapter 8 in the context of trend modeling.) The detrended data yields a set of 21 realizations $y_{i,n}$, $i = 1, \ldots, 21$, $n = 1, \ldots, 365$ of maximum daily temperatures. We assume that data to be normally distributed with mean zero and daily changing variance. That is $y_{i,n} \sim N(0, \sigma_n^2)$ and we can get a first coarse estimate of the instantaneous variance from the 21 detrended realizations. In the smoothness priors modeling of this data,

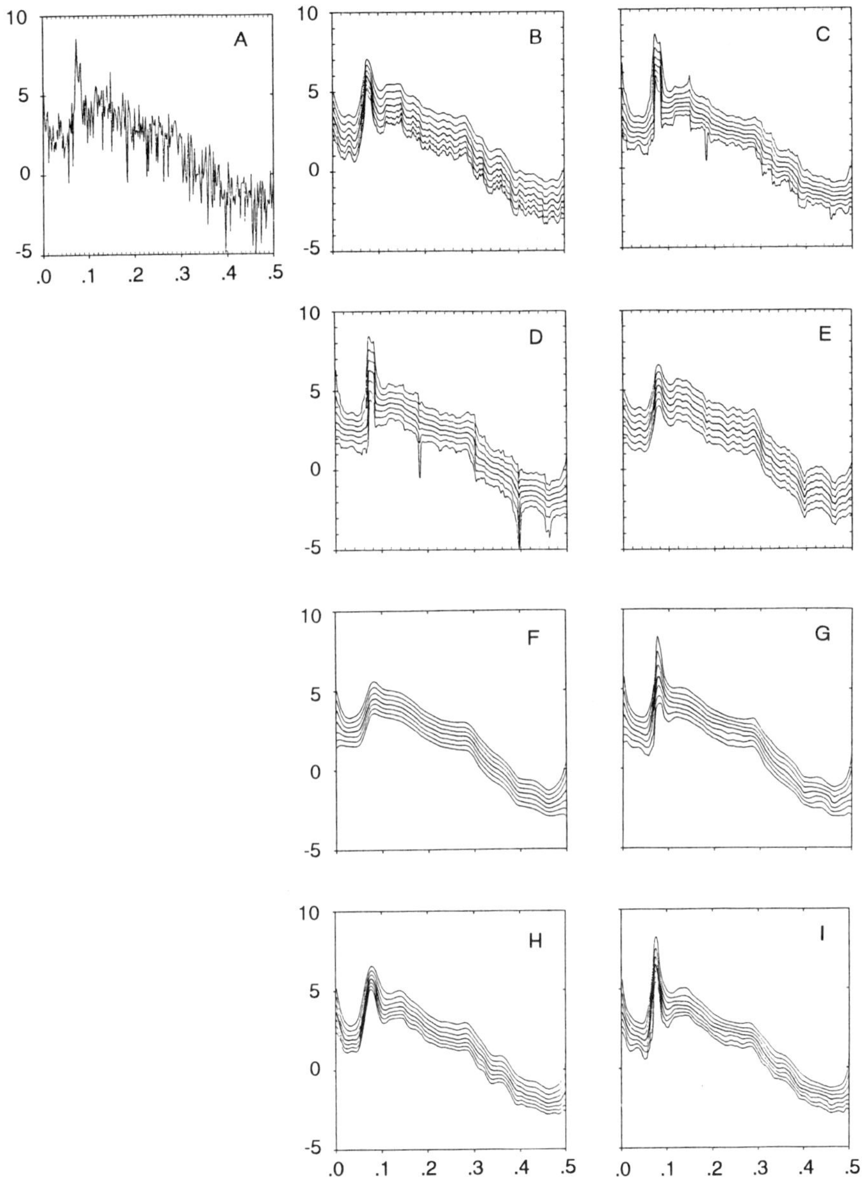

Figure 10.2: A: The log-periodogram. B-E: The log periodogram for k = 1. F-I: The log periodogram for k = 2.

Table 10.2 Summary Statisitcs. Three time-varying varaince models fitted to the maximum daily temerature data.

b	τ^2	log likelihood	AIC
∞	0.0030	-24324.66	48651.32
2	0.0033	-24325.15	48652.30
1	0.40×10^4	-24328.29	48658.58

assume a $k = 1$ order stochastic trend model for the log instantaneous variance data. That is let

$$\nabla^k \log \sigma_n^2 = w_n, \quad \sigma_n^2 = \sigma_{n-365}^2. \tag{10.8}$$

Then in the first order state space model identify the state $x_n = \log \sigma_n^2$ and the state space model is

$$x_n = x_{n-1} + w_n \tag{10.9}$$
$$y_{i,n} \sim N(0, \exp(x_n)), \quad i = 1, \ldots, 21. \tag{10.10}$$

Table 10.2 shows the summary statistics of three distinct models for the system noise. The AIC best model is $b = \infty$, namely the Gaussian distribution model.

Figure 10.3 shows the instantaneous log-variance computed as the logarithm of the sample variance of the 21 years of daily detrended data, the mean smoothed instantaneous log variance with the plus and minus 1, 2 and 3 sigma bands, and the estimated instantaneous variance. The estimated instantaneous variance is obtained by the transformation $\hat{\sigma}_n^2 = \exp(\hat{x}_n)$, where $\hat{x}_n = x_{n|N}$ is the smoothed estimate of the $k = 1$ trend model. The confidence bands around the smoothed state x_n are seen to be very tight.

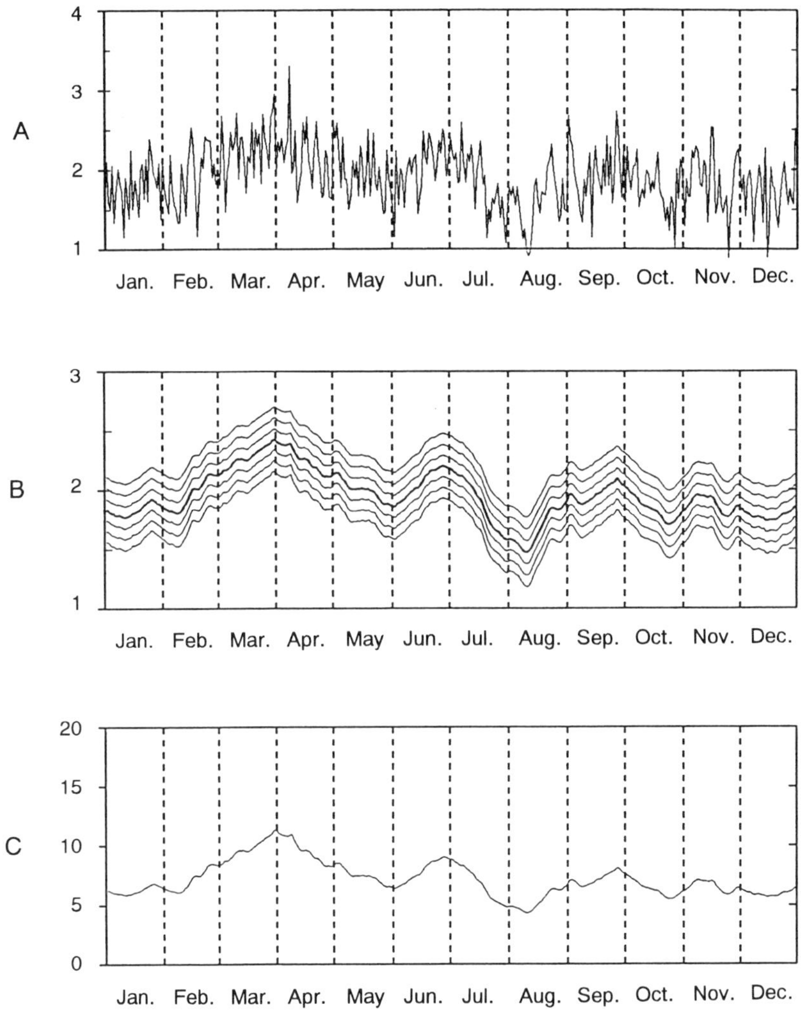

Figure 10.3: A: The instantaneous log-variance of the maximum daily temperature data. B: The mean smoothed and plus and minus 1,2 and 3 sigma instantaneous log-variance. C: The instantaneous variance.

11

Modeling Scalar Nonstationary Covariance Time Series

In this chapter scalar nonstationary covariance time series are modeled using time varying coefficient autoregressive models, as in equation 11.1. In such a model for N observations, if the order of the AR model is m there will be $N \times m$ AR coefficient parameters and as many as N innovations variance parameters. Fitting a time varying AR, (TVAR), model with smoothness priors constraints permits those parameters to be estimated implicitly in terms of only a small number of explicitly estimated hyperparameters. Several different mechanisms for implementing the smoothness priors constraints are possible. Our own approach to this important topic has evolved over several years, and two different methods for fitting the time varying AR coefficient model with smoothness priors constraints are shown here. (Each of the methods has implications for the modeling of multivariate nonstationary covariance data. That topic is treated in Chapter 12.) Our primary application for the scalar nonstationary covariance modeling is the evolution with time of the power spectrum. The estimated TVAR model yields what we refer to as an "instantaneous power spectral density".

Some of the background of our approach to the modeling of nonstationary covariance time series is discussed in Section 11.1. A scalar TVAR model, with Gaussian distributed smoothness priors constraints on the evolution with time of the AR parameters and its state space realization are shown in Sections 11.2 and 11.3 respectively. That particular model is well suited to nonstationary covariance time series with relatively slowly changing covariance structure. An alternate approach to fitting the scalar time varying AR model, is with a not necessarily Gaussian smoothness priors constraint on the evolution with time of the partial correlation coefficients, (PARCORs), of the AR model. That is shown in Section 11.4. Such a model is well suited to modeling nonstationary covariance time series with abruptly changing covariance structure. Examples of the modeling of both simulated data and of seismic data with both the Gaussian disturbance model and the non-Gaussian PARCOR model are shown in Section 11.5.

11.1 Introduction

The problem of modeling a time series with nonconstant covariance structure arises in several applications areas, including speech processing, image processing and automatic analysis of biomedical signals such as electroencephalograms and electrocardiograms. Several time domain approaches have been proposed for the modeling of such nonstationary covariance time series. A "classical" approach has been to partition the time series into stationary segments and to model each segment as an AR or ARMA model, (Ozaki and Tong 1975, Bodenstein and Praetorious 1977, Kitagawa and Akaike 1978, Ishii *et al.* 1979, and Segen and Sanderson 1980). Bohlin (1976) is an early application of the analysis of time series models with time-dependent coefficients. Kozin (1977) is an orthogonal polynomial expansion of the AR coefficients of a TVAR model with an application to earthquake data analysis. Gersch and Kitagawa (1983b), is an extension of Kozin's orthogonal expansion approach to vector time varying AR models in an application to econometric data analysis. Gersch (1985, 1987b) shows an application of that idea to the analysis of human electroencephalograms. Other notable research in TVAR coefficient models have been reported in Whittle (1965), Åstrom and Wittenmark (1973), Kozin and Nakajima (1980), Nicholls and Quinn (1985), and Grenier (1982). The latter is a lattice network time varying AR model.

Kitagawa (1983) introduced a smoothness priors state space approach for the modeling of TVAR models. We note that a Bayesian model for time varying AR coefficient modeling does not appear in the earlier literature. Subsequently variations of and improvements on the state space modeling and the computation of the likelihood of the model's hyperparameters appeared in Kitagawa and Gersch (1985b), Gersch and Kitagawa (1985), Kitagawa's program TVCAR in Akaike *et al.* (1985), Kitagawa (1986, 1987), and Gersch and Kitagawa (1988). Examples are shown in Section 11.5.

Extensions of these ideas to the modeling of multivariate nonstationary covariance time series by time varying vector AR, (TVVAR), models was shown in Gersch (1991, 1992), Stone (1993), and Jiang and Kitagawa (1993). The key idea in those realizations is the use of an instantaneous response-orthogonal innovations representation of the vector autoregressive, (VAR), model. That permits VAR modeling to be realized one scalar AR model at-a-time. TVVAR modeling of multivariate nonstationary time series is treated in Chapter 12.

The problem in modeling nonstationary covariance time series is to achieve an efficient parameterization to capture the local and global statistical relationships in the time series. That objective is achieved here via the TVAR coefficient model in two different ways. One is by imposing smoothness priors constraints in the form of stochastically perturbed difference equations directly on the evolution of the individual AR coefficients. The variances of the white noise stochastic perturbations are the hyperparameters of the AR coefficient distributions. An alternative approach is to impose smoothness priors constraints in the form of stochastically perturbed difference equations on the evolution of the PARCOR coefficient in a Levinson algorithm type lattice structure AR modeling. That in turn yields the time varying (instantaneous), AR coefficients. In this case, the variances of the white noise stochastic perturbations are the hyperparameters of the PARCOR distributions. In each approach, the relevant difference equations can be embedded into a state space representation. In the case of a slowly changing covariance, a Gaussian disturbance model and Kalman filter computation may be used for the computation of the likelihood. Numerical integration in a non-Gaussian modeling, is appropriate in cases of abrupt changes of the covariance structure. In either case, in the event that the magnitude of the envelope of the nonstationary covariance time series has a large dynamic range, the first step in the modeling will be to smooth the estimate of the nonstationary variance series of the time varying model by the method shown in Chapter 10. A byproduct of that computation is an estimate of the instantaneous innovations variance, which in conjunction with the estimates of the instantaneous AR coefficients, is used in the computation of the instantaneous spectral density.

11.2 A Time Varying AR Coefficient Model

The generic scalar time varying AR coefficient model of the observed data $y_1, \ldots, y_N$ is given by

$$y_n = \sum_{i=1}^{m} \alpha_{i,n} y_{n-i} + w_n, \quad w_n \sim \mathrm{dist}(0, \sigma_n^2) \tag{11.1}$$

In general, in (11.1), the innovations $w_n, n = 1, \ldots, N$, are constrained to be independent but not necessarily Gaussian distributed or necessarily with constant variance.

In our first approach, we assume that we are modeling a nonstationary covariance time series whose covariance structure changes slowly in time. In that case, the coefficients $a_{i,n}$, in (11.1), are assumed to change "gradually" with time and w_n is assumed to be a normally distributed white noise sequence with perhaps an instantaneous variance σ_n^2. Since there are $m \times N$ AR coefficients in the model in (11.1), an attempt to fit the parameters by the least squares method or any other ordinary means to the N observations $y_1, \ldots, y_N$, will yield poor parameter estimates. We consider the unknown AR coefficients to be random variables and impose Gaussian distributed stochastic constraints on those coefficients. Those constraints define a Gaussian smoothness prior distribution on the time history of the AR coefficients.

A simple and useful constraint for a time varying AR coefficient model is obtained by imposing the stochastically perturbed difference equation constraint model on the evolution of the AR model parameters,

$$\nabla^k \alpha_{i,n} = \delta_{i,n}, \quad i = 1, \ldots, m. \tag{11.2}$$

For convenience, in (11.2) $\delta_{i,n}$ is assumed to be a zero-mean Gaussian white noise sequence with variance τ_i^2 independent of i and n. That is, $\tau_i^2 = \tau^2, i = 1, \ldots, m$. In addition, the frequency domain constraints shown in Section 4.2 in the treatment of the smoothness priors long AR model, are also incorporated here.

The smoothness priors constraints on the AR coefficients mitigate the problem of overparameterization by permitting the AR coefficients to be expressed as the solution of the constrained least squares problem

$$\sum_{n=1}^{N}\Big(y_n - \sum_{i=1}^{m} \alpha_{i,n} y_{n-i}\Big)^2 + \zeta^2 \sum_{n=1}^{N}\sum_{i=1}^{m}\Big(\nabla^k \alpha_{i,n}\Big)^2 + \nu^2 \sum_{n=1}^{N}\sum_{i=1}^{m} \alpha_{i,n}^2 + \lambda^2 \sum_{n=1}^{N}\sum_{i=1}^{m} i^4 \alpha_{i,n}^2 \tag{11.3}$$

In (11.3) m and k are assumed known and the tradeoff parameters ζ^2, ν^2 and λ^2 which balance the infidelity of the model to the data and the infidelity of the model to the smoothness constraints are not known.

Similar to the analysis in Chapter 4, (11.3) yields a Bayesian interpretation of the least squares problem. Multiply (11.3) by $-1/2\sigma^2$ and exponentiate. Then, to within a constant term,

$$\exp\left\{-\frac{1}{2\sigma^2}\sum_{n=1}^{N}\left(y_n-\sum_{i=1}^{m}\alpha_{i,n}y_{n-i}\right)^2\right\}\exp\left\{-\frac{\zeta^2}{2\sigma^2}\sum_{n=1}^{N}\sum_{i=1}^{m}\left(\nabla^k\alpha_{i,n}\right)^2\right\}$$
$$\times\exp\left\{-\frac{\nu^2}{2\sigma^2}\sum_{n=1}^{N}\sum_{i=1}^{m}\alpha_{i,n}^2\right\}\exp\left\{-\frac{\lambda^2}{2\sigma^2}\sum_{n=1}^{N}\sum_{i=1}^{m}i^4\alpha_{i,n}^2\right\} \tag{11.4}$$

expresses the product of the conditional data distribution, the prior distribution of the smoothness of the AR parameters and the prior distribution of the smoothness of the spectrum. As in the development of smoothness priors in Chapter 3, the product of the distributions in (11.4) is proportional to the posterior distribution of the AR parameters. Also as before, integration of the posterior distribution of the AR parameters yields the likelihood for the smoothness tradeoff parameters.

A first version of this method of modeling time varying AR models was reported in Kitagawa and Gersch (1985a). The program TVCAR by Kitagawa in Akiake *et al.* (1985) uses this method, (reported in Gersch and Kitagawa 1988). That program incorporates a provision for human intervention which permits the modeling of nonstationary covariance time series with abruptly changing covariance structure as occurs for example in seismic events time series. At the specific instants of abruptly changing covariance structure, the observation variance is made very large. In effect, that artifice is equivalent to starting the program over again at those specified time points. In Section 11.4, we show a non-Gaussian method of time varying AR coefficient modeling which achieves such an effect automatically without human intervention.

11.3 A State Space Model

First, because of the complexity of the complete smoothness priors state space model, we show an unconstrained state space model. Define the $km \times 1$ state vector at time n to be $x_n = (a_{1,n}, \cdots, a_{m,n}, \cdots, a_{1,n-k+1}, \cdots, a_{m,n-k+1})^T$. The state space representation of the time varying AR coefficient model is

$$\begin{aligned} x_n &= Fx_{n-1} + Gw_n \\ y_n &= H_n x_n + \varepsilon_n. \end{aligned} \tag{11.5}$$

In (11.5), H_n, is an $(m+1) \times km$ observation matrix, w_n is the m vector, $w_n = (\delta_{1,n}, \ldots, \delta_{m,n})^T$ and the $m+1$ vector $(w_n^T, \varepsilon_n)^T$, is defined in (11.8). For the difference equation orders $k_1 = 1$ and $k_2 = 2$, the matrices F, G and H_n are

$$k=1: \quad F=(I_m),\ \ G=(I_m),\ \ H_n=H_{1,n}=[\ y_{n-1},\ldots,y_{n-m}\ ] \tag{11.6}$$

$$k=2: \quad F=\begin{bmatrix} 2I_m & -I_m \\ I_m & 0 \end{bmatrix},\ \ G=\begin{bmatrix} I_m \\ 0 \end{bmatrix},\ \ H_n=H_{2,n}=[\ H_{1,n},0,\cdots,0\ ].$$

In a more general notation the state space model can be written,

$$
\begin{aligned}
F &= F^{(k)} \otimes I_m, \quad G = G^{(k)} \otimes I_m, \quad H = H^{(k)} \otimes (y_{n-1}, \cdots, y_{n-m}) \\
x_n &= (a_{1n}, \cdots, a_{mn}, \cdots, a_{1,n-k+1}, \cdots, a_{m,n-k+1})^T
\end{aligned}
\tag{11.7}
$$

where $\otimes$ denotes the Kronecker product, (Graham 1981), and I_m is the m-dimensional identity matrix.

The system noise w_n, and the observation noise ε_n form the $m+1$ vector $(w_n^T, \varepsilon_n)^T$ that is assumed to be normally distributed and independent with time and independent of each other with mean and covariance matrix,

$$
\begin{bmatrix} w_n \\ \varepsilon_n \end{bmatrix} \sim N\left(\begin{bmatrix} 0 \\ 0 \end{bmatrix}, \begin{bmatrix} Q & 0 \\ 0 & \sigma_n^2 \end{bmatrix} \right). \tag{11.8}
$$

In (11.8), the $m \times m$ system noise covariance matrix Q is a diagonal matrix with $1/\zeta^2$ on the diagonal.

For $k = 1$, a complete state space model which incorporates the smoothness priors constraints is

$$
\begin{aligned}
x_n &= \begin{bmatrix} a_{1,n} \\ \vdots \\ a_{m,n} \end{bmatrix} = \begin{bmatrix} 1 & & \\ & \ddots & \\ & & 1 \end{bmatrix} \begin{bmatrix} a_{1,n-1} \\ \vdots \\ a_{m,n-1} \end{bmatrix} + \begin{bmatrix} w_{1,n} \\ \vdots \\ w_{m,n} \end{bmatrix} \\
\begin{bmatrix} y_n \\ 0 \\ \vdots \\ 0 \end{bmatrix} &= \begin{bmatrix} y_{n-1} & \cdots & y_{n-m} \\ 1 & & \\ & \ddots & \\ & & 1 \end{bmatrix} x_n + \begin{bmatrix} \varepsilon_n \\ u_{1,n} \\ \vdots \\ u_{m,n} \end{bmatrix} \\
\begin{bmatrix} w_{1,n} \\ \vdots \\ w_{m,n} \end{bmatrix} &\sim N\left(\begin{bmatrix} 0 \\ \vdots \\ 0 \end{bmatrix}, \begin{bmatrix} \zeta^{-2} & & \\ & \ddots & \\ & & \zeta^{-2} \end{bmatrix} \right) \\
\begin{bmatrix} \varepsilon_n \\ u_{1,n} \\ \vdots \\ u_{m,n} \end{bmatrix} &\sim N\left(\begin{bmatrix} 0 \\ 0 \\ \vdots \\ 0 \end{bmatrix}, \begin{bmatrix} \sigma^2 & 0 & \cdots & 0 \\ 0 & (\nu^2+\lambda^2)^{-1} & \ddots & \vdots \\ \vdots & \ddots & \ddots & 0 \\ 0 & \cdots & 0 & (\nu^2+m^4\lambda^2)^{-1} \end{bmatrix} \right).
\end{aligned}
\tag{11.9}
$$

For a fixed difference order k, the best fit of the state space smoothness priors constraints-time varying AR coefficient model to the data $y_1, \ldots, y_N$, is the one for which the likelihood of the hyperparameters is maximized. The likelihood is computed using the recursive formulas shown in Chapter 5.

11.3.1 Instantaneous Spectral Density

Primarily the applications we are concerned with here, in the fitting of time varying autoregressive models to data, involves the estimation of an instantaneous power spectrum. That particular term is due to Page (1952) in a pionering paper. Page's concept was that the instantaneous spectrum, roughly speaking, is the difference between the power distributions of the process over the interval $(0, t)$ and over the interval $(0, t+dt)$. An alternative concept introduced in Priestley (1965) and reviewed and summarized in Priestley (1981), is that of the evolutionary spectrum. (Also see Loynes 1968 and Brillinger 1993b.) The evolutionary spectrum represents the power distribution of the process within the interval $(t, t + dt)$. That is, the evolutionary spectrum corresponds roughly to what we would measure if the portion of time series lying between t and $t + dt$ were passed through a wave analyzer. Priestley's estimation of the evolutionary spectrum is done via an adaptation of windowed periodogram analysis. The analysis are only applicable to very slowly varying spectra and are not of great interest today. On the other hand, our concept of instantaneous spectrum is similar to the evolutionary spectrum concept used by Priestley. Another approach to the estimation of an instantaneous spectrum is the Wigner-Ville spectrum, (introduced in Ville 1948, other theoretical work and references are in Martin and Flandrin 1985). At any particular time instant, that spectrum is defined as the Fourier transform of the covariance function over a small interval time interval around that particular time instant. The Wigner-Ville spectrum does not have a meaningful physical interpretation and may in fact take on negative values for certain processes.

The evolutionary spectrum of Dahlhous (1993), motivated by the recent work in wavelets, allows for both the unique definition and consistent estimation of what might be the spectrum of a time series at a single time point. Currently nonlinear wavelet thresholding methods are very popular due to the statistical estimation breakthrough papers by Donoho and Johnstone, (1992, 1993). In those methods an orthonormal system of basis functions, localized in spatial position and frequency are capable of optimally compressing functions under a variety of smoothness assumptions. Some generalizations of that nonparametric regression method can be found in Brillinger (1994) and Johnstone and Silverman (1994). Dahlhous *et al.* (1995), is a related complex nonlinear wavelet threshold nonparametric estimation of the coefficients of a scalar TVAR process.

Our own approach is to define an instantaneous spectrum operationally. In a natural way we extend the definition of the power spectral density of a stationary process to that of an instantaneous power spectrum in terms of the time varying autoregressive model. That approach, while not mathematically formal, appears to yield meaningful computational results. Examples were shown for the scalar case in Kitagawa (1983) and Kitagawa and Gersch (1985a) in the modeling of scalar nonstationary covariance time series. Houacine and Demoment (1987) is an adaptive spectrum estimation variation of our approach which combines the smoothness priors long AR modeling (Chapter 4) and state space modeling.

The instantaneous power spectral density matrix is the distribution of energy in the time series as a function of frequency at an instant in time. The more conventional analysis of stationary time series by parametric or Fourier transform methods extended to a Priestly-type evolutionary spectrum do not have the time-frequency resolution properties necessary to capture the transitory characteristics of relatively rapidly changing nonstationary covariance data.

Therefore then, motivated by earlier work on spectrum estimation and the considerations discussed above, we define the instantaneous spectrum of a time varying coefficient AR process by

$$S_n(f) = \frac{\sigma_n^2}{\left|1 - \sum_{k=1}^{m} a_{k,n} \exp(-2\pi i k f)\right|^2}; \quad -1/2 \leq f \leq 1/2. \tag{11.10}$$

The value of the instantaneous spectrum is obtained by substituting the smoothed estimates of the TVAR coefficients and the smoothed estimate of the innovations variance σ_n^2 into (11.10).

We suggest that the method of estimation of instantaneous frequency introduced here is superior in time-frequency resolution properties than any of the methods reported in recent survey articles on the subject, (Boahash 1992).

11.4 PARCOR Time Varying AR Modeling

Alternatively, the coefficients of a stationary AR model can be estimated recursively via the estimation of the partial autocorrelation coefficients, (PARCORs), in a lattice AR structure model, (Goodwin and Payne 1977, Haykin 1991). This concept was extended to the time varying AR coefficient model in (11.1) in Kitagawa (1988). It is used here to model the PARCORs in terms of stochastic trend models. In contrast with the approach in Sections 11.2–11.3, in which stochastic contraints are placed on each of m time varying AR coefficients, in the time-varying PARCOR method, stochastic constraints are only placed on the PARCORs (in successively increasing model orders). In that case, models for smoothing time-varying partial autocorrelation coefficients are

$$\begin{aligned} f_n^{(m-1)} &= \alpha_{mn}^{(m)} b_{n-m}^{(m-1)} + f_n^{(m)} \\ \nabla \alpha_{mn}^{(m)} &= v_n^{(m)} \\ b_n^{(m-1)} &= \gamma_{mn}^{(m)} f_{n+m}^{(m-1)} + b_n^{(m)} \\ \nabla \gamma_{mn}^{(m)} &= u_n^{(m)} \end{aligned} \tag{11.11}$$

with $f_n^{(0)} = b_n^{(0)} = y_n$, $f_n^{(m)}$ and $b_n^{(m)}$ respectively the forward and backward prediction errors of the autoregressive model of order m. In the stationary case, ($v_n^{(m)} = u_n^{(m)} = 0$) and $\alpha_{mn}^{(m)} = \gamma_{mn}^{(m)}$ are identical to the partial autocorrelation coefficients. A somewhat brief description of the method is as follows.

The zero-th order forward and backward innovations are

$$f_n^{(0)} = y_n, \quad b_n^{(0)} = y_n. \tag{11.12}$$

The relations between the order update innovations and the instantaneous forward and backward PARCORs are in general given by

$$\begin{aligned} f_n^{(j)} &= f_n^{(j-1)} - \alpha_{j,n}^{(j)} b_{n-j}^{(j-1)} \\ \nabla \alpha_{jn}^{(j)} &= v_n^{(j)}, \qquad v_n^{(j)} \sim N(0, \tau_1^2) \\ b_n^{(j)} &= b_n^{(j-1)} - \gamma_{j,n}^{(j)} f_{n+j}^{(j-1)} \\ \nabla \gamma_{jn}^{(j)} &= u_n^{(j)}, \qquad u_n^{(j)} \sim N(0, \tau_2^2). \end{aligned} \tag{11.13}$$

The instantaneous updated forward and backward AR model parameters for model order $i = 1, \ldots, j-1$, then become,

$$\begin{aligned} \alpha_{i,n}^{(j)} &= \alpha_{i,n}^{(j-1)} + \alpha_{j,n}^{(j)}\gamma_{j-i,n}^{(j-1)} \\ \gamma_{i,n}^{(j)} &= \gamma_{i,n}^{(j-1)} + \gamma_{j,n}^{(j)}\alpha_{j-i,n}^{(j)}. \end{aligned} \tag{11.14}$$

In (11.13) the order updated instantaneous PARCORs, $\alpha_{j,n}^{(j)}$, $\gamma_{j,n}^{(j)}$ are scalar regression coefficients and the updated forward and backward innovations are the residuals of the regressions. The innovation in (11.13) is the stochastic difference equation representation for the order updated forward and backward instantaneous PARCORs. State space computations for the likelihood of the hyperparameter models corresponding to τ^2 is used, (for convenience we let $\tau_1^2 = \tau_2^2 = \tau^2$). Akaike's AIC is used to determine the AIC best order PARCOR model.

The distribution of the noise inputs may be either Gaussian or non-Gaussian. After filtering and smoothing the time-varying AR coefficients, the instantaneous spectrum of the nonstationary process is estimated by (11.9).

11.5 Examples

TVCAR Modeling: A Synthetic Data Set

The first example is the application of the constrained AR coefficient model to the estimation of the gradually changing spectrum of a simulated times series shown in Figure 11.1A. The corresponding evolution of the theoretical instantaneous power spectral density is shown in Figure 11.1B. The time series was generated by the model

$$y_n = \sum_{i=1}^{4} a_{i,n} y_{n-i} + w_n, \quad w_n \sim N(0,1). \tag{11.15}$$

The characteristic roots of the AR operator

$$s^4 - a_{1,n}s^3 - a_{2,n}s^2 - a_{3,n}s - a_{4,n} = 0 \tag{11.16}$$

are given by

$$0.83\cos\theta_{1,n} \pm \sqrt{-1} \times 0.83\sin\theta_{1,n}, \quad 0.88\cos\theta_{2,n} \pm \sqrt{-1} \times 0.88\sin\theta_{2,n}, \tag{11.17}$$

$$\theta_{1,n} = \frac{5}{9}\pi + \frac{\pi}{6}\sin\left\{\frac{7\pi(n+100)}{2700}\right\}, \quad \theta_{2,n} = \frac{\pi}{6} + \frac{\pi}{12}\sin\left\{\frac{7\pi(n+100)}{5400}\right\}. \tag{11.18}$$

The program TVCAR by Kitagawa in Akaike *et al.* (1985) was used to model this data. In that program for convenience, the trend model order k is "fixed at $k = 2$ and the AR model order is "fixed" internally at $m = 8$, (those parameters could of course be modified). As is apparent in Figure 11.1C, that model achieves a reasonable reproduction of the theoretical changing spectrum shown in Figure 11.1B.

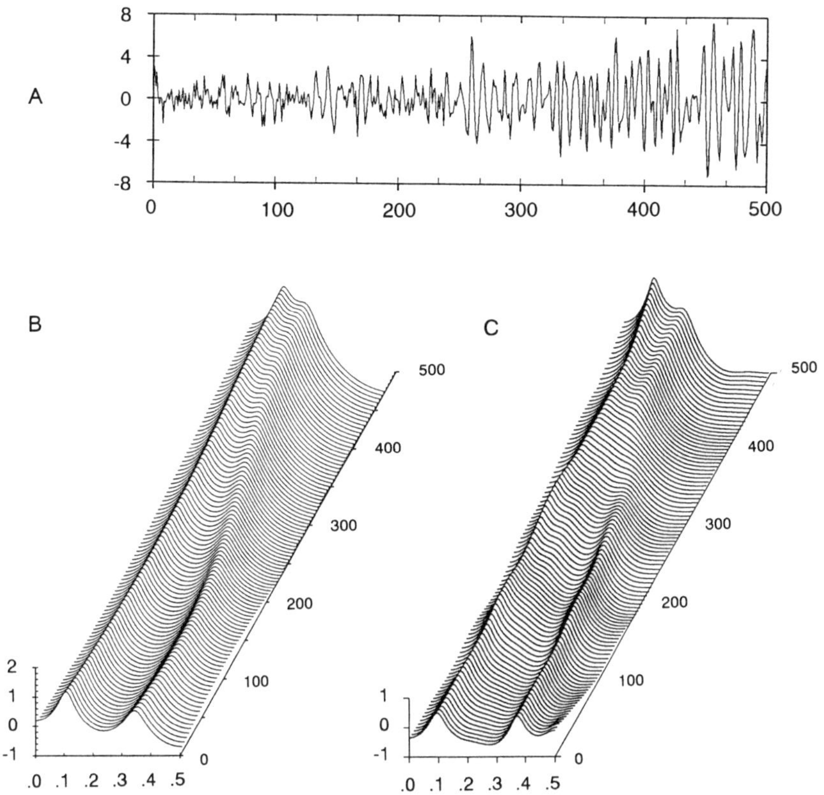

Figure 11.1 A: A $N = 500$ simulated nonstationary covariance data set with gradually changing spectrum. B: The time evolution of the theoretical instantaneous power spectral density. C: The time evolution of the estimated instantaneous power spectral density.

TVCAR and PARCOR Modeling: A Seismic Data Example

The MYE2F Hokkaido seismic data, $n = 2600$, was modeled using both the TVCAR program as well as the PARCOR method of fitting time varying AR coefficient models.

Two different runs of the TVCAR program were made with the AR model order $m = 8$. One was a linear Gaussian modeling, the other was a "human intervention" run in which the P-wave and S-wave arrival times estimated by using the locally stationary AR model were inputs to the program. The likelihoods for the Gaussian and human intervention runs were respectively were, -2792.35 and -2723.97. Correspondingly the AICs were, 5592.70 and 5455.95. The human intervention model was the better AIC model. Figure 11.2A respectively shows the original earthquake data,

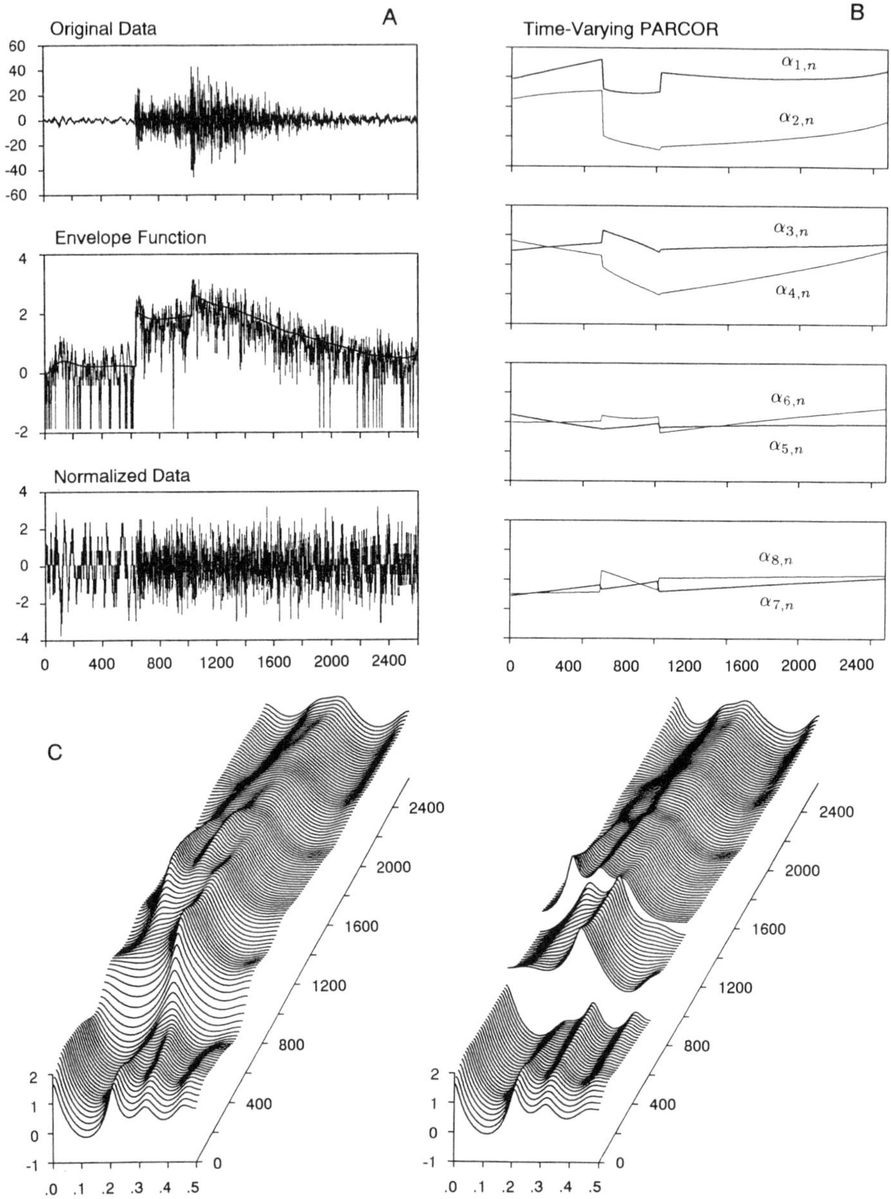

Figure 11.2 A: Original earthquake data, envelope function and variance normalized data. B: Evolution of the PARCORs. C: The evolution of the instantaneous spectra computed from the linear Gaussian model and the human intervention model.

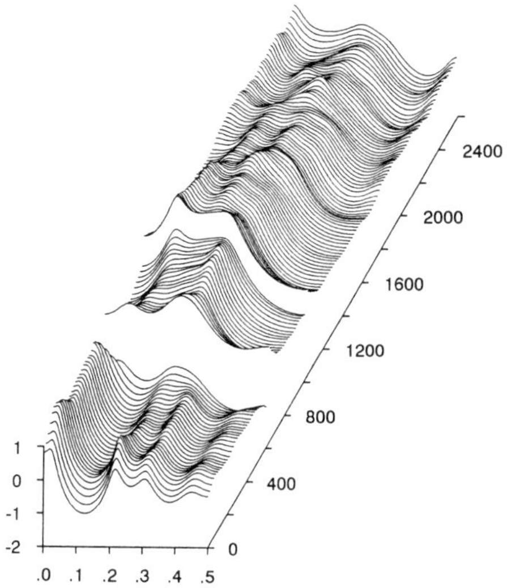

Figure 11.2 D: The evolution of the instantaneous spectra computed using the the non-Gaussian Cauchy distributed PARCOR model.

the envelope function and the variance normalized data for the human intervention run. Figure 11.2B illustrates the evolution of the PARCORs. The abrupt changes in the appearance of the envelope function and PARCORs is a direct consequence of the intervention. In Figure 11.2C the evolution of the instantaneous spectra computed from both the linear Gaussian model and the human intervention model are shown. The intervention model instantaneous spectrum captures the abrupt changes in the spectrum. Initially we observe the spectrum of the background or ongoing vibrations of the earth. The dramatic changes in the spectrum, coincident with the arrival of the P-wave and the S-wave are clearly discernible. After the S-wave, the spectrum tends to return to its original steady state form. The appearance of the evolutionary spectrum in the linear Gaussian non-intervention modeling, is seen to only change gradually and the transitions from background to P-wave to S-wave spectra are blurred.

Two computational runs using the PARCOR method of modeling were also done on this data using a first order trend model for the PARCORs. In one computation, the trend model variations on the PARCORs were Gaussian and the observation noise was also assumed to be Gaussian. In the other computation the assumption was that v_n and u_n are Cauchy and ε_n is Gaussian. The instantaneous spectra for the Gaussian distribution PARCORs modeling was similar to the results for the TVCAR modeling and is not shown here. The illustration of the instantaneous spectra for the Cauchy distribution PARCORs modeling is shown in Figure 11.2D. The point to be emphasized here is that using the non-Gaussian PARCOR method of modeling, the

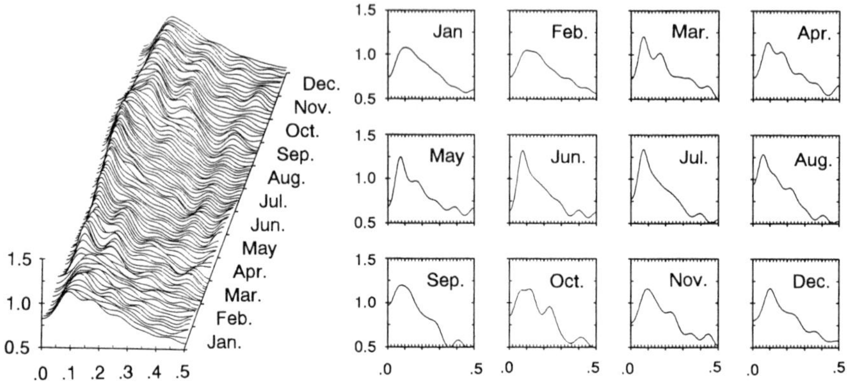

Figure 11.3 Evolutionary spectrum analysis of the deviations from the Tokyo maximum temperature data.

abrupt changes in both the PARCORs and the evolutionary spectrum at the times of occurrence of the P and S waves is automatic.

Deviations from the Maximum Temperature Common Trend

In this section we show the nonstationary covariance modeling of the deviations from the common mean of 21 years (1971–1991), of the Tokyo maximum daily temperature. For completeness, the time varying AR coefficient model is

$$y_n = \sum_{i=1}^{m} \alpha_{i,n} y_{n-i} + w_n, \quad w_n \sim N(0, \sigma_n^2). \tag{11.19}$$

The time domain constraint on the evolution of the individual AR components is

$$\nabla^k \alpha_{i,n} = \delta_{i,n}, \quad i = 1, \ldots, m. \tag{11.20}$$

In addition, an AR parameter constraint in the form,

$$\alpha_{i,n} = \alpha_{i,n-365} \tag{11.21}$$

was imposed. The evolutionary spectrum and graphs of individual years of spectra are shown in Figure 11.3.

The spectral peak with period length from one week to 10 days is clearly seen. The location of the peak shifts to the low frequency side in March to June. In this period, the increase of the power is also clearly seen in the figure.

Deviations from the Mid-Atlantic Ridge Data Trend

Here we show an example of a two-stage analysis of a time series, the first stage being the nonstationary mean or trend removal, the second stage being the analysis of the

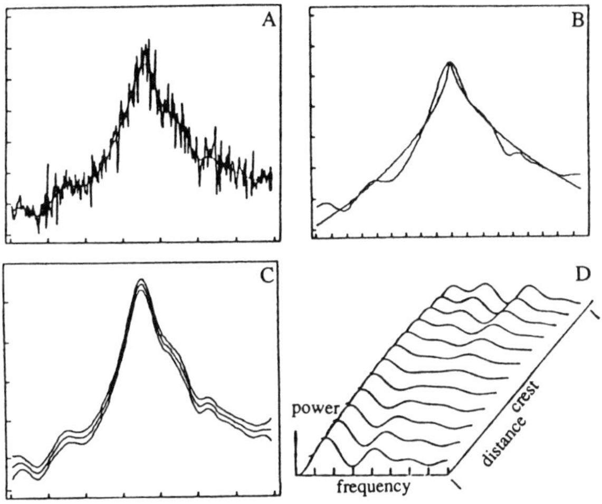

Figure 11.4 A: The mid-Atlantic ridge data and superimposed AIC best trend. B: The AIC best trend and superimposed $\sqrt{t}$ model. C: The AIC best trend and $\pm 2\sigma$. D: The evolutionary spectrum in db vs normalized frequency with distance.

residuals as a nonstationary covariance time series. This analysis, in the category of a nonparametric regression with nonstationary residuals is novel, we do not know of other analyses that treat this problem.

The bathymetric, (depth of the ocean), data along flowlines across the mid-Atlantic ridge that was analyzed in Gilbert and Malinverno (1988) is analyzed here, (also see Gersch 1990). There, the Parson and Sclater (1977) "deterministic" $\sqrt{t}$ component of ocean floor topography, (where t = time is measured in millions of years, and is related to thermal contraction), was removed from the data by a (parametric) least squares model analysis, prior to their stochastic model analysis of the residual. (The residual is the difference between the original data and the fitted trend component).

In our analysis, first the smoothness priors nonstationary mean analysis was applied to the data. The trend model order $k = 2$ was the AIC best model for the flowline data. Figures 11.4A shows (every 20th point of) the original data and the superimposed estimated mean value. Figure 11.4B shows the superimposition of the estimated mean value and the fit of the $\sqrt{t}$ model to the estimated mean value. The estimated trend is asymmetric around the crest and it deviates from the strict Parson and Sclater decrease of trend $\sqrt{t}$ relationship. Figure 11.4C shows the estimated mean value and a 95% (plus and minus two standard deviations) confidence limit around that mean value.

A TVCAR program analysis of the residuals, the difference between the flowline data and the best fitted trend, reveal them to have a nonstationary covariance. Graphical results of the power spectral density versus normalized frequency as a function of distance, with one spectral line every 15 (reduced rate) data points is shown in Figure 11.4D. The power spectrum is evolutionary or nonstationary, and not symmetrically nonstationary with distance away from the crest along the flowline.

Two potentially scientifically important results were revealed in this analysis. One is that the $\sqrt{t}$ deterministic trend model does not appear to be a statistically correct model of the (Gilbert-Malinverno data) ridge axis spread. Our analysis of the Gilbert-Malinverno data reveals asymmetry and anomalies in the spreading rates across the mid-Atlantic ridge axis that deviate from the Parson and Sclater theory. The scientific validity and the scientific importance of the Parson and Sclater theory is not an issue here. The anomalies in spreading rate revealed by the nonstationary trend analysis, remain to be compared to deviations from the more traditional $\sqrt{t}$ modeling and to be explained by marine geologists. This analysis does have implications on the generally accepted practice of removing "deterministic" geological features, (including trend surface), from the overall evidence in modeling seafloor data, even if those features are motivated by theory. The specification and removal of a deterministic or stochastic component in general time series modeling, is in general a statistically questionable practice, and it may lead to scientifically questionable results as well. The reason for this is that there is only one likelihood. Removing one component from the time series and then analyzing the remaining time series involves two likelihood maximizations. What is not sufficiently removed or what is excessively removed in the first stage of the analysis, contaminates the analysis in the second stage of the analysis. (Of course, under some circumstances it is possible to iterate the procedures and converge to a stable "correct" solution.)

The second result, revealed by the TVCAR analysis, yields a not-necessarily-homogeneous seafloor field description. Local variability of the ocean bottom roughness is revealed in the evolutionary spectral density of the field . The deviations from a homogeneous field are apparent, they are asymmetric and appear to be systematic. The structure in the evolution of the spectrum across the mid-Atlantic ridge axis also remains to be explained by marine geologists.

12

Modeling Multivariate Nonstationary Covariance Time Series

The modeling of multivariate nonstationary covariance time series is relevant to many econometric, human electroencephalogram, oceanographic, meteorological and seismic data situations. (See for example Brillinger and Hatanaka 1969, 1970.) In this chapter multivariate nonstationary covariance time series are modeled using time varying multivariate autoregressive or equivalently, time varying vector autoregressive, (TVVAR), models. This model is a natural extension of the scalar TVAR model shown in Chapter 11. Analogous to Chapter 11, two different modeling methods are presented. In one method, smoothness priors constraints are placed directly on the evolution of the individual AR parameters of the TVVAR model. In the other method, smoothness priors constraints are placed on the evolution of the PARCORs of AR models. The key idea which permits TVVAR modeling is a "one channel at-a-time" paradigm. This idea is a consequence of an instantaneous response-orthogonal innovations representation of multivariate AR models. That permits the multivariate time series to be modeled one scalar AR model at-a-time.

12.1 Introduction

An earlier, and in fact still used approach to the modeling of nonstationary covariance time series, is to segment the series into approximately stationary time series segments and to model each segment as a stationary time series model. The problem with this approach is that, in regions in which the covariance structure is slowly varying, the precise determination of segments becomes problematic and the method is unlikely to adequately capture the local and global statistical properties of the time series. On the basis of success with previous use of multivariate autoregressive models of stationary time series, Gersch and Kitagawa (1983b) considered the fitting of a time varying coefficient multivariate (or vector) autoregressive (TVVAR) model,

$$y_n = \sum_{m=1}^{p} A_{mn} y_{n-m} + w_n, \quad n = 1, \ldots, N. \tag{12.1}$$

to nonstationary covariance economic data. In (12.1), the y_n's are d-dimensional vectors and $w_n, 1, \ldots, N$ is a zero-mean uncorrelated normally distributed d-dimensional vectors sequence with instantaneous covariance matrix, W_n. That method was subsequently applied to the modeling of nonstationary covariance human epileptic episode electroencephalograms, (Gersch 1985, 1987a).

In the model above for a d channel N observation time series there are $N \times d$ observations and $d \times d \times p \times N$ autoregressive model parameters and $d \times (d+1) \times N/2$ innovations matrix parameters. In order to fit such a model with so many more parameters than observations, it is necessary to efficiently parametrize the time evolution of the parameters. The method that we used was to characterize the evolution of each of the $d \times d \times p$ autoregressive parameters in terms of a discrete Laguerre polynomial in a least squares computational solution. The order of the multivariate AR model and the degree of the Laguerre polynomial were determined using the AIC. The results obtained in the aforementioned papers seemed to be quite reasonable. On the other hand in that procedure, the Laguerre polynomial coefficients were determined by global statistical considerations. Ultimately it became clear that it was necessary to develop a multivariate time varying AR parameter modeling procedure that was more sensitive to local time variations. Each of the TVVAR modeling methods developed in this chapter have that property of being more sensitive to local time variations of the covariances.

The key idea in our approach to the realization of TVVAR models, is in an alternative representation of the conventional multivariate autoregressive model in an instantaneous response-orthogonal innovations representation. In that representation the original multivariate time series data is interleaved in such a way as to allow us to model the multivariate time series data in one scalar autoregressive channel model at-a-time. We refer to that approach as the "one channel at-a-time" paradigm.

Here too, our primary application for this modeling is in the evolution with time of the power spectral density matrix. Correspondingly, as a direct extension of the concept of an instantaneous power spectral density for scalar time series, we introduce an instantaneous power spectral density matrix. Examples of the modeling of both simulated and real physical multivariate seismic data and human electroencephalographic, (EEG) data are shown.

The instantaneous response-orthogonal innovations model idea is described in Section 12.2. The first variation, also described in Section 12.2, in an application of the one channel at-a-time paradigm, imposes smoothness priors trend-type constraints on the evolution of the AR parameters. In Section 12.3 a state space modeling of this method is shown. A more complete treatment of this material appears in Jiang and Kitagawa (1993). The second variation in which smoothness priors constraints are placed on the evolution of the PARCORs of the individual AR time series, is described in Section 12.4. More complete treatments of that material appears in Gersch and Stone (1994, 1995) and in the Ph.D thesis, Stone (1993). Examples of the modeling of multivariate nonstationary time series are in Section 12.5

12.2 The Instantaneous Response-Orthogonal Innovations Model

Here an instantaneous response-orthogonal innovations representation of multivariate AR time series is introduced and exploited to accomplish the modeling of multivariate AR time series, one scalar AR time series at-a-time. We have referred to this idea as the "one channel at-a-time" paradigm. The method is applicable to both multivariate stationary and nonstationary mean and nonstationary covariance time series. The one channel at-a-time instantaneous response-orthogonal innovations variance model is related to the VAR constant coefficient model with a general innovations variance by an invertible algebraic transformation. In the one channel at-a-time modeling, the order of each of the scalar AR models is determined using Akaike's AIC.

An instantaneous response-orthogonal innovations VAR parameterization was introduced in Akaike (1968) and Pagano (1978). The same parameterization was used in the program MULMAR by Kitagawa and Akaike (Akaike *et al.* 1978, Kitagawa and Akaike 1981), and also used in Newton (1982), Sakai (1982) and Takanami and Kitagawa (1991). Each of those papers dealt with the VAR modeling of stationary time series. Gersch (1991, 1992), and Jiang and Kitagawa (1993) exploit the one channel at-a-time paradigm in different ways to achieve the TVVAR modeling of multivariate nonstationary covariance data. In Stone (1993), the one channel at-a-time paradigm is exploited for parsimonious VAR modeling and enhanced one-step-ahead prediction of stationary time series, for the extension of the scalar smoothness priors long AR modeling that was described in Section 4.2 to the multivariate case, (see Chapter 16), and to TVVAR modeling of nonstationary covariance time series.

For clarity and simplicity of presentation of the one channel at-a-time paradigm and its application to TVVAR modeling, we separate the discussion into two parts and first present that idea in context of stationary VAR modeling.

Constant Coefficient VAR Time Series Modeling

Let $y_n = (y_{1,n}, y_{2,n}, \ldots, y_{d,n})^T$, $(n = 1, 2, \ldots, N)$ be a d-variate time series. The vector autoregressive, (VAR), model of that stationary time series is given by

$$y_n = \sum_{l=1}^{p} A_l y_{n-l} + w_n, \tag{12.2}$$

where p is the autoregressive order, A_l is the $d \times d$ autoregressive coefficient matrix and w_n is assumed to be a d-variate Gaussian white noise sequence with mean zero and covariance matrix W which is independent of y_m for $n > m$.

Premultiplying both sides of (12.2) by the unique unit diagonal lower triangular Cholesky factor of W so that the covariance matrix of w_n is reduced to a diagonal matrix U yields the instantaneous response-orthogonal innovations model,

$$y_n = B_0 y_n + \sum_{l=1}^{p} B_l y_{n-l} + \varepsilon_n. \tag{12.3}$$

Here the coefficient matrix of the instantaneous response, B_0, is in the form:

$$B_0 = \begin{bmatrix} 0 & 0 & \cdots & 0 \\ b_0(2,1) & 0 & \cdots & 0 \\ \vdots & \ddots & \ddots & \vdots \\ b_0(d,1) & \cdots & b_0(d,d-1) & 0 \end{bmatrix},$$

and ε_n is a d-variate white noise sequence with mean zero and diagonal covariance matrix $U = \text{diag}\ (\sigma_1^2,\ \sigma_2^2,\ \cdots,\ \sigma_d^2)$. Using simple algebraic operations, (12.3) can be re-expressed as

$$y_n = (I - B_0)^{-1} \sum_{l=1}^{p} B_l y_{n-l} + (I - B_0)^{-1} \varepsilon_n. \tag{12.4}$$

Thus the instantaneous response-orthogonal innovations model in (12.3) is related by an algebraic transformation to the usual VAR model in (12.2) by

$$\begin{aligned} A_l &= (I - B_0)^{-1} B_l, \qquad l = 1, 2, \cdots, m, \\ W &= (I - B_0)^{-1} U (I - B_0)^{-T}. \end{aligned} \tag{12.5}$$

The power spectral density matrix of the VAR process defined in (12.2), (Whittle 1963), is

$$P_f = A_f^{-1} W A_f^{-1*}, \tag{12.6}$$

where A_f^* is the complex conjugate transpose matrix of A_f and ,

$$A_f = I - \sum_{l=1}^{p} A_l \exp(-2\pi i f l), \tag{12.7}$$

is the frequency response matrix or, the whitening filter of the VAR model.

The model (12.3) also has a direct representation for the power spectral density matrix in the form

$$P_f = B_f^{-1} U B_f^{-T*}, \quad -\frac{1}{2} \le f \le \frac{1}{2}, \tag{12.8}$$

where

$$B_f = I - \sum_{l=0}^{p} B_l \exp(-2\pi i f l), \qquad B_0 = B_0 - I. \tag{12.9}$$

A very important property of the instantaneous response-orthogonal innovations VAR model is that it can be estimated by successively fitting the scalar models for each of the d orthogonal components ($y_{i,n},\ i = 1, 2, \cdots, d$) of the model in (12.3).

Time Varying Coefficient VAR Modeling

Next, we extend the model in (12.3) to the fitting of multivariate time-varying VAR models for the modeling of multivariate nonstationary covariance data. A time varying coefficient d-variate VAR model, a variation of the model in (12.3) is

$$y_n = B_{0n} y_n + \sum_{l=1}^{p} B_{l,n} y_{n-l} + \varepsilon_n, \tag{12.10}$$

where $\varepsilon_n = (\varepsilon_{1,n}, \varepsilon_{2,n}, \cdots, \varepsilon_{d,n})^T \sim N(0, U)$ with $U = \text{diag}\,(\sigma_1^2,\, \sigma_2^2,\, \cdots,\, \sigma_d^2)$. The time varying coefficient matrices are expressed by

$$\begin{aligned} B_{0n} &= \begin{bmatrix} 0 & 0 & \cdots & 0 \\ b_{0n}(2,1) & 0 & \cdots & 0 \\ \vdots & \ddots & \ddots & \vdots \\ b_{0n}(d,1) & \cdots & b_{0n}(d,d-1) & 0 \end{bmatrix}, \\ B_{l,n} &= \begin{bmatrix} b_{ln}(1,1) & \cdots & b_{ln}(1,d) \\ \vdots & \ddots & \vdots \\ b_{ln}(d,1) & \cdots & b_{ln}(d,d) \end{bmatrix}, \quad l = 1, 2, \cdots, p. \end{aligned} \tag{12.11}$$

As a direct extension of the scalar instantaneous power spectral density, for time varying VAR models, we use the concept of the instantaneous power spectral density matrix. The instantaneous power spectral density matrix corresponding to the time varying coefficient VAR model in (12.1) is, (Gersch and Kitagawa 1983),

$$P_{f,n} = A_{f,n}^{-1} W_n A_{f,n}^{-1*}, \quad -\frac{1}{2} \le f \le \frac{1}{2}, \tag{12.12}$$

with $A_{f,n}$ the instantaneous whitening filter given by

$$A_{f,n} = I - \sum_{l=1}^{p} A_{l,n} \exp(-2\pi i f l).$$

In terms of the instantaneous response-orthogonal innovations time varying AR model, the multivariate power spectral density matrix is given by

$$P_{f,n} = B_{f,n}^{-1} V_n B_{f,n}^{-1*}, \quad -\frac{1}{2} \le f \le \frac{1}{2}, \tag{12.13}$$

where the matrix $B_{f,n} = (b_{f,n}(i,j))_{d \times d}$ is defined by

$$B_{f,n} = I - \sum_{l=0}^{p} B_{l,n} \exp(-2\pi i f l). \tag{12.14}$$

12.3 State Space Modeling

Here we assume that in (12.11), the $b_{ln}(i,j)$'s change gradually with time and can be considered as random variables. As in the scalar TVAR coefficient modeling in

Chapter 11, stochastically perturbed difference equations with Gaussian noise are imposed as the smoothness prior distributions on the time history of the VAR model coefficients. That is, let

$$\nabla^q b_{ln}(i,j) = \nu_{ln}(i,j), \qquad i,j = 1,\cdots,d;\ l = 0,1,\cdots,p, \tag{12.15}$$

where ∇ is the difference operator defined by $\nabla b_{ln}(i,j) = b_{ln}(i,j) - b_{l,n-1}(i,j)$.

In (12.15), $\nu_{ln}(i,j)$ is assumed to be a Gaussian white noise sequence with mean zero and unknown variance $\tau_l^2(i,j)$, and independent of $\nu_{tn}(r,s)$ for any $\{r,s,t\} \neq \{i,j,l\}$.

Once fitting the instantaneous response model, we can obtain the more familiar time varying coefficient VAR model, (12.1),

$$y_n = \sum_{l=1}^{p} A_{l,n} y_{n-l} + w_n, \tag{12.16}$$

with $w_n \sim N(0, W_n)$ and uncorrelated in time. Similar to the stationary VAR model situation, this time varying VAR model can be related to the instantaneous response-orthogonal innovations time varying VAR model by the linear algebraic transformations

$$\begin{aligned} A_{l,n} &= (I - D_n)^{-1} B_{l,n}, \qquad l = 1,2,\cdots,p, \\ W_n &= (I - D_n)^{-1} U_n (I - D_n)^{-T}. \end{aligned} \tag{12.17}$$

A substantial difficulty with this model is that it contains excessively many hyperparameters $\tau_l^2(i,j)$, $i,j = 1,\cdots,d$; $l = 1,\cdots,p$ and $\tau_0^2(i,j)$, $1 \le j < i \le d$. Using the related idea in Kitagawa and Gersch (1985a) for the reduction of the number of hyperparameters in scalar time-varying AR modeling, we let $\tau_l(i,j)^2 = \tau_i^2$. That is, the p time varying coefficients in the scalar AR model for each of the d channels have the same hyperparameter and it is sufficient to merely determine the d hyperparameters $\tau_1^2, \cdots, \tau_d^2$.

Since $\varepsilon_n(i)$ is independent of $\varepsilon_n(j)$ for $j \neq i$, for $i = 1,\cdots,d$, each of the scalar time varying AR models

$$y_n(i) = \sum_{j=1}^{d} \sum_{l=0}^{p} b_{ln}(i,j) y_{n-l}(j) + \varepsilon_n(i), \tag{12.18}$$

with $\nu_{ln}(i,j) \sim N(0,\tau_i^2)$, $j = 1,\cdots,d$; $l = 1,2,\cdots,p$, $b_{0n}(i,j) = 0$ for $j \ge i$, can be fitted independently.

To estimate and identify the time varying coefficient VAR model efficiently, we introduce the state space form of the model and use the Kalman filtering and smoothing algorithms. The generic state space form of the i-th component model for the observations $\{y_n(i);\ n = 1,2,\cdots,N\}$ is

$$\begin{aligned} x_n &= F x_{n-1} + G w_n, \\ y_n(i) &= H_n x_n + \varepsilon_n(i), \quad i = 1,2,\cdots,d, \end{aligned} \tag{12.19}$$

where F, G, and H_n are $M \times M$, $M \times L$, and $1 \times M$ matrices, respectively. Also, x_n and $y_n(i)$ are the state vector and the observation at time n. In our case, the matrices F, G, and H_n are known and for convenience the initial state is assumed to be normally distributed with mean $x_{0|0} = 0$ and covariance matrix $V_{0|0}$ a diagonal matrix with arbitrarily large diagonal elements.

In particular, for the difference equation order $q = 1, 2$, and for the i-th component model, the matrices F and G are given by

$$\begin{aligned} q = 1: &\quad M = L, \quad F = [I_L], \quad G = [I_L], \\ q = 2: &\quad M = 2L, \quad F = \begin{bmatrix} 2I_L & -I_L \\ I_L & 0 \end{bmatrix}, \quad G = \begin{bmatrix} I_L \\ 0 \end{bmatrix}, \end{aligned} \tag{12.20}$$

where $L = dp + i - 1$ and I_L is the L dimensional identity matrix. The state vector x_n and the matrix H are respectively given by

$$\begin{aligned} x_n &= [\underbrace{b_{0n}(i,1), \cdots, b_{0n}(i,i-1)}_{i-1}, \underbrace{b_{1n}(1,1), \cdots, b_{pn}(1,1)}_{p}, \cdots, \underbrace{b_{1n}(1,d), \cdots, b_{pn}(1,d)}_{p}]^T \\ H_n &= [\underbrace{y_{1n}, \cdots, y_{i-1,n}}_{i-1}, \underbrace{y_{1,n-1}, \cdots, y_{1,n-p}}_{p}, \cdots, \underbrace{y_{d,n-1}, \cdots, y_{d,n-p}}_{p}]. \end{aligned} \tag{12.21}$$

We assume that the $L+1$ vector $(w_n^T, \varepsilon_{i,n})^T$ is a white noise sequence with

$$\begin{pmatrix} w_n \\ \varepsilon_n(i) \end{pmatrix} \sim N\left(\begin{bmatrix} 0 \\ 0 \end{bmatrix}, \begin{bmatrix} \tau_i^2 I_L & 0 \\ 0 & \sigma_i^2 \end{bmatrix} \right). \tag{12.22}$$

Given the observations $\{y_n(i); n = 1, 2, \cdots, N\}$ and the initial conditions $x_{0|0}$ and $V_{0|0}$, the one-step-ahead predictor $x_{n|n-1}$ and the filter $x_{n|n}$ of the state vector x_n are given by the Kalman filter algorithm described in Chapter 5.

Akaike's minimum AIC procedure is used to select the best of alternative parametric models from the set of time varying VAR models with VAR order p and difference order q.

As noted before, the state space representation and the Kalman filter yields an efficient algorithm for computing the likelihood of a time series model. The maximum likelihood estimate of the vector of model parameters, $\hat{\theta}_i$, is obtained by maximizing the log-likelihood, $\ell(\theta_i)$ with respect to those parameters. Correspondingly, the AIC of the i-th component model is defined by

$$\text{AIC}_i = -2\ell(\hat{\theta}_i) + 2(\text{dimension of } \theta_i). \tag{12.23}$$

The AIC for the entire model is then given by

$$\text{AIC} = \sum_{i=1}^{d} \text{AIC}(i), \tag{12.24}$$

and the best VAR order p and the difference order q in the priors are determined by minimizing the AIC of the overall model.

12.4 Time Varying PARCOR VAR Modeling

In this section we show a smoothness priors constraint on the time varying PARCOR VAR modeling method. It is an alternative to the smoothness constraints on the AR coefficients TVVAR model in Sections 12.2 and 12.3. First some background is offered.

Whittle (1963) was a seminal development in the fitting of multichannel autoregressive (AR) models to stationary time series data. In extending the Levinson algorithm for fitting AR models to scalar stationary time series to the multichannel case, Whittle found it necessary to employ both forward and backward multichannel AR models. The latter was critical in the implementation of a multichannel lattice-form algorithm, Wiggins and Robinson (1965). That realization and subsequent multichannel realizations by Lee *et al.* (1981), Friedlander (1982), and Porat *et al.* (1982) involved matrix computations. Scalar implementations of multichannel recursions, which required no matrix processing, were realized by Sakai (1982), Lev-Ari (1987), Lev-Ari and Kailath (1984), and Ling and Proakis (1984). Sakai's implementation, (motivated by Pagano 1978), transforms multichannel AR models into scalar AR models, by converting the multichannel Levinson algorithm into a scalar equivalent which Sakai calls the "circular lattice filter". Lev-Ari's (1987) realization emphasizes the modularity of the architecture and is the most elegant of the multichannel lattice algorithm developments. (The structure of Lev-Ari's algorithm and Sakai's algorithm are equivalent.) Sakai deals only with stationary time series modeling and Lev-Ari's concern was with a conceptual theoretical structure, not with implementation.

The development of the TVVAR algorithm shown here was directly influenced by Sakai's paper and his use of the Levinson algorithm type lattice structure. The variations in our modeling are that our algorithm is computed on the instantaneous data whereas Sakai's algorithm is based on sample correlation function computations. Another conceptual difference is that, as in Chapter 11, we use the PARCORs in a stochastic regression model context. Here too, a random walk smoothness priors trend model is used in the regression equation for the PARCORs. Instantaneous PARCORs are computed for each channel which in turn yields an instantaneous multichannel AR model. In this case the estimate of the instantaneous PARCORs is the mean of the posterior distribution of the PARCORs. In order to simplify the presentation of the time varying PARCOR VAR modeling, the development is first shown in Section12.4.1 in terms of multichannel stationary AR time series, independent of the development in Sections (12.1) and (12.2). The extension of stationary VAR time series modeling to TVVAR time series modeling is presented in Section 12.4.2. There, nonstationarity is invoked in computing the nonstationary PARCORs.

12.4.1 Constant Coefficient PARCOR VAR Time Series Modeling

The stationary VAR model is,

$$x_n = \sum_{m=1}^{M} A_m x_{n-m} + w_n, \quad n = 1, \ldots, N. \tag{12.25}$$

In (12.25) $w_n, n = 1, \ldots, N$ is a zero-mean time uncorrelated d-vector sequence with covariance matrix W.

Following Pagano (1978), interlace the d-variate x_n time series into the scalar series y_t using, $x_t(j) = y_{j+d(t-1)}$, $j = 1, \ldots, d$. Then organize the y_t sequence into a d-vector AR model according to

$$y_{k+d(t-1)} = \sum_{j=1}^{p_k} \alpha_{k,j} y_{k+d(t-1)-j} + \tilde{w}_{k+d(t-1)}, \quad k = 1, \ldots, d, \tag{12.26}$$

with $\tilde{w}_{k+d(t-1)} \sim N(0, \sigma_k^2)$ i.i.d. with p_k the order of the k-th channel AR model. Then the x_t sequence can be written in terms of the y_t in the matrix form

$$x_n = \sum_{m=0}^{M} B_m x_{n-m} + \tilde{w}_n \tag{12.27}$$

where the $d \times 1$ vector sequence $\tilde{w}_n$ has diagonal covariance matrix $U = \mathrm{diag}(\sigma_1^2, \ldots, \sigma_d^2)$ and B_0 is lower triangular. The relations between the matrix parameters in (12.25) and (12.27) and the covariance of the innovations sequence are:

$$A_j = -(I - B_0)^{-1} B_j, \quad W = (I - B_0)^{-1} U (I - B_0)^{-T}. \tag{12.28}$$

Next, the crucial idea of the LWWR algorithm (Levinson-Whittle-Wiggins-Robinson, Whittle 1963, Wiggens and Robinson 1965) is invoked to obtain a recursive order updating algorithm in terms of a reflection coefficient parameterization, (Morf *et al.* 1978). Toward that end, define the j-th order k-th channel forward and backward prediction errors,

$$\begin{aligned} \varepsilon_{k+nd}^{(j)} &= y_{k+nd} - \sum_{i=1}^{j} \alpha_k^{(j)}(i) y_{k+nd-i} \\ \eta_{k+nd}^{(j)} &= y_{k+nd-j} - \sum_{i=1}^{j} \beta_k^{(j)}(j+1-i) y_{k+nd-i+1} \end{aligned} \tag{12.29}$$

for $k = 1, \ldots, d$ and integer n respectively in terms of the forward model $\alpha_k^{(j)}(1), \ldots, \alpha_k^{(j)}(j)$ parameters where k is the channel number and j is the AR model order and backward model parameters $\beta_k^{(j)}(1), \ldots, \beta_k^{(j)}(j)$.

Now, the recursive order update relations are invoked to yield

$$\begin{aligned} \varepsilon_{k+nd}^{(j+1)} &= \varepsilon_{k+nd}^{(j)} - \alpha_{j+1}^{(j+1)}(k) \eta_{k-1+nd}^{(j)} \\ \eta_{k+nd}^{(j+1)} &= \eta_{k+nd}^{(j)} - \beta_{j+1}^{(j+1)}(k) \varepsilon_{k+nd}^{(j)} \end{aligned} \tag{12.30}$$

with

$$\sigma_j^2(k) = E[(\varepsilon_{k+nd}^{(j)})^2], \quad \tau_j^2(k) = E[(\eta_{k+nd}^{(j)})^2]. \tag{12.31}$$

Then as is conventional in LWWR type algorithms, parameter update equations for the $(j+1)$-st order model can be written in terms of the $(j+1)$-st order PARCORs and the parameters for the j-th order model. Sakai (1982) implemented a circular lattice filter algorithm using the methodology just shown for the modeling of stationary time series using covariance data. The TVVAR modeling version is shown next.

12.4.2 Time Varying PARCOR Coefficient VAR Modeling

In our realization, instantaneous data is used. The k-th channel zero-th order forward and backward innovations are

$$\varepsilon_{k+nd}^{(0)} = y_{k+nd}, \quad \eta_{k+nd}^{(0)} = y_{k-1+nd}. \tag{12.32}$$

Imposing the nonstationarity of the PARCORs, the relations between the order update innovations and the instantaneous PARCORs are given by

$$\begin{aligned}
\varepsilon_{k+nd}^{(j+1)} &= \varepsilon_{k+nd}^{(j)} + \alpha_{j+1,n}^{(j+1)}(k)\eta_{k-1+nd}^{(j)} \\
\nabla\alpha_{j+1,n}^{(j+1)}(k) &= v_n(k) \sim G(\tau_1^2) \\
\eta_{k+nd}^{(j+1)} &= \eta_{k+nd}^{(j)} + \beta_{j+1,n}^{(j+1)}(k)\varepsilon_{k+nd}^{(j)} \\
\nabla\beta_{j+1,n}^{(j+1)}(k) &= u_n(k) \sim G(\tau_2^2).
\end{aligned} \tag{12.33}$$

In (12.33), the notation G identifies an arbitrary distribution. In what follows we consider only $G(\tau) = N(0, \tau^2)$, which is satisfactory for modeling relatively slowly changing covariances. However, as in Chapter 11, the distribution G can be heavy tailed for the modeling of abruptly changing covariances.

Following from (12.33), the instantaneous order updated forward and backward model parameters then become,

$$\begin{aligned}
\alpha_{i,n}^{(j+1)}(k) &= \alpha_{i,n}^{(j)}(k) + \alpha_{j+1,n}^{(j+1)}(k)\beta_{j+1-i,n}^{(j)}(k-1) \\
\beta_{i,n}^{(j+1)}(k) &= \beta_{in}^{(j)}(k-1) + \beta_{j+1,n}^{(j+1)}(k)\alpha_{j+1-i,n}^{(j)}(k-1).
\end{aligned} \tag{12.34}$$

From (12.33) the order updated instantaneous PARCORs, $\alpha_{j+1,n}^{(j+1)}(k), \beta_{j+1,n}^{(j+1)}(k)$ are scalar regression coefficients and the updated forward and backward innovations are the residuals of the regressions. The key innovation in (12.33) is the stochastic difference equation representation for the order updated forward and backward instantaneous PARCORs. Kalman filter computations for the likelihood of the hyperparameter models corresponding to τ_1^2 and τ_2^2 are used. Akaike's AIC statistic is used to determine the AIC best order PARCOR model. Following the selection of the best order PARCOR model for each channel the instantaneous multichannel AR model parameters are computed one channel at-a-time.

12.5 Examples

In this section we show examples of the TVVAR modeling of simulated bivariate nonstationary covariance time series and of the TVVAR analysis of three components of a seismic event by the smoothness priors time varying AR coefficient method. An example of the analysis of three components of an epileptic event human electroencephalogram by the smoothness priors PARCOR method of analysis is also shown.

Table 12.1 The summary statistics of the models fitted to the simulated bivariate data.

	$q = 1$			$q = 2$		
p	AIC	σ_1^2	σ_2^2	AIC	σ_1^2	σ_2^2
1	16950	1.785	1.577	17073	1.799	1.595
2	15459	1.254	1.154	15666	1.280	1.166
3	14954	1.112	1.067	15220	1.130	1.068
4	14762	1.020	1.029	15037	1.037	1.033
5	14802	1.024	1.029	15153	1.032	1.025
6	14835	1.025	1.031	15267	1.039	1.017
7	14866	1.029	1.026	15376	1.036	1.006

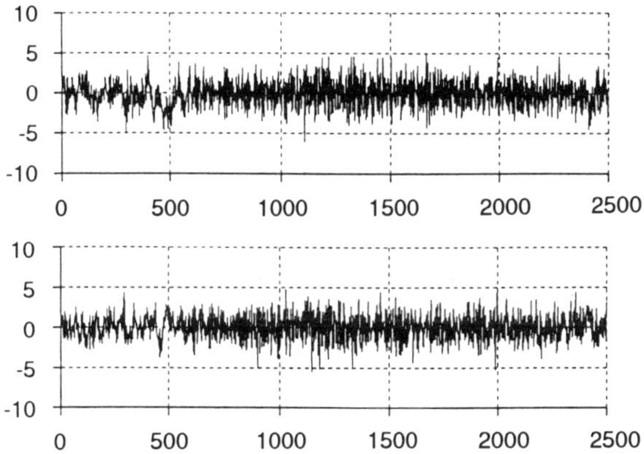

Figure 12.1 A: The simulated bivariate nonstationary covariance time series.

Example: An Analysis of Simulation Data

The first example that we consider is that of the simulation of a bivariate ($d = 2$), time varying VAR model of AR order 4 (Jiang and Kitagawa 1993). Figure 12.1A shows the bivariate time series data generated by the model

$$
\begin{aligned}
y_n(1) &= \sum_{j=1}^{2}\sum_{l=1}^{4} b_{ln}(1,j)y_{n-l}(j) + \varepsilon_n(1), && (12.35)\\
y_n(2) &= b_{0n}(2,1)y_n(1) + \sum_{j=1}^{2}\sum_{l=1}^{4} b_{ln}(2,l)y_{n-l}(j) + \varepsilon_n(2), && n = 1,2,\cdots,2500,
\end{aligned}
$$

with

$$
\begin{pmatrix} \varepsilon_{1,n} \\ \varepsilon_{2,n} \end{pmatrix} \sim N\left(\begin{bmatrix} 0 \\ 0 \end{bmatrix}, \begin{bmatrix} 1 & 0 \\ 0 & 1 \end{bmatrix}\right).
$$

The time varying coefficients were random and were themselves simulated. Table 12.1 shows the summary statistics of the variety of the models fitted to the simulated

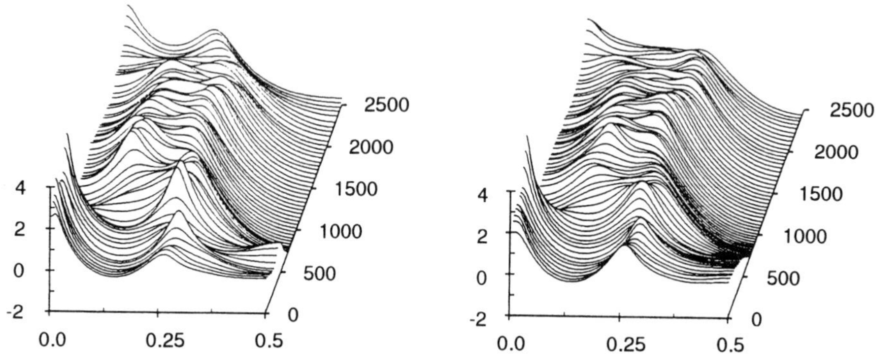

Figure 12.1 B: The evolution of the theoretical and estimated instantaneous spectra $p_{f,n}(1,1)$.

data. The time varying coefficient VAR model of order 4 ($p = 4$) with first order difference smoothness prior ($q = 1$) attains the minimum of the AIC's. This model was used to estimate the time varying VAR coefficients and the instantaneous spectra. The evolution of the theoretical instantaneous spectra $P_{f,n}(1,1)$ and the corresponding estimated instantaneous spectra are illustrated in Figure 12.1B. The estimate of the instantaneous spectrum by our model nicely resembles the theoretical spectrum.

Example: Analysis of a Seismic Event

The performance of the time varying AR coefficient VAR model is illustrated here with an analysis of the three components of the seismograms of 1982 Urakawa-Oki earthquake data in the AISM data library, (Takanami 1991). Viewing from top-to-bottom, the left hand side of Figure 12.2A illustrates the 3-variate seismic data with sample size 2500. The code names of the data are MYE2F (east-west component), MYN2F (north-south component) and MYU2F (vertical component), respectively. The sampling interval of the original data is $\Delta t \approx 0.01$ second. The data analyzed was sampled at every other observation. The "background noise", P-wave (after the first abrupt change) and S-wave (after the second abrupt change) are clearly distinguishable. Since the variances of the original data change over 1000 times in magnitude, the data are first transformed so that the series become roughly homogeneous in variance. Viewing from top-to-bottom, the right hand side of Figure 12.2A illustrates the normalized series of the original seismograms obtained by the procedure described in Chapter 9.

Table 12.2 shows the values of the AIC for various values of the VAR order p and the difference order q. The minimum AIC is attained at $p = 9$ and $q = 1$. The maximum likelihood estimates of the AIC best model are $\tau_1^2 = 0.361 \times 10^{-3}$, $\tau_2^2 = 0.152 \times 10^{-3}$, $\tau_3^2 = 0.349 \times 10^{-3}$, $\sigma_1^2 = 0.594$, $\sigma_2^2 = 0.598$, $\sigma_3^2 = 0.714$.

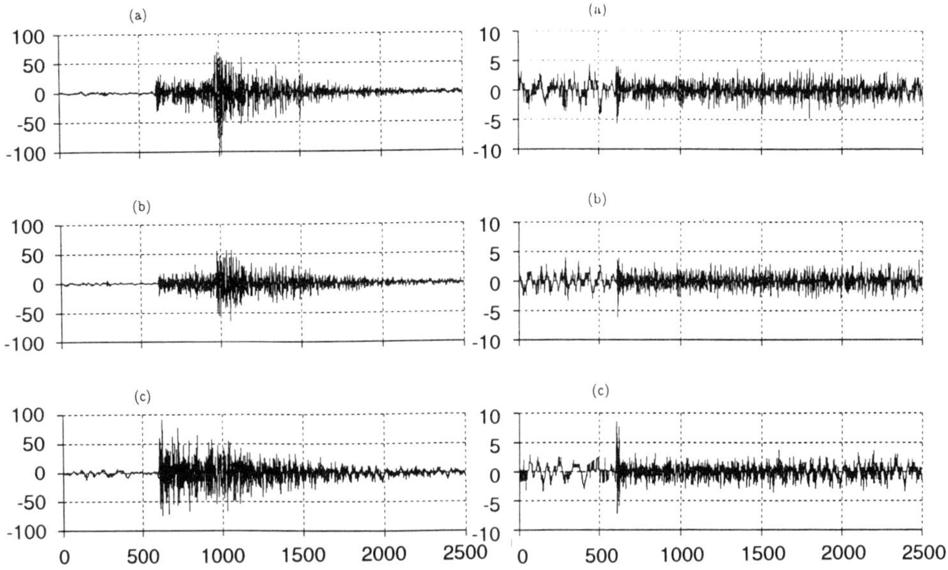

Figure 12.2 A: On the left-hand side, from top-to-bottom: The 3-component seismogram data. On the right-hand side, from top-to-bottom: The corresponding normalized data.

Table 12.2 The summary statistics of the models fitted to the seismic data.

	AIC			AIC	
p	$q = 1$	$q = 2$	p	$q = 1$	$q = 2$
1	22131	22475	6	19041	20073
2	20201	20745	7	19019	20194
3	19231	19909	8	19015	20365
4	19071	19788	9	18988	20522
5	19019	19907	10	19006	20689

The time evolution of the VAR coefficients $b_{kn}(i,i)$ $(i = 1,2,3,\ k = 1,2,3)$ and $b_{0n}(i,j)$ $(1 \le j < i \le 3)$ are illustrated in Figure 12.2B. The coefficients show significant step changes at $n = 600$ and $n = 960$. Those time instants which correspond to the arrival of the P-wave and the S-wave respectively, can be determined either by visual inspection of the original data or by an automatic procedure described in Takanami and Kitagawa (1991). The arrival time information is used in the program, to specify the positions of "outliers" with very large system noise variances. That in turn, relaxes the priors constraint and permits the VAR coefficients and subsequently derived quantities to change abruptly at those instants. We note that the appear-

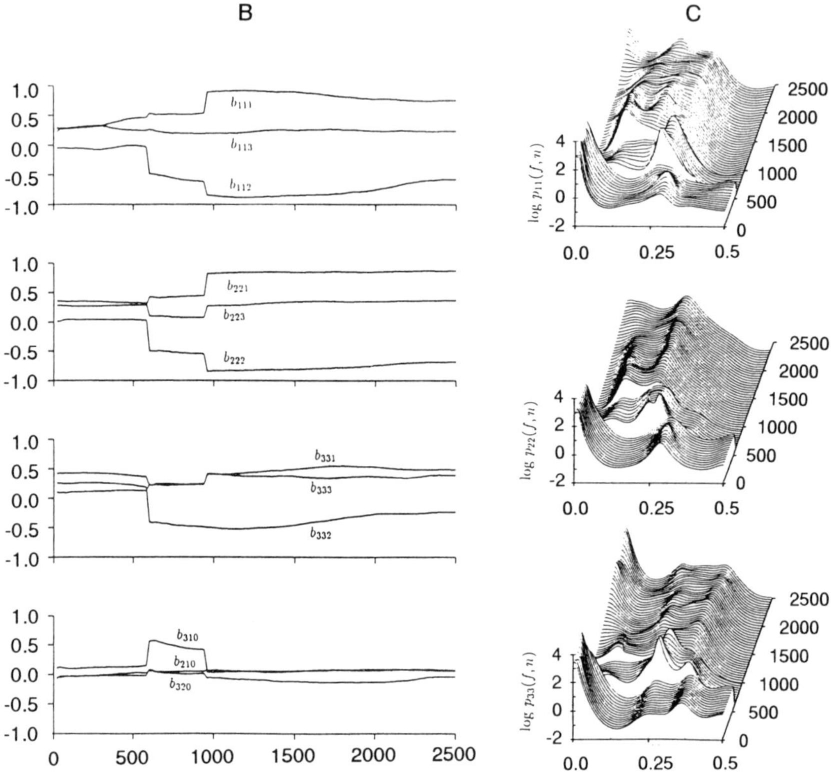

Figure 12.2 B: The time evolution of some of the model parameters. C: The time evolutionary power spectral density of the the three seismic components.

ance of the two horizontal waves are quite similar and differ from the vertical wave's appearance particularly during the S-wave.

The time evolution of the power spectra of the three components is shown in Figure 12.2C. The power spectra of the two horizontal components resemble each other during the background noise interval prior to the P-wave and during the P-wave. The vertical component peaks at a higher frequency than the other components during the background noise interval. Early in the S-wave interval, the spectrum is dominated by a peak at a normalized frequency of approximately $f = 0.125$. Subsequently the spectrum peaks at about $f = 0.2$. The spectrum of the vertical motion is different from that of the horizontal components, with more pronounced power at the higher frequencies. A more comprehensive discussion and interpretation of the seismic phenomena associated with this earthquake is in Takanami and Kitagawa (1991).

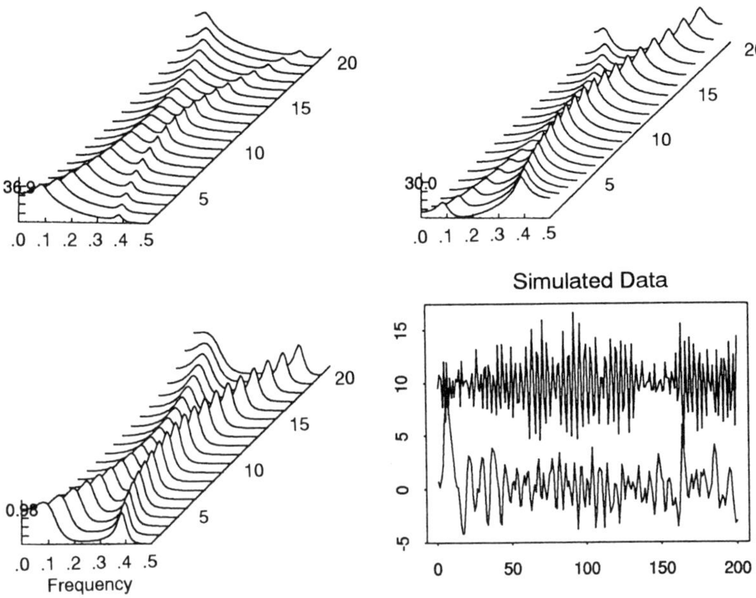

Figure 12.3 A: Theoretical power spectra and spectral coherences of bivariate nonstationary covariance data.

Example: PARCOR Analysis of Simulated Data

Here we illustrate the performance of the PARCOR TVVAR modeling method on bivariate simulated data. The theoretical and estimated spectral properties of a bivariate nonstationary covariance time series are shown in Figures 12.3A and 12.3B. The estimated evolutionary power spectra and spectral coherence functions are seen to be reasonable replicas of the corresponding theoretical quantities.

Example: EEG Data Analysis

An example of the data analysis of a three-channel data episode of an "aura" or pre-epileptic seizure EEG event, by the PARCOR TVVAR analysis method is shown here. Some previous work in the study of human epileptic EEGs appears in Gersch (1985, 1987).

Figure 12.4A illustrates data from a 7-second 3-site data record of the electrical activity at the onset of a spontaneous seizure event. (In fact, 16 channels of data were continuously monitored by telemetry.) The recorded data was sampled at the rate of 200 samples per second. Every 4th data point, (Nyquist frequency = 25 Hertz), was used for analysis. (J. Gotman of the Penfield Neurological Laboratory, Montreal in a personal communication said that EEG seizure events have negligible power above 25

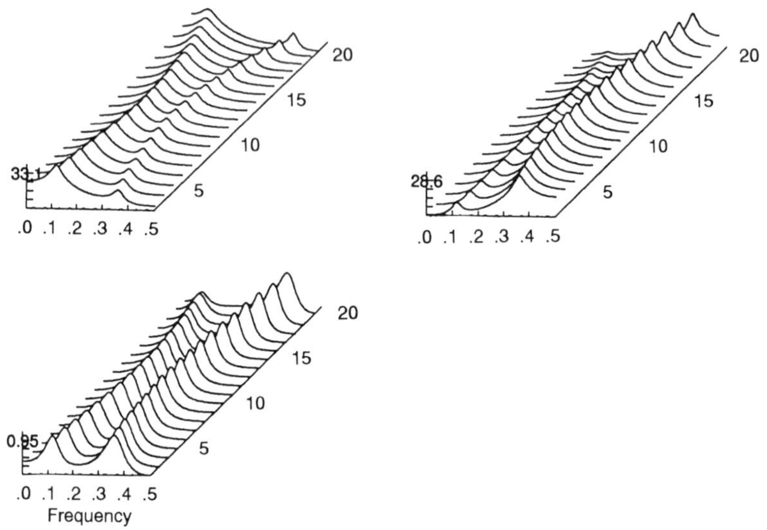

Figure 12.3 B: Estimated power spectra and spectral coherences of bivariate nonstationary covariance data.

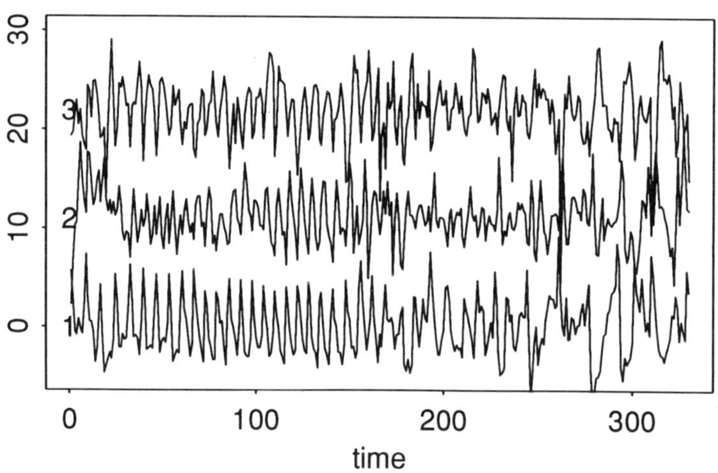

Figure 12.4 A: Three channels of human epileptic episode EEG data.

Hertz.) The appearance of each individual EEG trace reveals that the local structure of the time series changes with time. Our analysis involved the fitting of the PARCOR TVVAR model to the data and a subsequent spectral analysis using the concept of an instantaneous power spectral density matrix,

$$P_{f,n} = A_{f,n}^{-1} W_n A_{f,n}^{-T} \tag{12.36}$$

where the multichannel AR model instantaneous system function $A_{f,n}$ is given by $A_{f,n} = I - \sum_{m=1}^{M} A_{m,n} \exp(-2\pi imf)$.

A Concept of Driving or Causality in Seizure EEG Analaysis

The objective of the analysis was to determine whether the electrical activity at any one of the observed anatomical recording sites could be interpreted as "driving", or causing the electrical activity at the other sites. The data in Figure 12.4A is complex, and there is no obvious visual clue that might unambiguously identify an initiating or driving site or identify when driving is present. We identify causality or driving at an instant in time by a frequency domain analysis via the instantaneous power spectral density matrix. The concept of causality that we use was introduced for stationary time series in Gersch and Goddard, (1970) in a study of epilepsy induced in cats and used in Gersch and Tharp (1976) in a study of epilepsy in humans. Our concept of causality differs from the "feedback free" definition of causality in econometrics (Geweke 1982), or the time precedence definition previously and conventionally used in the location of epileptic foci from EEG time series, (Brazier 1972, Gotman 1983). (The technical problems in using the time precedence definition of driving arise because "canned" windowed periodogram estimates are used for the estimation of delay as if the time series were stationary. Much of the data segments of interest for analysis are not stationary and frequently considerable expertise is required to obtain reliable estimates of delay from windowed periodogram analyses.)

Let $W_{f,n}^2(x, y)$ denote the value of the spectral coherence between channels x and y at frequency f at time n. Similarly, let $W_{f,n}^2(x, y|z)$ denote the value of the partial spectral coherence between channels x and y conditioned on channel z at frequency f at time n. Then, our definition of causality is equivalent to the following conditions on spectral coherences and partial coherences

$$W_{f,n}^2(x, y) \neq 0, \qquad W_{f,n}^2(x, w) \neq 0, \qquad W_{f,n}^2(y, w) \neq 0,$$

$$W_{f,n}^2(x, y|w) \neq 0, \quad W_{f,n}^2(x, w|y) \neq 0, \quad W_{f,n}^2(y, w|x) \equiv 0.$$

That is, the identification of causality at frequency f at time n is determined by the detection of zero partial instantaneous coherence. (Tests for zero partial coherence are in Brillinger 1975 and Koopmans 1974.) The instantaneous coherences and partial coherences are computed from the instantaneous power spectral density matrix. The TVVAR model yields estimates of the spectral coherence and partial spectral coherence at every instant in time. Computational results of both the evolving spectral coherence and evolving partial spectral coherence for each of the three possible pairs of three different data channels are shown in Figure 12.4B in row 2 and

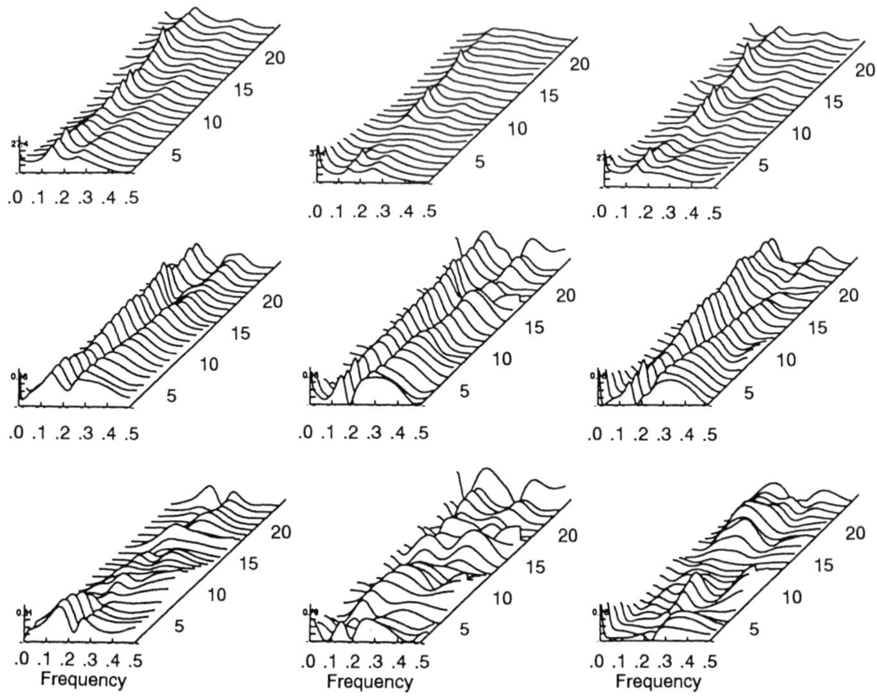

Figure 12.4 B: The evolutionary power spectral densities, spectral coherences and partial spectral coherences.

row 3. The computational results are used to determine whether or not and when one channel drives the other two.

The second row and third rows of Figure 12.4B respectively illustrate the time-changing structure of pairwise coherence versus frequency between pairs of channels and the corresponding time-changing partial coherence versus frequency. The partial coherence between channels 1 and 2 and between channels 1 and 3 with the partialing done on the excluded third channel remain significantly correlated. On the other hand, the partial coherence between channels 2 and 3 with the influence of channel 1 removed is statistically indistinguishable from zero over the frequency interval between 7 and 11 Hertz. We conclude that the pairwise coherence between the three channels is a consequence of the fact that channel 1 was driving channels 2 and 3 for the first 2.5 seconds of the record in the vicinity of 7.25 Hertz. Channels 2 and 3 have very little explanatory power.

The relatively low partial coherence between channels 2 and 3 partialed on channel 1 throughout the epoch, (third row of Figure 12.4B), results in the appearance of a relatively flat partial coherence surface and suggests that this kind of illustration may be a potentially powerful diagnostic aid to identify candidate driving channels.

The covariance structure and hence the ability to identify the location of the focal driving site of the epileptic episode changes over a very short time interval. That fact makes a proper method of analysis of multichannel nonstationary covariance data critical for the analysis of human focal epilepsy EEG data.

13

Modeling Inhomogeneous Discrete Processes

In this chapter we illustrate the application of non-Gaussian modeling method to the analysis of inhomogeneous discrete random processes data. Two data analytic examples, Tokyo rainfall data, a nonstationary binary process and the analysis of a simulated nonstationary Poisson process data set are shown. The latter simulation is based on research of the count of X-rays from the star Cygnus X1 data. Kashiwagi and Yanagimoto (1992), is an application of the non-Gaussian state space methodology in the analysis of nonstationary Poisson process analysis of medical data.

13.1 Nonstationary Discrete Process

Consider an inhomogeneous discrete process

$$y_n \sim \mathrm{dist}(\lambda_n)$$

where the parameter of the process λ_n evolves with time. The problem considered here is the estimation of the mean value function λ_n from the observation $y_1, \ldots, y_n$. For this purpose, we consider λ_n as a state and use the smoothness prior model

$$\nabla^k \lambda_n = \omega_n.$$

In the formulation of our state space model, for simplicity we assumed that the random variables have density functions. But actually we only need to compute the convolution involved in the one step-ahead prediction formula for arbitrary distributions and Bayes formula. Therefore we can also handle discrete distributions.

From Chapter 6, the recursive formulae for obtaining one step ahead prediction and filtering densities are as follows;

One step ahead prediction (time update)

$$p(x_n|Y_{n-1}) = \int_{-\infty}^{\infty} p(x_n|x_{n-1})p(x_{n-1}|Y_{n-1})dx_{n-1}, \tag{13.1}$$

where $p(x_n|x_{n-1})$ is the density of x_n given the previous state vector x_{n-1}.

Filtering (observation update)

$$\begin{aligned}
p(x_n|Y_n) &= p(x_n|Y_{n-1}, y_n) \\
&= \lim_{\nabla x \to 0} \frac{\Pr(x \le x_n < x + \nabla x|Y_{n-1}, y_n)}{\nabla x} \\
&= \lim_{\nabla x \to 0} \frac{\Pr(x \le x_n < x + \nabla x, y_n = y|Y_{n-1})}{\nabla x \Pr(y_n = y|Y_{n-1})} \\
&= \lim_{\nabla x \to 0} \frac{\Pr(x \le x_n < x + \nabla x|Y_{n-1})}{\nabla x} \times \frac{\Pr(y_n = y|x \le x_n < x + \nabla x)}{\Pr(y_n = y|Y_{n-1})} \\
&= \frac{p(x_n|Y_{n-1})\Pr(y_n|x_n)}{\Pr(y_n|Y_{n-1})},
\end{aligned} \tag{13.2}$$

where $P_r(y_n|x_n)$ is the probability of y_n given x_n and $P_r(y_n|Y_{n-1})$ is obtained by the computation $\int P_r(y_n|x_n)p(x_n|Y_{n-1})dx_n$.

13.2 Nonstationary Binary Processes

Here, we show a problem concerning estimating the time varying mean of a nonstationary (inhomogeneous) binary process. This problem was considered earlier in Ishiguro and Sakamoto (1983) under Gaussian assumptions and a quadratic approximation to the likelihood function. That assumption is not required in our analysis, and we do not need to approximate the likelihood function.

To illustrate what the data is like, Table 13.1 shows the number of occurrences of rainfall over 1mm in Tokyo in each day in the 2 years starting January 1, 1983 through December 31, 1984. The general problem in the modeling of such nonhomogeneous binary process data is to estimate the probability, p_n of the occurrence of rainfall on a specific calendar day. The probability of rain is believed to be gradually changing with time.

The left hand side of Figure 13.1, illustrates the number of rainy days in Tokyo on each calendar day, successively for 1, 2, 4, 8 and 17 years, starting in 1975. Dividing those numbers by the number of observations, (1, 2, 4, 8, 17 respectively) yield a coarse and hopelessly irregular estimate of the probability of rain on a calendar day. Our objective here is to smooth those coarse probabilities, and get a feeling for the consistency of the annual rain patterns in Tokyo.

Our model for the probability of occurrence is given by

Table 13.1 Number of Occurrences of Rainfall Over 1 mm in Tokyo, 1983–1984

January	0 0 1 1 0 1 1 0 0	0 0 0 0 0 1 0 0 1 1	0 1 1 0 0 0 0 0 0 0 0 1
February	0 1 0 0 0 0 0 0 0	0 0 1 1 0 0 0 2 1 0	0 0 0 1 1 0 1 0 0 0
March	0 1 1 0 0 0 0 0 0	2 0 0 1 1 0 2 1 0 1	1 1 0 1 2 0 0 1 1 1 1 1
April	2 0 0 1 1 0 1 2 0	1 1 1 0 0 1 2 1 0 2	1 0 1 0 0 0 0 0 1 0 0
May	1 1 0 0 0 1 1 0 0	0 0 0 1 1 1 2 2 0 0	0 0 0 1 1 1 0 0 0 1 0 0
June	0 0 0 1 0 1 0 0 0	2 1 1 2 1 0 1 2 2 0	2 2 1 1 1 2 2 2 1 1 0
July	0 0 1 0 1 1 1 2 1	0 1 1 0 0 2 1 1 1 1	2 2 0 1 0 0 0 0 1 1 0 0
August	0 0 0 0 0 0 0 0 0	0 0 0 1 1 1 1 1 1 0	1 2 1 0 0 0 0 0 0 0 1 0
September	1 0 0 0 1 0 1 1 1	2 0 0 0 1 2 2 0 1 1	2 2 1 0 1 1 1 1 1 1 0
October	0 0 0 0 1 0 0 0 1	0 2 1 1 0 1 1 1 0 2	1 1 1 1 0 0 0 1 0 0 0 0
November	0 0 0 0 0 1 1 0 1	1 0 0 0 0 1 1 0 0 1	1 0 0 0 1 0 0 0 0 0 0
December	0 0 0 0 0 0 0 0 0	0 1 0 0 0 0 1 1 1 0	0 0 0 1 0 0 0 0 0 0 1 1

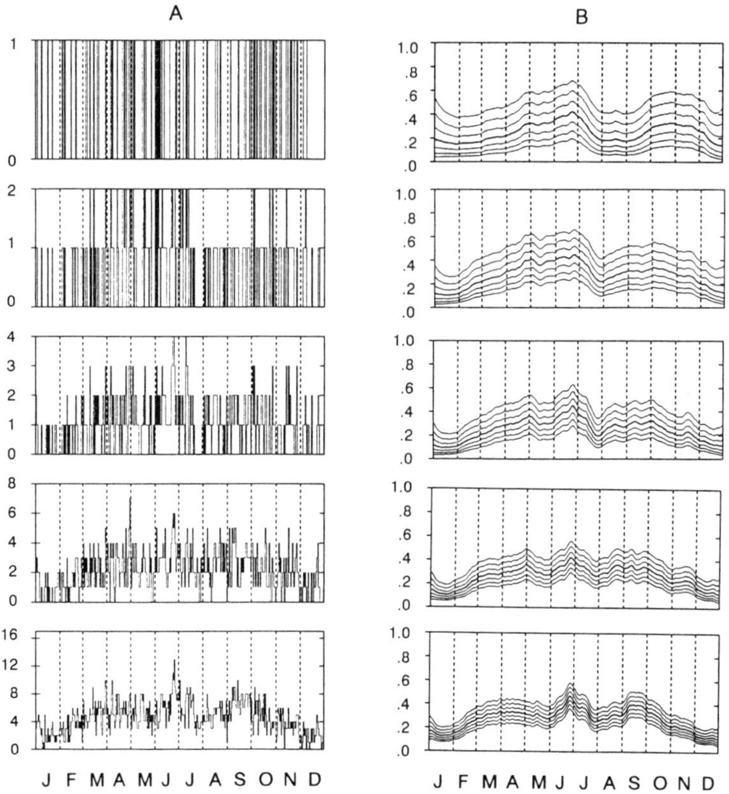

Figure 13.1 A: The ratio of the number of rainy days in Tokyo to the number of observations successively for the years, 1975, 1975-1976, 1975-1978, 1975-1982 and 1975-1991. B: The smoothed values of the probability of rain and $\pm 1, 2, 3\sigma$ in Tokyo successively for the years, 1975, 1975-1976, 1975-1978, 1975-1982 and 1975-1991.

$$\begin{aligned} z(m_n|\ell_n, p_n) &= \binom{\ell_n}{m_n} p_n^{m_n}(1-p_n)^{\ell_n - m_n} \\ q_n &= \log[p_n/(1-p_n)] \\ \nabla^k q_n &= v_n, \end{aligned} \tag{13.3}$$

Here ℓ_n is the number of observation at the n-th time point, m_n is the number of occurrences of an event at the n-th time point and $z(m_n|\ell_n, p_n)$, is the probability mass function of the binomial distribution. Also $q_n = \log[p_n/(1-p_n)]$, the logit transformation is applied for compatibility with the analysis by Ishiguro and Sakamoto (1983). In their treatment, the transformation from p_n to q_n was applied to guarantee that $0 < p_n < 1$. That is not an essential requirement in our method, since the state space is easily restricted. As indicated in (13.2), we apply a trend model computation to the logit of the daily probability of rain and in (13.3) we assume that $v_n \sim \text{dist}(0, \tau^2)$.

For this system with discrete observations, the analogue of the filtering equation is given by

$$p(q_n|Y_n) = C^{-1} z_{\ell_n}(m_n|\ell_n, p_n) p(q_n|Y_{n-1}) \tag{13.4}$$

with $C = \int_{-\infty}^{\infty} z_{\ell_n}(m_n|\ell_n, p_n)p(q_n|Y_{n-1})dq_{n-1}$ and $p_n = e^{q_n}/(e^{q_n}+1)$, where $z_{\ell_n}(m_n|\ell_n, p_n)$ is the binomial probability mass function that is now conditioned on the state at time n, q_n.

The right hand side of Figure 13.1, shows the smoothed mean and the ± 1, 2, 3 sigma intervals of those different data sets. As expected, the widths of $\pm$ sigma intervals decrease with an increasing number of observations. The relatively long term 1975–1978, 1975–1982 and 1975–1991 estimated probability of rain patterns are consistent with the concensus of opinion about weather in Tokyo; a dry winter, a possibly unsettled spring, little rain in May, a rainy season in late June to mid July, a stable hot summer from late July through August, and generally fine weather but with occasional typhoon associated rain in September and October.

13.3 Nonstationary Poisson Process

The same method of analysis as was done for the nonhomogeneous binary process data is applied here to a nonstationary Poisson process. (For an alternative approach see Cox 1981.) In this case, the data was simulated from a model based on research on the count of X-rays from the X-ray star Cygnus X1. The original Cygnus X1 data, obtained by a collimator pointed at a particular angle aboard a satellite was made available by Professor Oda, formerly of the Institute of Planetary and Space Science, Tokyo. The original data was regularly sampled every millisecond and 10^6 observations were available.

For the simulated data case, the theoretical or model non-homogeneous Poisson process mean value is shown in Figure 13.2A and a realization of $n = 2000$ observations is shown in Figure 13.2B. This data too, appears to be almost hopelessly

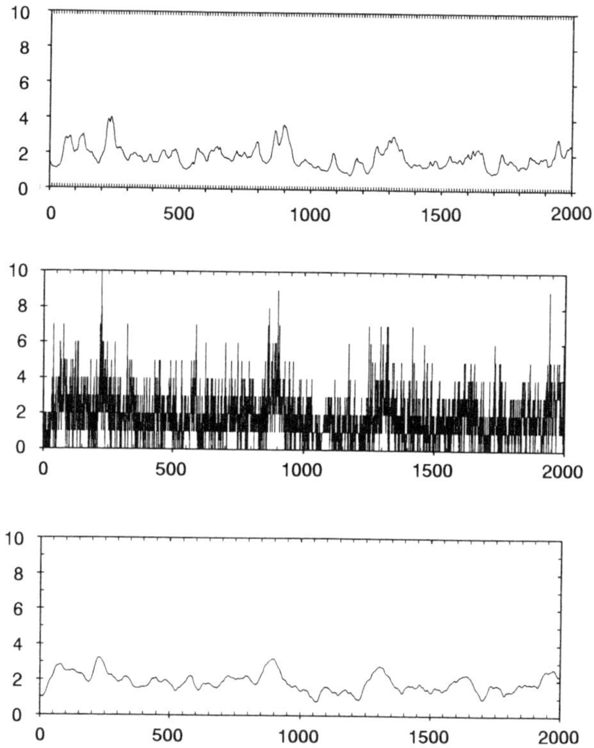

Figure 13.2 A: The theoretical mean of the number of counts per millisecond. B: The number simulated counts per millisecond. C: The estimated mean of the number of counts per millisecond.

irregular. The objective of our analysis is to estimate the mean value function of the inhomogeneous Poisson process data. The method of analysis is very similar to the analysis of the inhomogeneous observed daily rainfall binary distributed data. Here again numerical integration was used to compute the convolution in the state space analysis prediction step.

Consider the sequence of numbers, $y_1, \ldots, y_N$ that is assumed to come from a nonhomogeneous Poisson process, $P_{\lambda_1}, \ldots, P_{\lambda_N}$ with probability mass function for the n-th observation is

$$P(y_n|\lambda_n) \;=\; \frac{e^{-\lambda_n}\lambda_n^{y_n}}{y_n!}, \tag{13.5}$$

and $E[Y_n] = \lambda_n$. The objective of our analysis is to estimate the time varying mean value function $\lambda_1, \ldots, \lambda_N$.

The Kashiwagi and Yonemoto Analysis

Such a problem was in fact analyzed in a smoothness priors approach due to Yanagimoto and Kashiwagi (1990) and Kashiwagi and Yanagimoto (1992) that is analytically more extensive, (and computationally more complex), than our own approach. Mindful of Akaike's emphasis on the importance of the choice of model, it is shown here for illustrative purposes. A description of our own approach follows.

The authors assume a distribution on the logarithm of the λ_n, $n = 1, \ldots, N$ via a k-th order stochastic trend model. That is, let

$$\nabla^k \log \lambda_n \sim N(0, \tau^2). \tag{13.6}$$

Then, as indicated in Chapter 2, it is possible to get a closed form expression for the likelihood of the hyperparameter τ^2. In fact, from Yanagimoto and Kashiwagi (1990), a closed form expression for the likelihood of all of the model parameters is available under the following assumptions: i) the successive observations are independent, ii) the distribution of $\nabla^k \log \lambda_n$ is Gaussian and, iii) the trend order model $k = 1$. Then, for a given mean value parameter λ_n, and the probability mass function for the n-th observation as given in (13.5), and trend model order $k = 1$, the conditional distribution of the parameter λ_n given the values of the preceding λ_{n-1} and the hyperparameter τ^2 is

$$g(\lambda_n|\lambda_{n-1}, \tau^2) = (2\pi)^{-1/2}(\tau\lambda_n)^{-1} \exp\left\{-\frac{1}{2\tau^2}(\log\ \lambda_n - \log\ \lambda_{n-1})^2\right\},\ n = 2, \ldots, N \tag{13.7}$$

Combining these expressions, the conditional likelihood for the parameters $\lambda_2, \ldots, \lambda_N$ given the data $y = \{y_1, \ldots, y_N\}$ is

$$L(\lambda_2, \ldots, \lambda_N|y, \lambda_1, \tau^2) = \prod_{i=1}^{N} f(y_i|\lambda_i) \prod_{n=2}^{N} g(\lambda_n|\lambda_{n-1}, \tau^2). \tag{13.8}$$

The unknown parameters are estimated in a two-stage procedure. In the first stage the estimates $\hat{\lambda}_1$ and $\hat{\tau}^2$ are obtained by maximizing the marginal likelihood of y,

$$L(\lambda_1, \tau^2) = \int_0^\infty \cdots \int_0^\infty L(\lambda_2, \ldots, \lambda_N|y, \lambda_1, \tau^2) d\lambda_2 \cdots d\lambda_N. \tag{13.9}$$

The parameters $\lambda_2, \ldots, \lambda_N$ are estimated in the second stage by taking the expectations with respect to the smoothing densities

$$p(\lambda_n|y, \hat{\lambda}_1, \hat{\tau}^2) \propto \int_0^\infty L(\lambda_2, \ldots, \lambda_N\ |y, \hat{\lambda}_1, \hat{\tau}^2) d\lambda_2\ \cdots\ d\lambda_{n-1} d\lambda_{n+1}\ \cdots\ d\lambda_N. \tag{13.10}$$

A closed form expression for the likelihood of all the model parameters, could also be obtained for the case of trend order model, $k = 2$.

A Simpler Analysis

Our own approach is simpler. Since $E[Y_n] = \lambda_n$ the sequence of X-ray count data $y_1, \ldots, y_N$, is taken as a coarse estimate of the sequence of estimates, $\lambda_1, \ldots, \lambda_N$. The corresponding state space model is,

$$\begin{aligned} \nabla^k \lambda_n &= w_n \sim N(0, \tau^2) \\ y_n \sim P(y_n|\lambda_n) &= \frac{e^{-\lambda_n} \lambda_n^{y_n}}{y_n!}. \end{aligned} \tag{13.11}$$

A diffuse prior is assumed for λ_1 and also with a trend order $k = 1$ model with Gaussian system noise, using the non-Gaussian filtering algorithm we obtain the likelihood for the unknown hyperparameter τ^2.

For this system of discrete observations, the analogue of the filtering equation is

$$p(\lambda_n|Y_n) = C^{-1} P(y_n|\lambda_n) p(\lambda_n|Y_{n-1}) \tag{13.12}$$

with $C = \int_{-\infty}^{\infty} P(y_n|\lambda_n) p(q_n|Y_{n-1}) dq_{n-1}$. The resultant smoothed nonhomogeneous mean value of the X-ray count data is shown in Figure 13.2C. We note that the estimated smoothed mean is a very reasonable replica of the theoretical or model mean.

14

Quasi-Periodic Process Modeling

The early development of time series has been related to a quest for an understanding of cyclical phenomena. For example, Shuster's periodogram (1898, 1906), and Yule's (1927) introduction of autoregressive models, were devoted to the analysis of cyclical sunspot numbers and Whittle's (1954) analysis of the water level in a rock channel on the Wellington coast of New Zealand is also related to a cyclical phenomenon. In fact many time series exhibit cyclical behavior in the sense there appears to be an approximate repetition of a pattern with a not very well defined period or amplitude. Rather both the period and amplitude appear to change gradually. Typically ecological data, air pollution data and several other physical phenomena exhibit such behavior. Two of the most familiar examples of time series which exhibit such behavior and which have been extensively analyzed are the Canadian lynx data, (for example see Campbell and Walker 1977, Tong 1977, Bhansali 1979 and Priestly 1981), and the Wolfer sunspot series, (Morris 1977, Tong 1983). Such series are frequently modeled by AR, ARMA or AR plus sinusoidal models. However, none of these modeling methods are very satisfactory for the prediction of more than one lead time, (Tong 1982). As pointed out in Akaike (1977b) in his discussion of several analyses of the Canadian lynx data, those analyses were unconvincing and that the critical issue in modeling time series is the selection of a proper model.

A smoothness priors quasi-periodic modeling procedure is shown here. Several examples including the Wolfer sunspot data, the Canadian lynx data, the maximum daily Tokyo temperature data and synthesized phase-wrap data are shown to illustrate the behavior and to suggest the potential richness of the applicability of our quasi-periodic model.

In addition, a simulation study to examine the predictive performance of the quasi-periodic model is shown. It appears that the quasi-periodic model introduced here, has the flexibility to be an appropriate model for a large number of quasi-periodic time series.

14.1 The Quasi-periodic Model

The observation process of the quasi-periodic model is expressed by

$$y_n = c_n h(\theta_n) + \varepsilon_n, \tag{14.1}$$

where $h(\theta_n)$ is a cyclic or periodic function, θ_n is time varying phase function and c_n is a time varying amplitude (modulating) function. Our approach to modeling such quasi periodic time series is expressed in the form

$$\begin{aligned} \nabla^{k_1}\theta_n &= v_n \sim N(0,\tau_1^2) \\ \nabla^{k_2}a_n &= u_n \sim N(0,\tau_2^2) \\ c_n &= \exp(a_n) \\ y_n &= c_n \left\{ \sum_{j=0}^{m} \alpha_j \cos(2\pi j\theta_n) + \sum_{j=1}^{m} \beta_j \sin(2\pi j \theta_n) \right\} + \varepsilon_n. \end{aligned} \tag{14.2}$$

That is, we allow the phase θ_n to vary in accordance with a k_1-th order stochastic trend model and assume $v_n \sim N(0,\tau_1^2)$. Also the logarithm of the amplitude fluctuations c_n vary in accordance with a k_2-th order stochastic trend process with $u_n \sim N(0,\tau_2^2)$. From (14.2), $\varepsilon_n = y_n - e^{a_n}h(\theta_n)$ is nonlinear, so that despite the fact that we let the distribution of ε_n be Gaussian, i.e. $r(\varepsilon) \sim N(0,\sigma^2)$, this model is in the class of general state space models and the computations are realized by numerical integration, (Chapter 6). Note that the periodic component model in (14.1) is both cyclically and amplitude modulated to form the output. We also note that since this model is only nonlinear in the observation process the modifications to the filtering formula are rather simple and that no modifications are required for the smoothing computations.

This model yields estimates of the posterior distributions of phase and amplitude. Quasi-periodic analyses of both the Wolfer sunspot series and the Canadian Lynx data, as well as the analysis of synthetic phase-unwrapping data and the analysis of quasi-periodicity in 17 years of rainfall data are shown.

14.2 The Wolfer Sunspot Data

Schwabe (1838) was very likely the first to identify the sunspot cycle. Rudolf Wolf introduced a formula for calculating the sunspot number and using historical data reconstructed the annual mean sunspot numbers for the previous 100 years (Izenman 1985). As a result of considerable work by solar astronomers, records of annual means are available dating back to 1700. The continued interest in the sunspot data reflects the assumption that data is a good indicator of the evolution of the magnetic oscillation of the sun and there are a variety of physical theories to account for that phenomenon, (Yoshimura 1979, for example).

The earliest linear model to account for this data is probably due to Yule (1927). Some of the many other time series analysis studies of the Wolfer sunspot series data have been for example by Moran (1954), Craddock 1967, Bloomfield (1976), Morris (1977), Akaike (1977b) and Tong (1983).

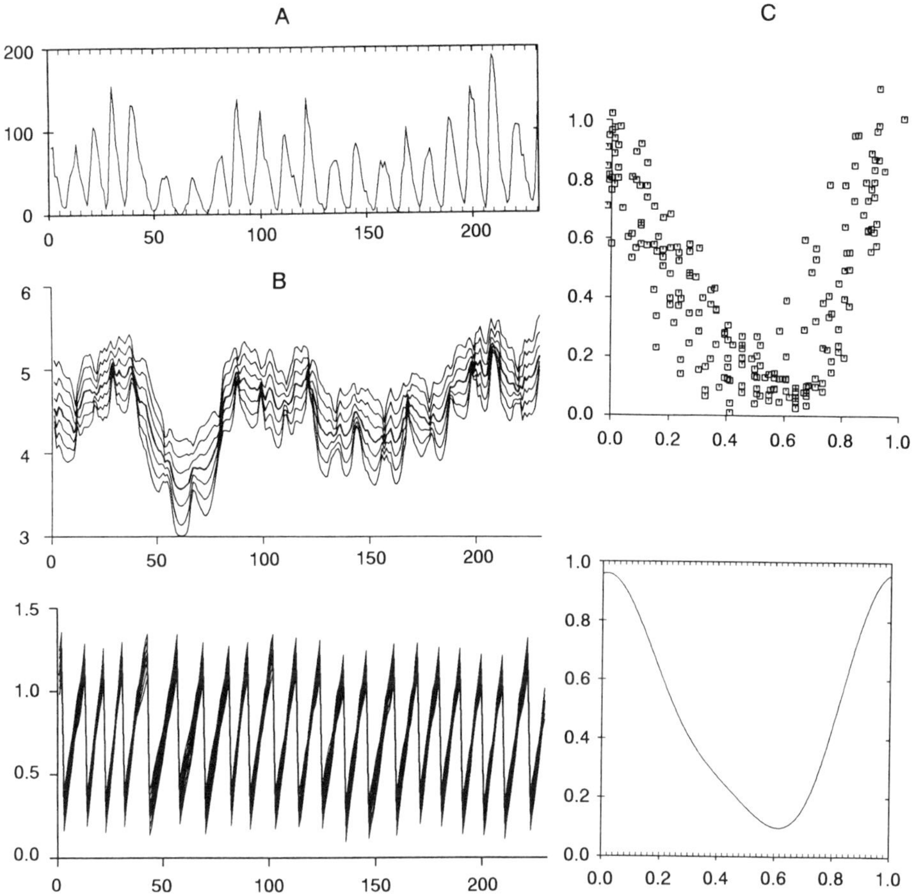

Figure 14.1 A: The Wolfer sunspot series data, 1749-1979. B: The estimated amplitude and phase. C: The estimated cyclic function and corresponding scatter diagram.

Our own approach is to model this data as a quasi-periodic process by the model in (14.1) and (14.2). We modeled that data with a first order trend model, $k_1 = 1$, and a first order amplitude modulation trend model, $k_2 = 1$. Both the phase noise and the observation noise were assumed to be Gaussian distributed. Since the observation equation is nonlinear, the model is in the class of general state space modeling and the computations were realized by numerical integration.

The graphical results obtained are as shown in Figure 14.1. The annual $N = 231$ data from 1749-1979 is shown in Figure 14.1A. Also shown in Figure 14.1 are the estimated amplitude and phase and the cyclical or periodic function as well as a scatter diagram which illustrates the observed variability of the cyclical function.

The computed results clearly exhibit the amplitude modulated and phase modulated structure of the series.

For completeness, and possible comparison with future modeling of this data, we report the estimated values of the fitted model. The estimated variances of the phase and amplitude noise process are $\tau_1^2 = 0.017$ and $\tau_2^2 = 0.0011$, respectively and the observation variance is $\hat{\sigma}^2 = 18.5$. The cyclical function h_n was a trigonometric sum of components up to order $m = 2$, with parameters $\alpha_0 = 0.47893$, $\alpha_1 = 0.39369$, $\alpha_2 = 0.08451$, $\beta_1 = 0.11371$, $\beta_2 = -0.021336$. These parameters were estimated by a Householder transformation least squares procedure, (as described in Chapter 2), and the AIC was used to determine the best fitting model order.

14.3 The Canadian Lynx Data

The Canadian lynx data consists of the annual record of the numbers of Canadian lynx trapped in the Mackenzie river district of North-west Canada for the period 1821-1934, $N = 114$. The data first appeared in Elton and Nicholson (1942) and since then, has been the subject of numerous studies. We cite for example Moran (1953a), Hannan (1960), Bulmer (1974), Tong (1977), and most comprehensively in the discussion paper by Campbell and Walker (1977). It is interesting to note that in the discussion of that paper, Akaike pointed out that the analyses of Campbell and Walker (1977) and Tong (1977) were inconclusive.

Here too, our approach was to model this data as a quasi-periodic process by the model in (14.1) and (14.2). The approach to modeling the Canadian lynx data was the same as that used for the modeling of the Wolfer sunspot series data. That is we also modeled that data with first order trend models for both phase and amplitude modulation. Again, each of the phase noise, amplitude noise and observation noise were assumed to be Gaussian distributed.

The graphical results obtained here are as shown in Figure 14.2. The $N = 114$ Canadian lynx data from 1821-1934 is shown in Figure 14.2A. The estimated amplitude and phase and the cyclical or periodic function as well as a scatter diagram which illustrates the observed variability of the cyclical function are also shown in Figure 14.2. The computed results clearly exhibit the amplitude modulated and phase modulated structure of the fitted model.

Again, for completeness we report the estimated values of the fitted model. The variances of the phase and amplitude system noises were $\tau_1^2 = 0.090$ and $\tau_2^2 = 0.0011$, respectively and the observation variance was $\hat{\sigma}^2 = 5000.0$. The cyclic function parameters were again estimated by a Householder transformation least squares procedure, and the AIC was used to determine the best fitting model order. The estimated parameters in the trigonometric cyclical function model, h_n, were $\alpha_0 = 0.35197$, $\alpha_1 = 0.41072$, $\alpha_2 = -0.04080$, $\beta_1 = 0.14777$, $\beta_2 = 0.00558$, $\beta_3 = 0.05218$.

We point out that Kooperberg *et al.* (1995) considered a logspline function estimation of both the Wolfer sunspot series and the Canadian lynx data. They concluded that the Wolfer series did not contain a spectral atom while the Canadian lynx data did contain an atom at an angular frequency near $\frac{24}{114}\pi$ which explained 56% of the variability in the data. We did not attempt to reconcile our own analyses with those

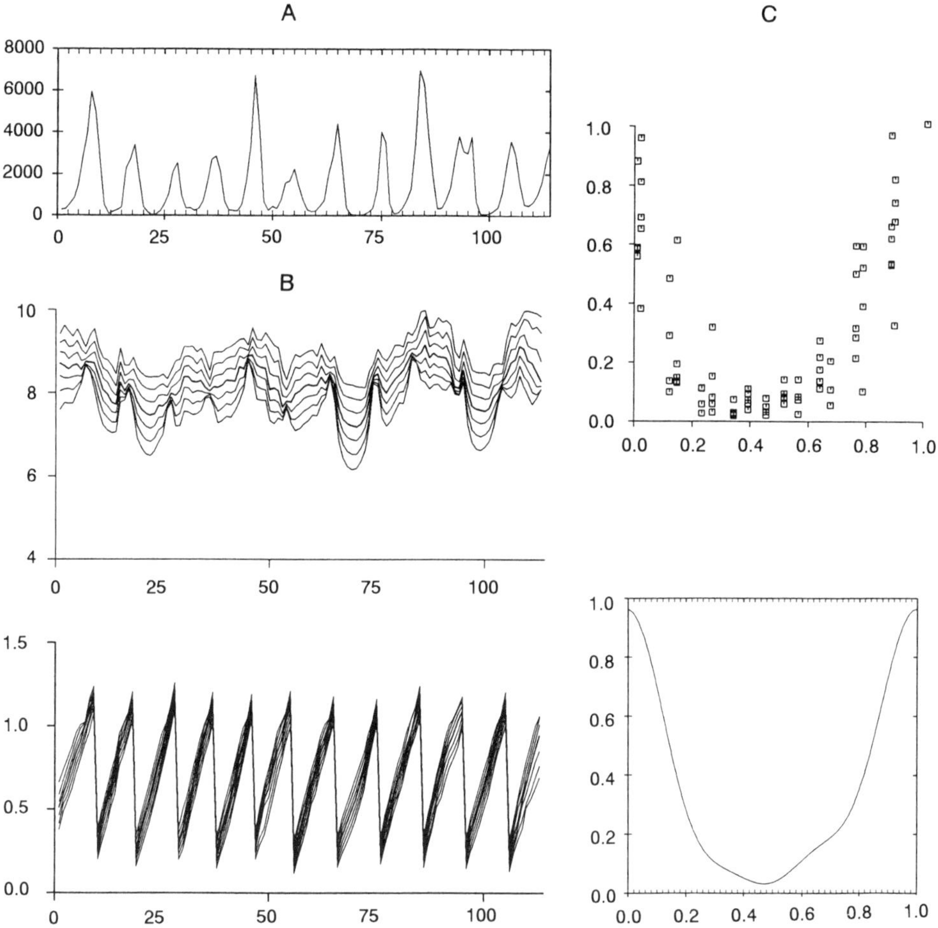

Figure 14.2 A: The Canadian lynx series data, 1821-1934. B: The estimated amplitude and phase. C: The estimated cyclic function and corresponding scatter diagram.

of Kooperberg *et al.*

14.4 Other Examples

In this section we briefly show results of the state space quasi-periodic modeling of simulated phase-wrap data and of the 17 years of Tokyo rainfall data.

14.4.1 Phase-unwrapping

Phase-unwrapping is a problem that has applications in speech analysis, geophysics, underwater acoustics and spectrum analysis. An interesting phase-unwrap problem in underwater acoustics involving communication under the arctic ice flows (Moura and Baggeroer 1988) motivated this study. For a variety of physical reasons there are discontinuities in the phase of that communications carrier signal. By virtue of the discontinuities, the problem of phase-wrap is nonlinear. Moura and Baggeroer model the phase-unwrap with a sum of Gaussian models. They did not obtain smoothed estimates, so their analysis is incomplete. Here we merely illustrate phase-unwrap as a state space quasi-periodic process modeling, and obtained smoothed estimates, to illustrate the fact that such a modeling is appropriate for this problem. Only simulated data is considered.

The model considered here is slightly different and simpler than the one in (14.1)-(14.2). The observation process of this quasi-periodic model is given by

$$y_n = a\cos\theta_n + \varepsilon_n \tag{14.3}$$

and θ_n is modeled by

$$\nabla^2\theta_n \;=\; w_n \sim N(0,\tau^2). \tag{14.4}$$

That is, we take the phase function to be $\theta_n = 2\theta_{n-1} - \theta_{n-2} + w_n$, and in this example the system noise and observation noise were assumed to be Gaussian distributed. For convenience, the fitted model was assumed to have the same structure as the model that generated the data.

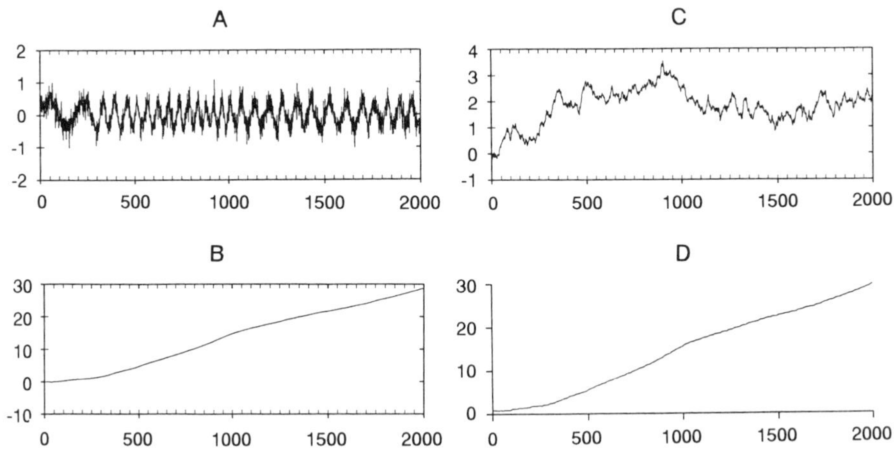

Figure 14.3 A: The simulated phase-wrap data, N = 2000. B: The estimated almost linear phase function. C: Successive differences of the fitted phase function. D: True phase function.

Graphical results are shown in Figure 14.3. The $N = 2000$ simulated data is in Figure 14.3A. The estimated phase by the second order stochastic trend model is shown in Figure 14.3B. The successive differences of the phase, shown in Figure 14.2C, more clearly suggest the stochastic nature of this problem.

For the modeling of real physical phase-wrap data, we would very likely consider fitting the same model as indicated in (14.3)-(14.4) but allow for abrupt discontinuities in the phase function by letting the system noise be heavy tailed. The general state space modeling would be realized by numerical analysis.

14.4.2 Quasi-periodicity in the Rainfall data

In this problem, we re-examined the seventeen years of Tokyo rainfall data from the perspective of thinking of it as quasi-periodic process. The objective of the analysis is to examine the relative phase behavior of the annual rainfall cyclical phenomenon. The model is

$$\begin{aligned} x_n &= a x_{n-1} + w_n \\ z(m_n|\ell_n) &= \binom{\ell_n}{m_n} p(n,x_n)^{m_n}(1-p(n,x_n))^{\ell_n-m_n} \\ p(n,x_n) &= p_{[\mathrm{mod}(n+x_n,365)]} \end{aligned} \tag{14.5}$$

where $p_{[\mathrm{mod}(n+x_n,365)]}$ is the rainfall probability for each day that is obtained as the common trend over the computed 17 years of smoothed estimates described in Chapter 13. $[x]$ denotes the integer part of the real number x. As before $z(m_n|\ell_n)$ is the Binomial distribution for the probability of rain given m_n rainy days out of the l_n observation days. As seen in (14.5), the probability of rain is now allowed to have an AR(1) model phase variation with $w_n \sim N(0, \sigma^2)$.

The results obtained by this modeling are shown in Figure 14.4. Figure 14.4A illustrates the mean and $\pm 1, 2, 3\sigma$ of the phase of the rain over the seventeen year period. Figure 14.4B is an expanded version of Figure 14.4A which permits us to understand the annual relative phase shifts of the rain pattern. For example, in the early months of 1975 there was a slight phase lag or relatively late arrival of the rainy period. That late arrival pattern was more pronounced over the summer months and it lasted through the fall. A similar patter is seen for the early months of 1977, however the rainy period arrival was relatively early in the spring of 1977.

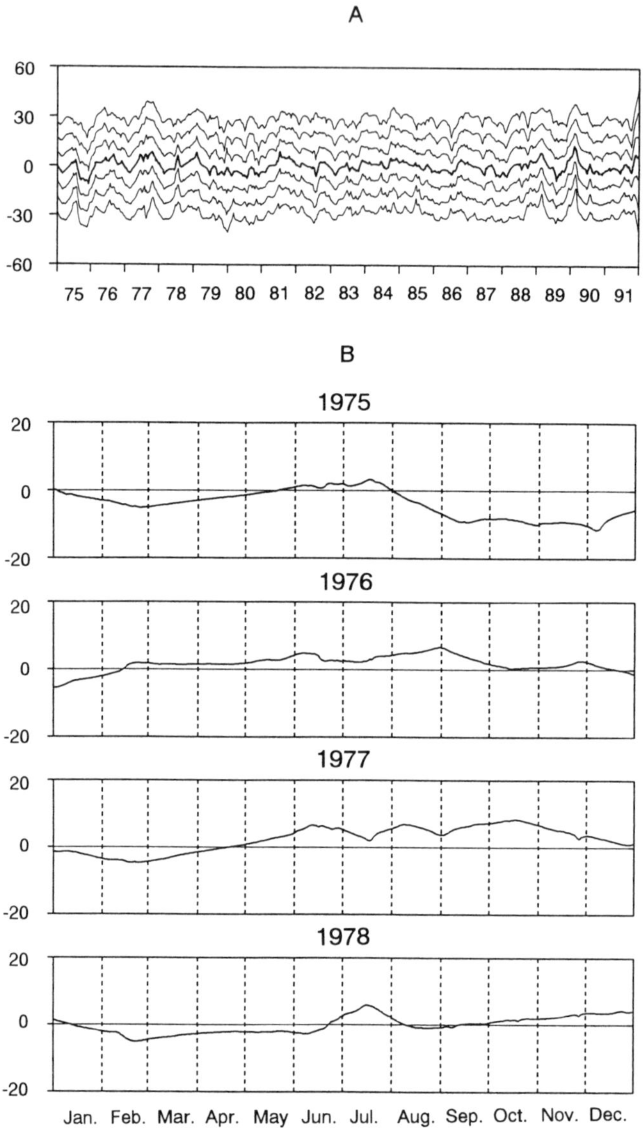

Figure 14.4 A: The mean and $1, 2, 3\sigma$ of the phase of rain over 17 years. B: The relative phase for several selected years.

14.5 Predictive Properties of Quasi-periodic Process Modeling

Finally, motivated by the remark in Tong (1982), that previous modeling of the sunspot and lynx data were not satisfactory for the prediction of more than one lead time, we constructed a simulated data example to investigate the predictive properties of our quasi-periodic model. The data was simulated from the following constant amplitude model:

$$
\begin{aligned}
t_n &= 1.85t_{n-1} - 0.855t_{n-2} + w_n \\
\theta_n &= \theta_{n-1} + 0.2\pi(1 + 0.01t_n) \\
y_n &= \cos(\theta_n) + \varepsilon_n \\
\varepsilon_n &\sim N(0, 0.04), \quad w_n \sim N(0, 1).
\end{aligned} \tag{14.6}
$$

N = 160 observations were generated from this model.

The model parameters and the state were estimated using the first 120 observations. Figure 14.5A and B respectively show $\nabla\theta_n$, the successive differences $\theta_n - \theta_{n-1}$ of the randomly generated phase data and the $y_n, n = 1, \ldots, 160$, observation data. The two different models considered for fitting this data were:

$$\text{Model 1:} \quad \theta_n = \theta_{n-1} + \Delta\theta + w_n, \qquad w_n \sim N(0, \tau_1^2) \tag{14.7}$$

$$\text{Model 2:} \quad \theta_n = 2\theta_{n-1} - \theta_{n-2} + w_n, \quad w_n \sim N(0, \tau_2^2) \tag{14.8}$$

The results obtained for Model 1 were, $\hat{\tau}_1^2 = 0.00065$, $\hat{\sigma}^2 = 0.0325$, $\Delta\theta = 11.9$ with an AIC $= 2.52$. The results obtained for Model 2 were, $\hat{\tau}_2^2 = 0.000001$, $\hat{\sigma}^2 = 0.0380$ with an AIC $= -0.72$.

It is clear that the second model is an AIC better model than the first model. The mean of the fitted phase function is shown in Figure 14.5C. Figure 14.5D and E respectively show the increasing horizon predictive distributions, of the state $p(\theta_n|Y_{120})$ and of the observation process $p(y_n|Y_{120})$ for $n = 121, \ldots, 153$, obtained by the prediction formulae for the state and the observation

$$
\begin{aligned}
p(x_{n+k}|Y_n) &= \int p(x_{n+k}|x_{n+k-1})p(x_{n+k-1}|Y_n)dx_{n+k-1} \\
p(y_{n+k}|Y_n) &= \int p(y_{n+k}|x_{n+k})p(x_{n+k}|Y_n)dx_{n+k}.
\end{aligned} \tag{14.9}
$$

In contrast with the observation by Tong (1982), that prior methods of modeling quasi-periodic time series were not very satisfactory for the prediction of more than one lead time, Figures 14.5D and E show very reasonable predictive distributions for a considerable number of leads. As expected, with increasing horizon, the predictions do become less reliable. That is evidenced by the fact that the appearance of successive predictive distributions "flatten out", indicating an increase in prediction variance.

Finally in Figure 14.5F the increasing horizon prediction performance of an AR model fitted to this data is contrasted with that of the quasi-periodic model over the interval $n = 121, \ldots, 160$. Shown in the figure for the AR model are the mean

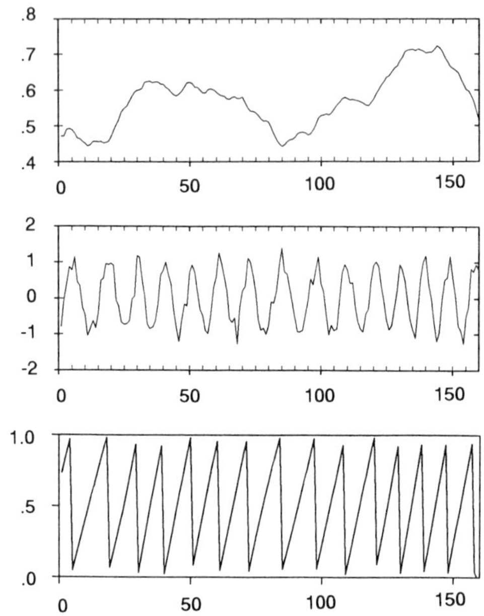

Figure 14.5 A: Successive phase differences, simulated data. B: The observed simulated data. C: The mean of the fitted phase function.

and plus and minus one, two and three σ prediction bands and for the quasi-periodic model the median and equivalent percentage prediction bands. The circles identify the true values of the data. The quasi-periodic model clearly is the better predictor model and the confidence interval bands of the quasi-periodic model are not as wide as those of the AR model. It appears that satisfactory predictive performance of the quasi-periodic model has been clearly illustrated.

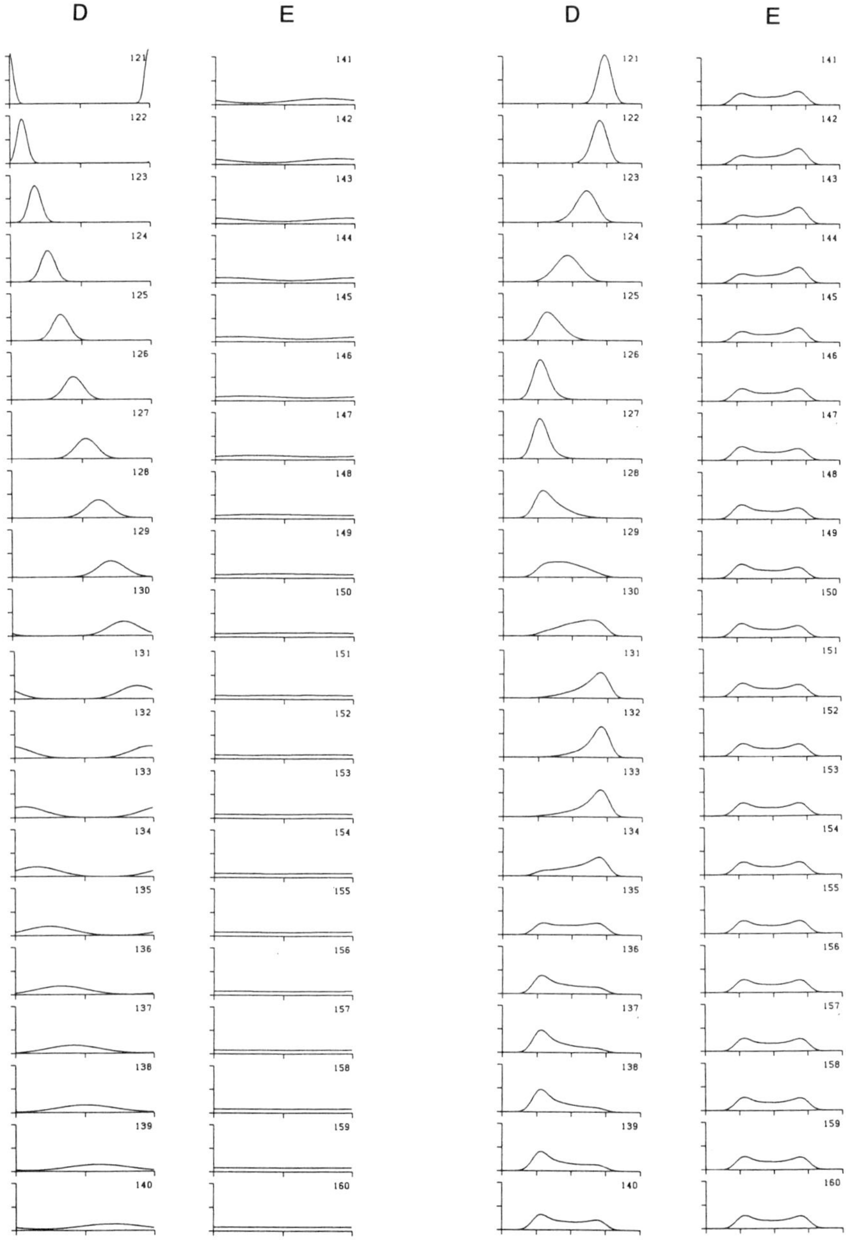

Figure 14.5 D: Increasing horizon state predictive distributions.
E: Increasing horizon observation predictive distributions.

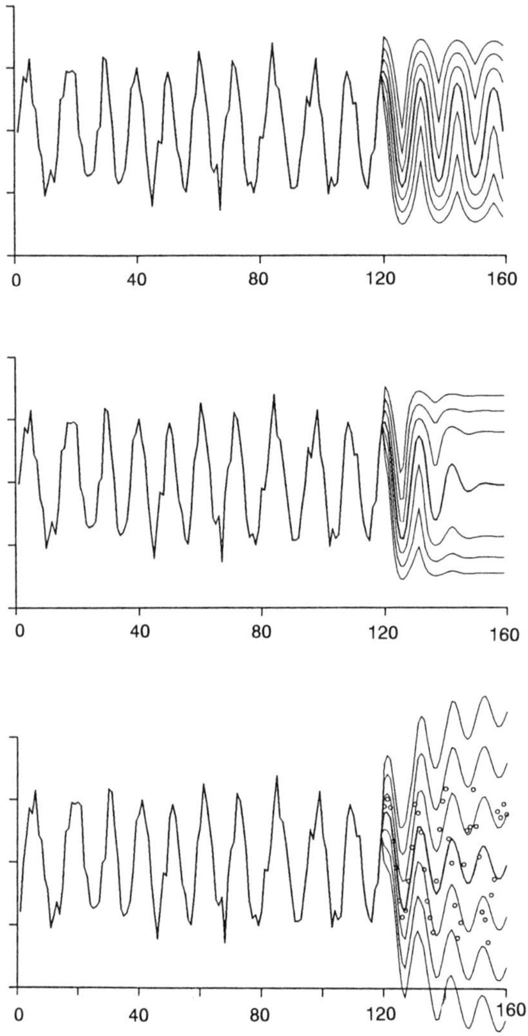

Figure 14.5 F: Predictions, $1, 2, 3\sigma$ intervals for the quasi periodic models (Model 1 and Model 2) and the AR model.

15

Nonlinear Smoothing

A nonlinear state space approach to the smoothing of time series is developed. The time series is expressed in state space model form where the system model or the observation model contains nonlinear functions of the state vector. Recursive formulas for prediction, filtering and smoothing for the nonlinear state space model are shown. The performance of the method is illustrated by the analyses of both a one dimensional and a two dimensional example that have been previously considered in the literature. These examples require numerical approximations of the relevant densities and numerical computations for the nonlinear transformations of variables, the convolution of two densities, Bayes formula, and normalization. Results for the one dimensional example are compared and contrasted with those obtained by the extended Kalman filter method, by a second moment approximation method and also by the Monte Carlo filter method. An empirical study of the numerical accuracy is also shown. A more complete treatment of this material is in Kitagawa (1991).

15.1 Introduction

A nonlinear smoothing methodology for time series analysis shown here is based on the general state space model treated in Chapter 6. The method is particularly useful for time series that can not be analyzed satisfactorily by the standard linear time series models or by any of a large variety of linear approximation techniques.

Of the well known approaches to state space modeling of nonlinear systems we mention the extended Kalman filter, (Sage and Melsa 1971), the second order filter and the Gaussian sum filter (Alspach and Sorenson 1972). These procedures approximate the non-Gaussian distribution by one or several Gaussian distributions and are known to be satisfactory in various nonlinear problems (Anderson and Moore 1979, Sorenson 1985). These methods do however have several drawbacks. The methods based on single Gaussian density such as the extended Kalman filter may yield disastrous results when the true density is not unimodal. The conventional Gaussian sum filter has several technical difficulties in its implementation. (The Gaussian mixture-two filter method, Section 6.4, does overcome some of those difficulties.) Typical phenomena related to these problems are illustrated here with numerical examples.

In view of these technical problems, we were motivated to develop a smoothing method that can handle general density functions. Here, we first show recursive filtering and smoothing formulas for a nonlinear state space model. These formulas can be easily derived from the formulas for a general state space model, (Chapter 6). Each of the probability density functions is approximated by a step function or a continuous piecewise linear function. The necessary operations on the densities are realized by numerical computations. This kind of direct method was attempted in an early stage of the development of nonlinear filters (Bucy and Senne 1971, de Figueiredo and Jan 1971.) The recent development of fast computing facilities now makes it practical to rely on such direct numerical methods, at least for lower order systems. In return for the intensive numerical computations, our method is free from Gaussian or linearity assumptions, and we also do realize a smoothing algorithm.

Other related works on state space nonlinear systems are, West *et al.* (1985), Harrison and Stevens (1976), West and Harrison (1989), Harvey and Fernandes (1989) and Carlin *et al.* (1992). Other interesting nonlinear state space work includes Ettore (1987), in the use of state space models for nonlinear continuous time point processes, applied to neurophysiological and seismological data, Brillinger (1988).

15.2 State Estimation

Consider a system described by a nonlinear state space model

$$\begin{aligned} x_n &= g(x_{n-1}) + w_n \\ y_n &= h(x_n) + \varepsilon_n, \end{aligned} \tag{15.1}$$

where y_n and x_n are ℓ-dimensional observation and m-dimensional state vectors, respectively. w_n and ε_n are m-dimensional and ℓ-dimensional white noise sequences having densities $q(w)$ and $r(\varepsilon)$, respectively, which are independent of the past history of x_n and y_n. The initial state vector x_0 is assumed to be distributed according to the density $p(x_0)$.

The notation used here is as follows: The collections of the states and the observations up to time n are denoted by X_n and Y_n, namely, $X_n \equiv \{x_0, x_1, \ldots, x_n\}$ and $Y_n \equiv \{y_1, \ldots, y_n\}$. The conditional density of x_n given X_i and Y_j is denoted by $p(x_n|X_i, Y_j)$.

Here too, the general problem is the evaluation of $p(x_n|Y_j)$, the conditional density function of the state x_n given the observations, Y_j. As with other state space modeling, the problems of prediction, filtering and smoothing are respectively, identified with the cases $n > j$, $n = j$ and $n < j$.

The nonlinear system (15.1) can be expressed in the general evolution and observation equation form as

$$\begin{aligned} x_n &\sim p(x_n|x_{n-1}) \\ y_n &\sim p(y_n|x_n), \end{aligned} \tag{15.2}$$

for which the conditional densities $p(x_n|x_{n-1})$ and $p(y_n|x_n)$ are

$$\begin{aligned} p(x_n|x_{n-1}) &= q(x_n - g(x_{n-1})) \\ p(y_n|x_n) &= r(y_n - h(x_n)). \end{aligned} \tag{15.3}$$

The general state space model implies that the conditional distributions satisfy the following Markov properties:

$$\begin{aligned} p(x_n|X_{n-1},Y_{n-1}) &= p(x_n|x_{n-1}) \\ p(y_n|X_n,Y_{n-1}) &= p(y_n|x_n). \end{aligned} \tag{15.4}$$

Then, from the general state space treatment in Chapter 6, we have that the density of x_n conditional on x_{n+1} and the entire set of observations Y_N is

$$p(x_n|x_{n+1},Y_N) = p(x_n|x_{n+1},Y_n). \tag{15.5}$$

Similarly from the general state space model in Chapter 6, the recursive formulas for obtaining one step ahead prediction, filtering and smoothing densities are:

One step ahead prediction:

$$p(x_n|Y_{n-1}) = \int_{-\infty}^{\infty} p(x_n|x_{n-1})p(x_{n-1}|Y_{n-1})dx_{n-1}. \tag{15.6}$$

Filtering:

$$p(x_n|Y_n) = \frac{p(y_n|x_n)p(x_n|Y_{n-1})}{p(y_n|Y_{n-1})}, \tag{15.7}$$

where $p(y_n|Y_{n-1})$ is obtained by $\int p(y_n|x_n)p(x_n|Y_{n-1})dx_n$.

Smoothing:

$$p(x_n|Y_N) = p(x_n|Y_n)\int_{-\infty}^{\infty} \frac{p(x_{n+1}|Y_N)p(x_{n+1}|x_n)}{p(x_{n+1}|Y_n)}dx_{n+1}. \tag{15.8}$$

As observed in Chapter 6, for linear Gaussian systems, the conditional densities $p(x_n|Y_{n-1})$, $p(x_n|Y_n)$ and $p(x_n|Y_N)$ are characterized by the mean vectors and the covariance matrices and hence (15.6)–(15.8) are equivalent to the well known Kalman filter and the fixed interval smoothing algorithms (Kalman 1960, Anderson and Moore 1979). (Note the derivation of the Kalman filter from a general state space model in Chapter 6.) For nonlinear state space models, however, due to the nonlinear transformation of the state variables, the conditional density $p(x_n|Y_j)$ is non-Gaussian even when both w_n and ε_n are Gaussian and cannot be specified by using the first two moments.

It should be noted that only the structure in (15.2) and the Markovianess in (15.4), are the critical model assumptions. Therefore, the additive error structure in (15.1) is not essential, and the method presented here can be applied to a wider class of nonlinear models than the one formulated in (15.1).

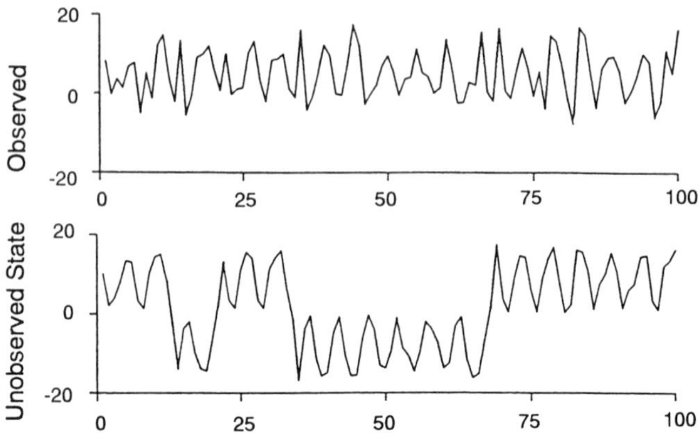

Figure 15.1: True signal x_n, $n = 1, ..., 100$ and the observations y_n, $n = 1, ..., 100$

15.3 A One Dimensional Problem

In this section the performance of our nonlinear smoother is compared with that of the extended Kalman filter, (EKF), a second moment approximation method as well as the Monte Carlo filter method, by computations on a well known model. We consider the data artificially generated by the nonlinear model originally considered by Andrade Netto *et al.* (1978).

$$\begin{aligned} x_n &= \frac{1}{2}x_{n-1} + \frac{25x_{n-1}}{1+x_{n-1}^2} + 8\cos(1.2n) + w_n \\ y_n &= \frac{x_n^2}{20} + \varepsilon_n. \end{aligned} \tag{15.9}$$

The x_n and y_n shown in Figure 15.1 are generated by independent Gaussian random numbers $x_0 \sim N(0,5)$, $w_n \sim N(0,1)$ and $\varepsilon_n \sim N(0,10)$. The problem is to estimate the true signal x_n from the sequence of observations $\{y_n\}$ assuming that the model (15.9) is known. Our nonlinear filter and smoother were applied to the problem. For comparison, the well-known extended Kalman filter, the second order filter and the linearized fixed interval smoother associated with these filters were also applied (Sage and Melsa 1971). In filtering and the smoothing, the following discretization was (arbitrarily) used: $k = 400$, $z_0 = -30$, $z_k = 30$ and $p(x_0) = N((z_0 + z_k)/2, (z_k - z_0)^2/16)$. Figure 15.2 shows the posterior densities $p(x_{17}|Y_m)$, $m = 16, \ldots, 20$ and 100. From the left to the right each column of the figure shows the results obtained by the extended Kalman filter, by the second order filter and by our nonlinear filter and smoother, respectively. This figure illustrates a quite typical situation where these algorithms yield substantially different results. Using our nonlinear filter, the one step ahead predictive density $p(x_{17}|Y_{16})$ is very broad and bimodal, and this bimodality extends to the filtered density $p(x_{17}|Y_{17})$ and to the smoothed density $p(x_{17}|Y_{18})$. On the other hand, the extended Kalman filter approximates each density $p(x_{17}|Y_m)$ by

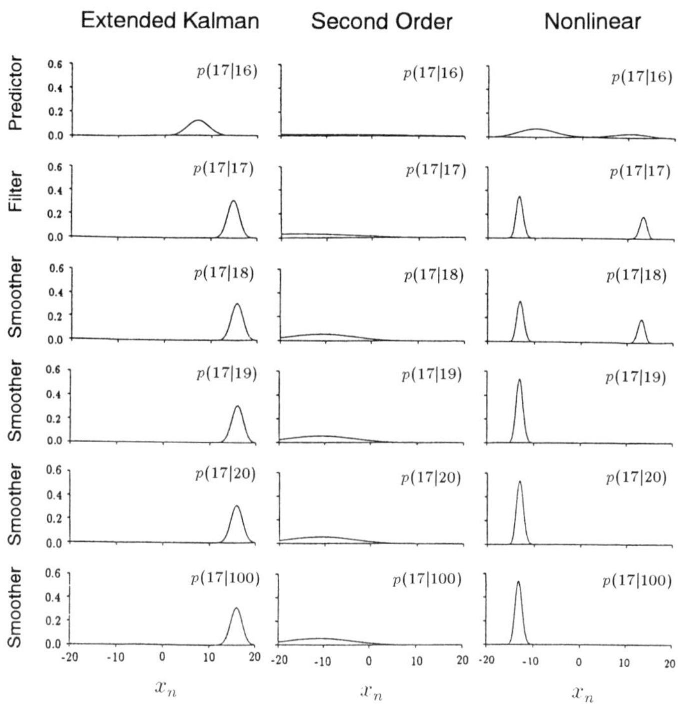

Figure 15.2: Posterior densities $p(x_{17}|Y_m)$, $m = 16, \ldots, 20$ and 100 obtained by the extended Kalman filter based smoother (left), the second order filter based smoother (middle) and our nonlinear smoother (right).

a single Gaussian density. Although for $m \geq 19$ the smoothed density obtained by our nonlinear smoother, $p(x_{17}|Y_m)$, also becomes unimodal and resembles a Gaussian density, its location is completely different from the one of the linearized smoother and is actually on the other side of the origin. The second order filter shown in the middle column also approximates the posterior density by a single Gaussian density. The second order filter posterior density estimates are very conservative in that they have large variances.

Figure 15.3 shows the smoothed posterior density $p(x_n|Y_N)$ versus time, n, obtained by our nonlinear smoother using numerical integration. In that illustration, the bold curve shows the 50% point of the posterior density and two fine curves express the 2.3% and 97.7% points which correspond to the two standard error intervals of the Gaussian densities. + indicates the true value of x_n. Figure 15.4 shows the plot of the median error of smoothing, $x_{n|N}$ minus the true x_n with two standard error intervals. Figure 15.5 shows the results realized by the extended Kalman filter based linearized smoother. In comparison with the the performance of our nonlin-

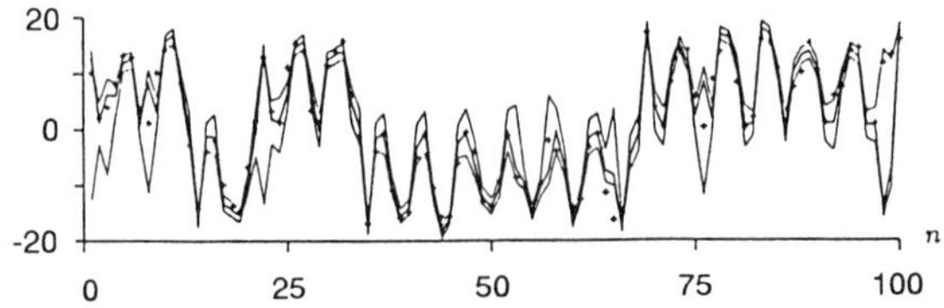

Figure 15.3: Posterior density $p(x_n|Y_N)$ obtained by our nonlinear smoother. The bold curve shows the median and the fine curves show two standard error interval. + indicates the true value.

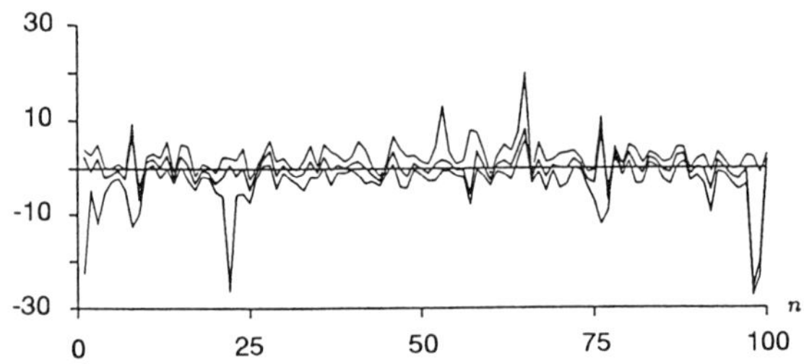

Figure 15.4: Posterior median minus true values with two standard error intervals for the nonlinear smoother.

ear smother, the EKF performance is very poor. The superiority of of the nonlinear smoother is evident.

The smoothed estimates of the state x_n obtained by the Monte Carlo method is shown in Figure 15.6. The results compare quite nicely with those obtained by the "exact" nonlinear smoother in Figure 15.3 and are also clearly superior to those obtained by the extended Kalman filter method.

An empirical study on the effect of the selection of the number of nodes was also performed using the same example. Table 15.1 shows the effect of the number of nodes, k, on the computing time and on the accuracy of the obtained posterior densities. The first column of the table shows the CPU-time in seconds spent for the computation by a main frame computer, HITAC-M682H with internal array processors (about 60MFLOPS). From this table, it can be seen that the expended CPU-time is less than the order of k^2. The convergence of the nonlinear filter and the smoother as the number of the nodes increases was checked by the four comparison of likelihoods type "performance criteria" defined as follows:

$$\nabla F(k) = \sum_{n=1}^{N} \log p_k(y_n|Y_{n-1}) - \sum_{n=1}^{N} \log p_{4096}(y_n|Y_{N-1}) \qquad (15.10)$$

$$E_1(k) = \sum_{n=1}^{N} I(p_{4096}(x_n|Y_n); p_k(x_n|Y_n)) \qquad (15.11)$$

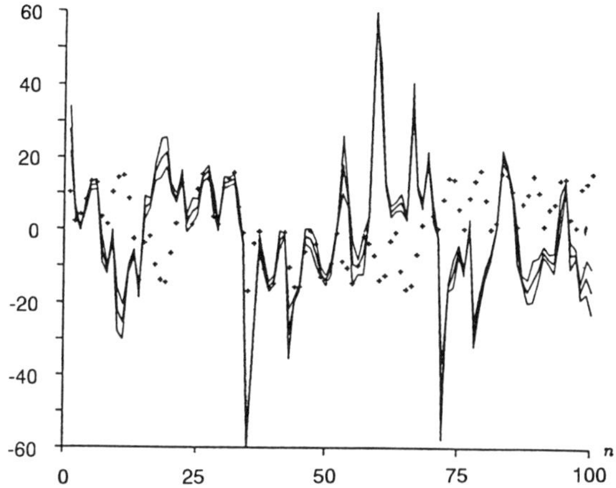

Figure 15.5: Posterior medians minus true values with two standard error intervals for the extended Kalman filter/smoother.

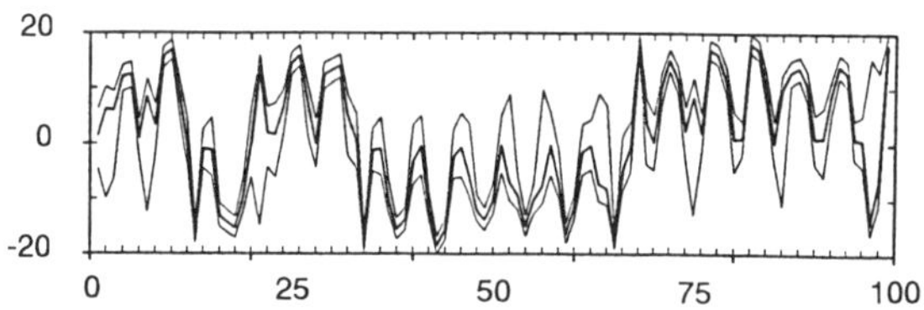

Figure 15.6: Posterior density $p(x_n|Y_N)$ obtained by the Monte Carlo method. Posterior median plus and minus two standard error intervals.

$$E_2(k) = \sum_{n=1}^{N} I(p_{4096}(x_n|Y_N); p_k(x_n|Y_N)) \tag{15.12}$$

$$E_3(k) = \sum_{n=1}^{N} I(p_{4096}(x_n|Y_N); p_k(x_n|Y_n)). \tag{15.13}$$

Here $p_k(x_n|Y_j)$ denotes the posterior density of x_n given the observation Y_j obtained by using the approximation with k nodes, and $I(p(x); q(x))$ is the Kullback-Leibler information number of the density $p(x)$ with respect to the density $q(x)$ defined by $I(p(x); q(x)) = \int \log\{p(x)/q(x)\}p(x)dx$. (See Section 2.1.1.) Since the true density is unknown, the one obtained by the finest mesh ($k = 4096$) is used as the "true" density. $\nabla F(k)$ is the difference of the log-likelihood values between the approximation and the "true" density. $E_1(k)$ and $E_2(k)$ are the summation over time of the information numbers of the filtered and the smoothed densities, respectively. $E_3(k)$ measures the

Table 15.1. Monte Carlo study on the effect of the number of nodes (k), on the CPU time and the accuracy.

k	CPU time	∇F	E_1	E_2	E_3
32	0.10	58.4	80.7	144.0	483.9
64	0.20	6.21	16.1	46.8	64.6
128	0.42	3.59	4.26	12.5	51.7
256	0.94	1.42	1.02	3.30	47.5
512	2.33	0.39	.234	.758	43.9
1024	6.55	0.21	.048	.152	42.3
2048	20.69	0.02	.007	.022	41.7
4096	71.80	—	—	—	—
EKF	.005	650.5	2114.7	2572.9	2238.1

difference between the "true" smoother and the approximating filter.

From the table, it can be seen that the difference between the log-likelihoods of the "true" and approximating densities is less than 1.0 for $k \geq 512$. However, from the experience gained in comparing several models, and observing similar fluctuations in the approximated log likelihood, it appears that a coarser mesh with k=128 might be sufficient for model identification. Also, $E_1(k)$ and $E_2(k)$ both converge to zero apparently with order $O(k^2)$. On the other hand, the $E_3(k)$ seems to converge to a constant. This indicates that there is a substantial advantage in doing smoothing, in comparison with only doing filtering. It is worth mentioning that the estimates obtained by smoothing with a coarse mesh (k=64), has an accuracy equivalent to that of estimation by filtering. An evaluation of the extended Kalman filter performance is also shown in the table. It is clear that the EKF performance is much worse than our nonlinear filter or smoother even with the coarsest mesh.

In summary, this example reveals two important points in the nonlinear filtering problem:

1. The extended Kalman filter and any other filter that approximates the density by a single Gaussian density may produce disastrous results when the true density is not unimodal.

2. The information from future observations is quite important to identify the location of the state. The difference of $p(x_{17}|Y_{19})$ and $p(x_{17}|Y_{17})$ and the values of $E_3(k)$ clearly demonstrate this. Thus the role of smoothing is essential to get a good estimate of the state.

15.4 A Two Dimensional Problem

The second example is a passive receiver problem. A similar problem was considered by Bucy and Senne (1971) and Alspach and Sorenson (1972). In this example, the

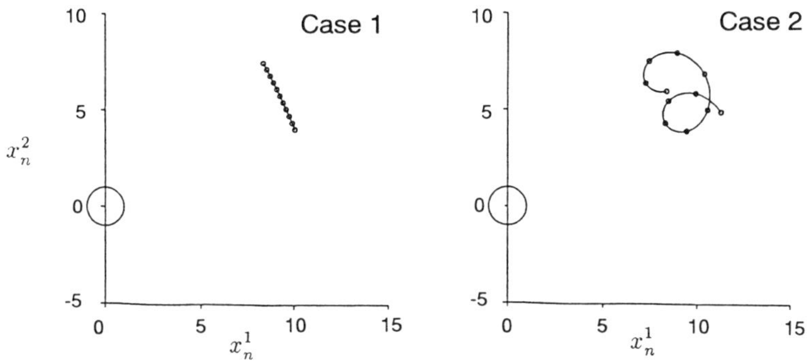

Figure 15.7: Trajectory of the moving object and the receiver: Case 1 and Case 2.

problem is to locate a slowly moving target in two dimensional space. This target is observed in accordance with the scalar nonlinear measurement model

$$y_n = h(x_n^1, x_n^2) + \varepsilon_n \tag{15.14}$$

where

$$\begin{aligned} h(x_n^1, x_n^2) &= \tan^{-1}\left\{\frac{x_n^2 - \sin\beta_n}{x_n^1 - \cos\beta_n}\right\}, \\ \beta_n &= \beta_{n-1} + \nabla\beta. \end{aligned} \tag{15.15}$$

Here β_0 and $\nabla\beta$ are given constants and w_n is a Gaussian white noise with known variance σ^2. This is a simple example of the vector tracking problem of locating a moving object by observing the relative angle observed on a rotating observatory. At time n, the target is at position (x_n^1, x_n^2) and $h(x_n)$ is the bearing angle from a sensor which proceeds in a unit circle around the origin. Figure 15.7 shows two examples of the trajectory. (For Case 1, $x_n^1 = 22\cos t_n - 9$, $x_n^2 = 30\sin t_n - 11$, and for Case 2, $x_n^1 = 1.5\cos s_n + 22\cos t_n - 9$, $x_n^2 = 1.5\sin s_n + 30\sin t_n - 11$ with $t_n = (30 + 0.08n)/180\pi$, $s_n = (30 + 6n)/180\pi$). Figure 15.8 shows two artificially generated series y_n which are obtained by observing these trajectories according to the measurement model (15.14) and (15.15) with $\beta_0 = 0$, $\nabla\beta = 1$, $\varepsilon_n \sim N(0, 0.02^2)$ for Case 1 and $\varepsilon_n \sim N(0, 0.01^2)$ for Case 2. For the estimation of this moving object, we consider the following smoothness prior model for the movement of the target in the two-dimensional state space.

$$\begin{aligned} \nabla^d x_n^1 &= w_n^1 \\ \nabla^d x_n^2 &= w_n^2. \end{aligned} \tag{15.16}$$

Here the difference order d is either 1 or 2 and w_n^1 and w_n^2 are mutually independent Gaussian white noise sequence with variances, τ_1^2 and τ_2^2, respectively. The smoothness prior model (15.16) with the observation model (15.14), (15.15) constitutes our nonlinear state space model for estimating the location of the object. It should be

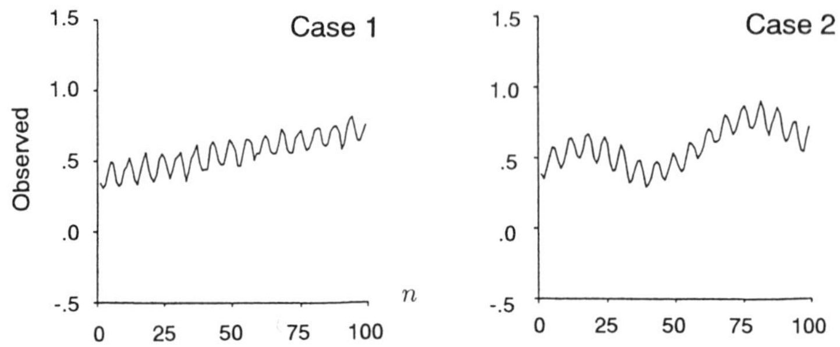

Figure 15.8: Observed angle y_n: Case 1 and Case 2.

noted that the Gaussianity of neither w_n^1 nor w_n^2 are essential in our model. The value of τ_1^2 and τ_2^2 are estimated by maximizing the log-likelihood as defined for the general state space model in in Section 6.2.2. To reduce computational complexity in the maximum likelihood computations, we let $\tau_1^2 = \tau_2^2 = \tau^2$ for the second order model. Table 15.2 shows the maximum likelihood estimates of τ^2, τ_1^2 and τ_2^2 and the associated log-likelihoods and the AICs. The discretization used in the computation are summarized in Table 15.3. The initial state density, $p(x_0)$, was arbitrarily set to the two or four dimensional Gaussian density with the i-th mean $(z_0^i + z_{k_i}^i)/2$ and the diagonal covariance matrix with the i-th diagonal element given by $(z_{k_i}^i - z_0^i)^2/16$. Figure 15.9 shows the contours of the posterior density $p(x_n^1, x_n^2|Y_N)$ for n=20, 40, 60 and 80 for Case 2. Figure 15.10 shows the trace of (p_{1n}^j, p_{2n}^j), where p_{1n}^j and p_{2n}^j $(j = 1, \ldots, 7)$ are the 0.13%, 2.27%, 15.87%, 50%, 84.13%, 97.73%, 99.87% points of the marginal posterior density of $p(x_n^1|Y_N)$ and $p(x_n^2|Y_N)$, respectively. In the non-Gaussian case, except for the 50% point (namely $j = 4$), these points do not have any particular meaning. The movement of the posterior density on the 2-D space can be appreciated by keeping the contour lines in Figure 15.9 in mind.

The CPU-time for the first order trend model $(m = 2, k_1 = k_2 = 200, n = 100)$, was 66 seconds. On the other hand, for the second order trend model $(m = 4)$, filtering with a coarse mesh $(k_1 = k_3 = 50, k_2 = k_4 = 7)$, $(k_1 = k_3 = 70, k_2 = k_4 = 11)$, and $(k_1 = k_3 = 100, k_2 = k_4 = 15)$, took 98, 590 and 2617 seconds, respectively.

Table 15.2. Summary statistics of the fitted models to the 2-D tracking problem

CASE	MODEL	Log-Like.	AIC	τ_1^2	τ_2^2
1	2D	223.36	-442.71	0.032	0.014
	4D	231.69	-461.39	0.12×10^{-5}	
2	2D	239.39	-474.78	0.041	0.0013
	4D	260.65	-518.30	0.031	

Table 15.3. Summary of the discretizing parameters used in the nonlinear filter and smoother

Case	Model			
1	2D	$k_1 = k_2 = 200$	$z_0^1 = 6, z_{k_1}^1 = 14$	$z_0^2 = 2, z_{k_2}^2 = 10$
	4D	$k_1 = k_3 = 100$	$z_0^1 = 6, z_{k_1}^1 = 14$	$z_0^3 = 2, z_{k_2}^3 = 10$
		$k_2 = k_4 = 15$	$z_0^2 = -0.013, z_{k_2}^2 = 0.021$	$z_0^4 = 0.031, z_{k_4}^4 = 0.039$
2	2D	$k_1 = k_2 = 200$	$z_0^1 = 5, z_{k_1}^1 = 15$	$z_0^2 = 2, z_{k_2}^2 = 12$
	4D	$k_1 = k_3 = 100$	$z_0^1 = 5, z_{k_1}^1 = 15$	$z_0^2 = -0.013, z_{k_2}^2 = -0.021$
		$k_2 = k_4 = 15$	$z_0^3 = 2, z_{k_3}^3 = 13$	$z_0^4 = -0.2, z_{k_4}^4 = 0.2$

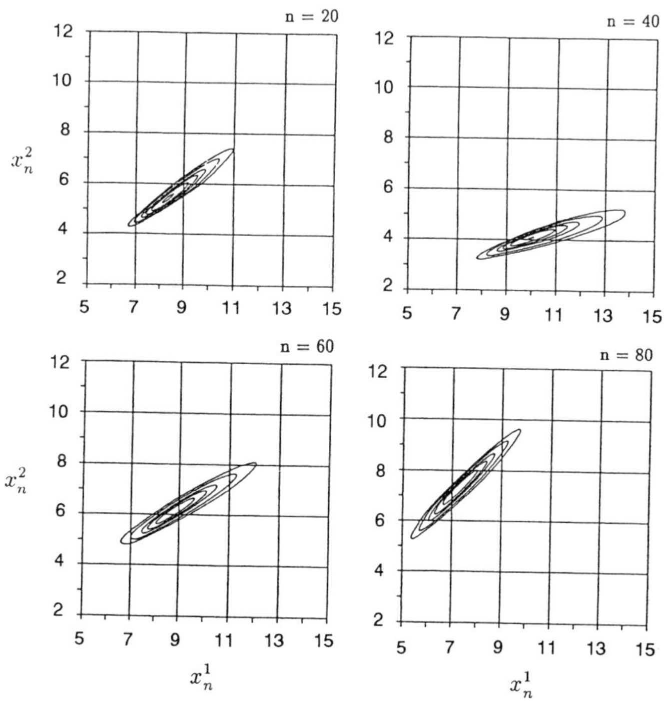

Figure 15.9: Contour of the posterior densities, $p(x_n^1, x_n^2|Y_N)$, n=20,40,60 and 80 for Case 2.

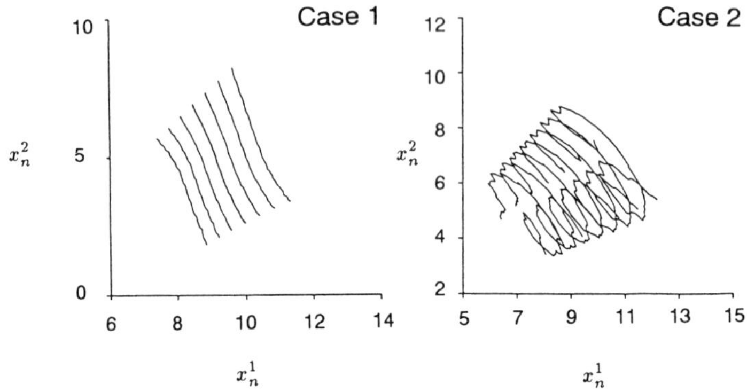

Figure 15.10: Trace of (p_{n1}^j, p_{n2}^j) where p_{n1}^j and p_{n2}^j are 0.13, 2.27, 15,87, 50, 84,13, 97,73 and 99.87 percentile points of the marginal posterior density of $p(x_n^1|Y_N)$ and $p(x_n^2|Y_N)$.

16

Other Applications

In this chapter three additional applications of smoothness priors time series modeling are addressed which for a variety of reasons, were not included in other chapters. The first application is a study of the modeling of a very large data set, (500,000 observations), with missing data and outliers in a complex stochastic trend and regression on covariates modeling (Kitagawa and Matsumoto 1996). The objective of the analysis is to decompose the data into its component parts. The second application is a Markov state classification problem in which each observed state corresponds to a different time series process and the states are switched at random times. An illustrative analysis is done on simulated data. The third application involves an extension of the smoothness priors long AR model for spectral estimation in scalar stationary time series, (discussed in Chapter 4), to the multivariate case.

16.1 A Large Scale Decomposition Problem

In this section we treat the analysis of the very long, ($N = 500{,}000$), regularly sampled underground water level data set observed at the Haibara well, Shizuoka in central Japan. Some of the data is missing and there are outliers. The analysis is a decomposition of the scalar ground water level observations into a nonstationary trend and several other observed covariate components. The objective of the modeling is to determine if, after the removal of the influence of the covariates, coseismic effects (effects which are a consequence of earthquakes), can be discerned in the underground water trend level. The analysis is prototypic of large data set analysis problems and is also quite important in the sense that a successful analysis can potentially yield insight into coseismic and perhaps even preseismic effects in the measurement of underground water levels.

Smoothness priors state space methods are used to analyze the underground water level data. The observation of the underground water level is assumed to be characterized by a nonstationary trend and be influenced by barometric pressure, tidal and precipitation effects. Earlier, Igarashi and Wakita (1991), fit a least squares computational smoothness priors ground water level model to observed barometric pressure and tidal influences. That analysis did not include the effects of precipitation. Matsumoto (1992) did include the effects of precipitation in a non Bayesian regression analysis of the data.

Our own analysis is a smoothness priors stochastic trend model with regression covariates. Specifically, the underground water level data is to be analyzed by the model

$$y_n = t_n + P_n + E_n + R_n + \varepsilon_n, \tag{16.1}$$

where t_n, P_n, E_n, R_n and ε_n are the trend, barometric pressure, earth tide, and precipitation effect components and the observation noise, respectively. (Additional details are shown in Kitagawa and Matsumoto 1996.)

16.1.1 Data Preparation and a Strategy for the Data Analysis

The nature of the modeling problem can be appreciated from Figures 16.1A, and B. In Figure 16.1A simultaneously observed segments of the water level, (measured in meters), precipitation, (measured in millimeters), barometric pressure, (measured in millibars), and earth tide (in μ gals, computed from theoretically known earth tides) are shown. The measurements were taken regularly at 10 minute intervals. The envelopes of the water level and barometric pressure are "drawn in" for convenience of visual interpretation. The gaps in the water level and barometric pressure data indicate data that is missing, (because of equipment problems), and which must be filled in order for the decomposition model (16.1), to be fitted. Figure 16.1B shows a small segment of the water level data. Occasionally that data is seen to be contaminated by positive noise spikes or outliers.

The Strategy

The first problem addressed in this data set, is that of filling the missing barometric and the missing water level data. Secondly, for several technical reasons, we can not fit the entire model to the long data set at once. Our approach, which we shall justify, will be to fit the observation model without the precipitation effect in a first modeling stage and in a second modeling stage, add in the precipitation effect.

Preparing the data

First we address the problem of filling in the missing data. The treatment of missing barometric data is different than the treatment of missing water level data and they are discussed separately.

Filling in the barometric pressure data

The envelope function of the barometric pressure in Figure 16.1 is suggestive. We consider fitting a first order trend model to the barometric pressure and filling in the missing barometric pressure data using the Kalman filter technique presented in

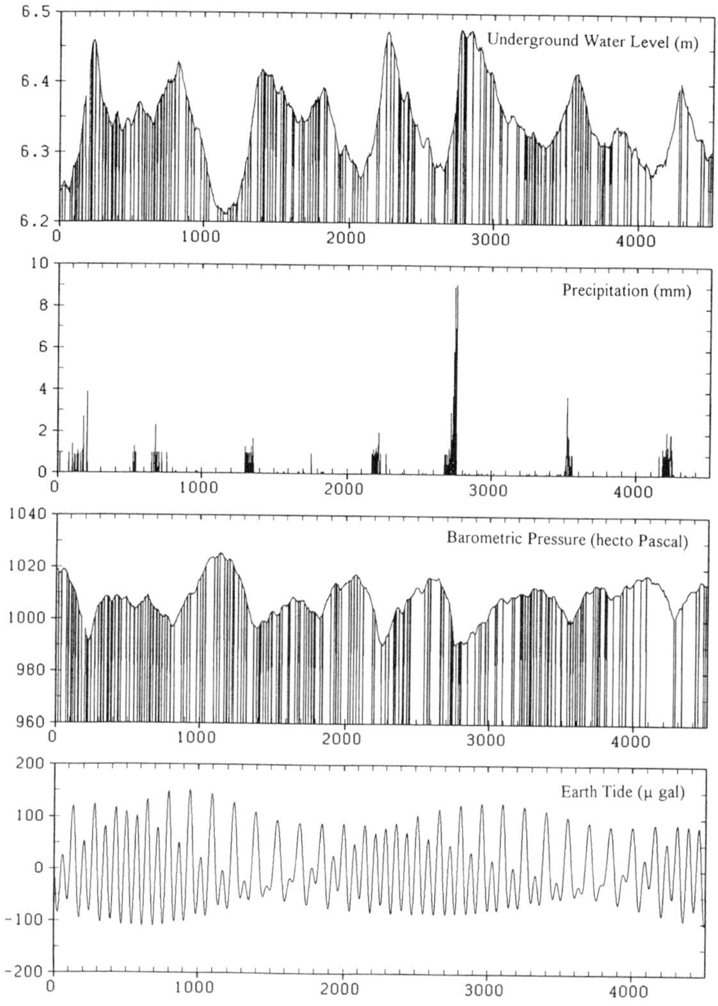

Figure16.1 A: Simultaneously observed water level, precipitation, barometric pressure, and theoretically known earth tides.

Chapter 5. A model for that procedure is

$$\begin{aligned} t_n &= t_{n-1} + w_n \\ y_n &= t_n + \varepsilon_n, \end{aligned} \tag{16.2}$$

where $w_n \sim N(0, \tau^2)$ and $\varepsilon_n \sim N(0, \sigma^2)$. Let $t_{n|m}$ and $V_{n|m}$ denote the mean and the variance of t_n given the observations up to time m, $Y_m \equiv \{y_1, \cdots, y_m\}$. Then one step ahead prediction $t_{n|n-1}$, the filter $t_{n|n}$, and the smoothed trend estimates $t_{n|N}$,

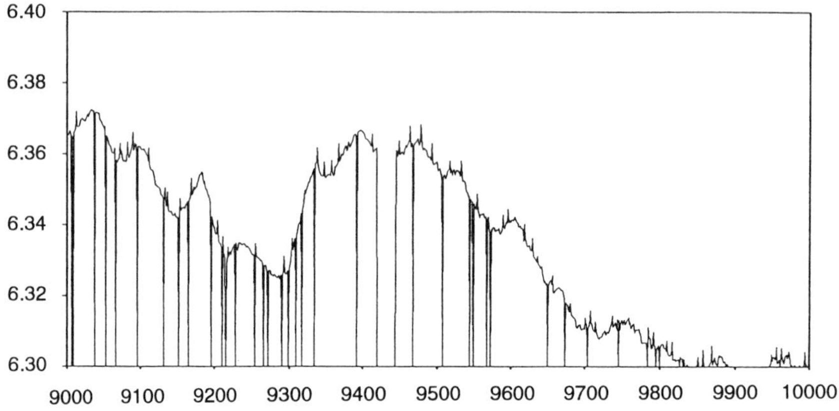

Figure 16.1 B: A segment of the water level data.

with their variances can be obtained by the method shown in Chapter 5 for filling in missing data.

Filling in the water level data

The fact that there are outliers as well as missing data in the water level observations implies that we are obliged to fill in that data with the non-Gaussian state space model. Here again, we consider the first order trend model

$$\begin{aligned} t_n &= t_{n-1} + w_n \\ y_n &= t_n + \varepsilon_n, \end{aligned} \tag{16.3}$$

where $w_n \sim N(0, \tau^2)$ and where we now consider using probability density functions with heavy tails for the observational noise. Such densities allow the occurrence of large deviations with a low probability.

Specifically, in the analysis, we tried the following three observation noise models.

1. Gauss model $\quad r(\varepsilon) \sim N(0, \sigma^2)$,
2. Cauchy model $\quad r(\varepsilon) = \dfrac{\sigma}{\pi(\varepsilon^2 + \sigma^2)}$,
3. Gaussian mixture model $\quad r(\varepsilon) \sim (1-\alpha)N(0, \sigma^2) + \alpha N(\mu, \zeta^2), \qquad \mu > 0.$

For the filtering and smoothing of the non-Gaussian state space model, we applied the non-Gaussian predictor, filter and smoother, (Chapter 6),

$$\begin{aligned} p(x_n|Y_{n-1}) &= \int_{-\infty}^{\infty} q(t_n - t_{n-1}) p(t_{n-1}|Y_{n-1}) dt_{n-1} \\ p(t_n|Y_n) &= \frac{r(y_n - t_n) p(t_n|Y_{n-1})}{p(y_n|Y_{n-1})} \\ p(t_n|Y_N) &= p(t_n|Y_n) \int_{-\infty}^{\infty} \frac{q(t_{n+1} - t_n) p(t_{n+1}|Y_N)}{p(t_{n+1}|Y_n)} dt_{n+1}. \end{aligned} \tag{16.4}$$

Table 16.1 AIC's of different observation noise models for water level data

model	AIC
1	−8743.0
2	−8656.8
3	−8832.4

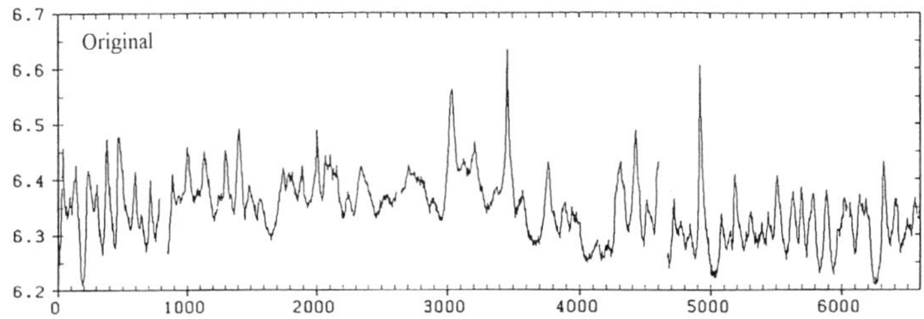

Figure 16.1 C: The interpolated underground water level data.

Table 16.1 shows the AIC's of three observation noise models. The Gaussian mixture model is the best observation noise model.

Figure 16.1C shows the interpolated underground data realized by the non-Gaussian smoother. As can be seen in Figure 16.1C, the underground water level data fluctuates within approximately 45cm and the effects of earthquakes on that data are not apparent. In the analysis that follows, the interpolated data are sampled hourly.

16.1.2 The Data Analysis

Next, with the missing data filled in, in an attempt to account for the effect of the covariates on the underground water level, we considered the following model,

$$y_n = t_n + P_n + E_n + \varepsilon_n, \tag{16.5}$$

where t_n, P_n, E_n and ε_n are the trend, the barometric pressure, the earth tide, and the observation noise components, respectively. We assumed that those components follow the following models

$$\begin{aligned} \nabla^k t_n &= w_n \\ P_n &= \sum_{i=0}^{m} a_i p_{n-i} \\ E_n &= \sum_{i=0}^{\ell} b_i et_{n-i}. \end{aligned} \tag{16.6}$$

Here p_n is the observed barometric pressure at time n and et_n is the theoretical earth tide at time n. In (16.6), the effects of the barometric pressure and the tidal

Table 16.2 AIC's of barometric pressure and tide effects regression model orders

m	ℓ = 0	1	2	3
23	−57819	−59368	−59498	−59488
24	−57815	−59374	−59536	−59526
25	−57830	−59393	−59580	−59569
26	−57836	−59386	−59575	−59563
27	−57832	−59379	−59566	−59554

component on the underground water level are expressed in terms of regressions on covariates. To estimate the trend t_n, the barometric pressure coefficients a_i, $i = 0, 1, \ldots, m$ and the earth tide coefficients b_i, $i = 0, 1, \ldots, l$, we use the state space representation of (16.5) and (16.6).

$$\begin{bmatrix} t_n \\ \hline a_0 \\ \vdots \\ a_m \\ \hline b_0 \\ \vdots \\ b_\ell \end{bmatrix} = \begin{bmatrix} 1 & & \\ \hline & I_{m+1} & \\ \hline & & I_{\ell+1} \end{bmatrix} \begin{bmatrix} t_{n-1} \\ \hline a_0 \\ \vdots \\ a_m \\ \hline b_0 \\ \vdots \\ b_\ell \end{bmatrix} + \begin{bmatrix} 1 \\ \hline 0 \\ \vdots \\ 0 \\ \hline 0 \\ \vdots \\ 0 \end{bmatrix} w_n, \qquad (16.7)$$

$$y_n = [1, p_n, \cdots, p_{n-m}, et_n, \cdots, et_{n-\ell}] x_n + \varepsilon_n$$

In the equation above, the trend order, k, is 1. The state vector x_n, is defined by the left hand side of (16.7).

Application of the Kalman filter and the fixed interval smoother yields the estimate of the state x_n and thus t_n, and the regression coefficients a_i, $i = 0, 1, \ldots, m$, and b_i, $i = 0, 1, \ldots, l$. Since the a_i and b_i are assumed to be time invariant, they are estimated by Kalman filtering and the fixed interval smoothing is necessary only for the trend component t_n.

The values of the AIC computed for several models with different values of barometric pressure and tidal effect regression model orders m and l are listed in Table 16.2. The AIC best model is the one for which $m = 25$ and $l = 2$.

Figure 16.1D shows the estimated trend, barometric pressure and the earth tide components. The annual variation of the trend is only about 6cm and the effect of the earthquake with magnitude M=4.8, at a distance D=42km from the ground water well, is clearly detected. Most of the range of about 45cm trend variations in Figure 16.1C can be now be seen to be the effect of barometric presure. The effect of the earth tide is also present, but is miniscule.

Evaluation of the Rainfall Effect

In Figure 16.1D, the effect of an earthquake on the underground water level is

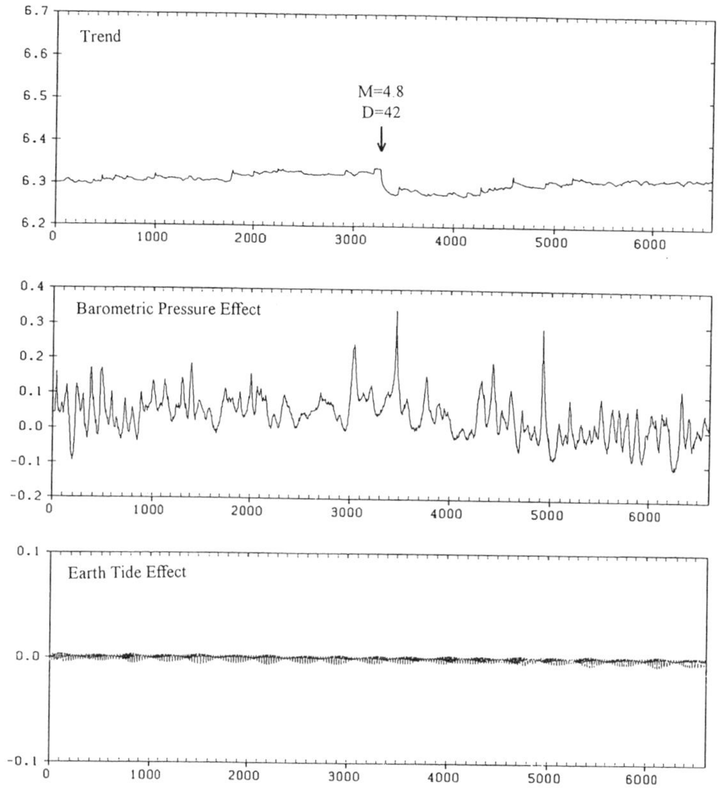

Figure 16.1 D: The estimated trend, barometric pressure and earthtide effects.

clearly discernible. Since it is known, (Matsumoto 1992), that the underground water level is strongly correlated with rainfall, we next tried to take the rainfall effect on the ground water level, into account. Therefore, we considered the decomposition

$$y_n = t_n + P_n + E_n + R_n + \varepsilon_n, \tag{16.8}$$

where t_n, P_n, E_n and ε_n are as above and R_n is the rainfall effect component. That is modeled here by the mixed autoregressive moving average model with a general driving force covariance structure.

$$R_n = \sum_{i=1}^{k} c_i R_{n-i} + \sum_{i=1}^{k} d_i r_{n-i} + v_n, \tag{16.9}$$

where r_n is the observed precipitation. This model can be expressed in the state space model form,

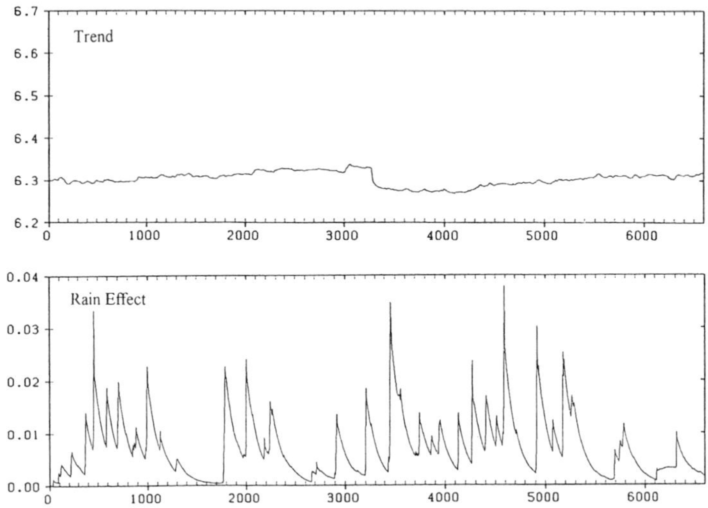

Figure 16.1 E: The estimated trend and rainfall effect.

$$
\left[\begin{array}{c} t_n \\ \hline R_{n1} \\ R_{n2} \\ \vdots \\ R_{nk} \\ \hline a_0 \\ \vdots \\ a_m \\ \hline b_0 \\ \vdots \\ b_\ell \end{array}\right]
=
\left[\begin{array}{c|cccc|c|c} 1 & & & & & & \\ \hline & c_1 & 1 & & & & \\ & c_2 & & \ddots & & & \\ & \vdots & & & 1 & & \\ & c_k & & & & & \\ \hline & & & & & I_{m+1} & \\ \hline & & & & & & I_{\ell+1} \end{array}\right]
\left[\begin{array}{c} t_{n-1} \\ \hline R_{n-1,1} \\ R_{n-1,2} \\ \vdots \\ R_{n-1,k} \\ \hline a_0 \\ \vdots \\ a_m \\ \hline b_0 \\ \vdots \\ b_\ell \end{array}\right]
+
\left[\begin{array}{c} 0 \\ \hline d_1 \\ 0 \\ \vdots \\ 0 \\ \hline 0 \\ \vdots \\ 0 \\ \hline 0 \\ \vdots \\ 0 \end{array}\right] r_n
+
\left[\begin{array}{c} u_n \\ \hline v_n \\ 0 \\ \vdots \\ 0 \\ \hline 0 \\ \vdots \\ 0 \\ \hline 0 \\ \vdots \\ 0 \end{array}\right].
\tag{16.10}
$$

Table 16.3 lists the AIC values for the different model orders. The fifth order rainfall effect model is the AIC best model.

Figure 16.1E shows the estimated rainfall effect and the trend. The barometric pressure and earth tide effects on the trend are visually indistinguishable from the ones in Figure 16.1D and are not shown here. It can be seen that a considerable part of the difference in the trends in Figures 16.1D and 16.1E can be explained as the effect of rainfall.

The trend is now smoother and the coseismic effect is more apparent. Figure 16.1F shows the estimated trend for the years 1981 to 1984. The arrows in Figure 16.1F

Table 16.3 AIC's for model orders versus covariance noise types

n	1	2	3	4	5	6
AIC	−61675	−61734	−61803	−61800	−61810	−61809

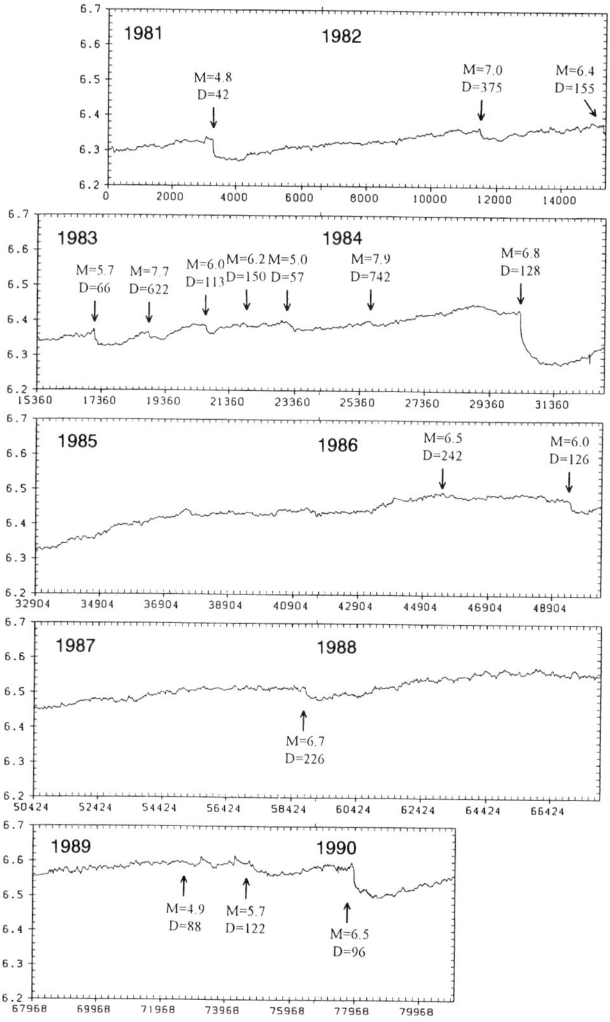

Figure 16.1 F: Ten years of estimated trend and indicated significant earthquakes.

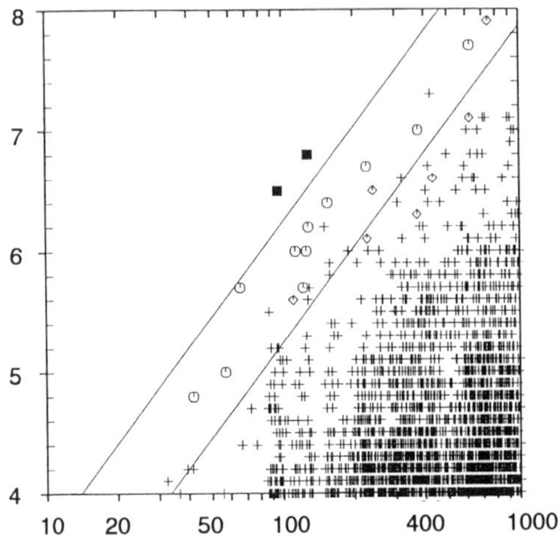

Figure 16.1 G: Scatter plot of earthquake occurrences, with distance from the hypocenter and magnitudes and indicated detected coseismic effects.

identify the time of occurrence of earthquakes of magnitude $M \geq 4.8$, and it can be clearly seen that the drop of the trend component corresponds to the occurrence of earthquakes. From the appearance of the estimated trend, we make the following important observations:

1. The drop of water level of about $5cm$ is seen as a coseismic effect.
2. Exclusive of the coseismic effect drop, the trend regularly increases at the rate of about $12cm$ per year.

Figure 16.1G shows the scatter plot of the earthquakes with distance in the horizontal axis and magnitude in the vertical axis. The symbols $\circ$, $\square$ and $\triangle$ identify earthquakes with detected coseismic effects larger than $15cm$, $5cm$ and $1cm$. The $+$ labeled events indicate the ones without significant (or less than $1cm$) coseismic response. The regression line in the figure which separates larger magnitude earthquakes from smaller ones is defined by

$$M > 2.62 \log_{10} D + 0.69. \tag{16.11}$$

It is clear that most of the larger earthquakes exhibit a coseismic effect.

16.2 Markov State Classification

There are numerous situations in which it appropriate to think that a time series may evolve in accordance with a Markov chain property. In that case, we assume that there is an unknown stationary transition probability matrix (of known dimension), which identifies the probability of a transition between states and in which each state is characterized by a different unknown stationary time series. The objective of the analysis is to identify the unknown individual state time series models and to assign a probability of belonging to a specific state at each time instant. In what follows we consider an artificial data example that involves a four state Markov chain with each state represented by an AR(2) model. The analysis is accomplished via the non-Gaussian state space modeling method.

16.2.1 Introduction

There is considerable interest in the development of appropriate models for those situations in which the dynamic structure of a time series under observation, changes abruptly under an assumed but unknown discrete Markov chain regime. The subject is referred to as hidden Markov models in the engineering literature and received significant attention because of its success in speech recognition, (Rabiner and Juang 1986). Other applications are to nonlinear time series analysis (Tong 1978), biology, (Churchill 1989, 1992 and Katzoff and Shumway 1993), medicine (Gordon and Smith 1988), and economics (Goldfield and Quandt 1973, Hamilton 1988). Also see Raftery (1995). There is a fairly intense interest in applications to macroeconomics with particular attention to the concern over whether an economy is contracting or expanding, (see McCulloch and Tsay 1994 and references therein).

Shumway and Stoffer (1991), in a noteworthy paper, embedded the unknown time series systems into a state space representation with the Markov switching accomplished by allowing the measurement matrix, (H_n in the usual state space notation), to be random with the possibility of assuming any of the possible values identified with the assumed set of states. Also noteworthy, Komaki (1993) used numerical integration in a non-Gaussian state space modeling of a hidden Markov state model application in endocrinology.

16.2.2 A Markov Switching Model

Here we consider a hidden four-state Markov chain autoregressive model. Denote the set of states by $S = \{1, 2, 3, 4\}$. Then the random variable X_n that takes a value in S constitutes a four-state Markov chain. Denote the stationary transition matrix by P, namely, $P = (p_{ij})$ and $p_{ij} = \Pr\{X_n = i \,|X_{n-1} = j\}$. In the particular case that we considered, the following transition matrix was assumed,

$$P = \begin{bmatrix} 1-3\alpha & \alpha & \alpha & \alpha \\ \beta & 1-\beta-2\gamma & \gamma & \gamma \\ \beta & \gamma & 1-\beta-2\gamma & \gamma \\ \beta & \gamma & \gamma & 1-\beta-2\gamma \end{bmatrix}. \tag{16.12}$$

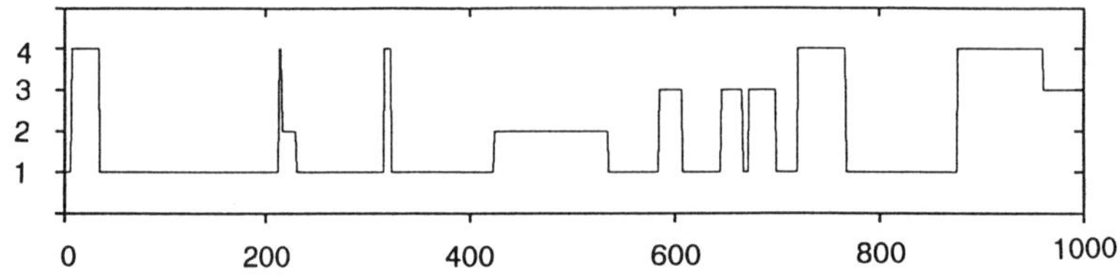

Figure 16.2 A: The Markov states sequence.

We also assume that at any time instant n, the observed time series is generated by one of the four AR(2) models listed below, where the model choice is specified by the state.

$$\begin{aligned} y_n &= 1.785y_{n-1} - 0.903y_{n-2} + \varepsilon_{1,n} && \text{if } X_n = 1, \\ &= 1.344y_{n-1} - 0.903y_{n-2} + \varepsilon_{2,n} && \text{if } X_n = 2, \\ &= 1.386y_{n-1} - 0.640y_{n-2} + \varepsilon_{3,n} && \text{if } X_n = 3, \\ &= 0.800y_{n-1} - 0.640y_{n-2} + \varepsilon_{4,n} && \text{if } X_n = 4. \end{aligned} \tag{16.13}$$

In (16.13), the $\varepsilon_{i,n}, i = 1, \ldots, 4$ are sequences of independent and identically distributed Gaussian random variables which for convenience are each with zero-mean and identical variance σ^2.

The four AR(2) time series models were chosen to have the zeros of their characteristic polynomials in a fairly compact cluster to make distinguishing between them nontrivial. The particular values of the parameters in the transition matrix P that were used were $\alpha = 0.0033$, $\beta = 0.016$ and $\gamma = 0.002$.

The sequence of states is illustrated in Figure 16.2A. The sequence was determined by selecting a starting state from the equilibrium solution of the transition matrix, P, and determining state transitions using the matrix P with random selections from a uniform number generator.

16.2.3 Analysis and Results

The state prediction and filtering formula are obtained as follows.
[Prediction]

$$\begin{aligned} \Pr(X_n = i\,|Y_{n-1}) &= \sum_{j=1}^{4} \Pr(X_n = i, X_{n-1} = j|Y_{n-1}) \\ &= \sum_{j=1}^{4} \Pr(X_n = i\,|X_{n-1} = j)\Pr(X_{n-1} = j\,|Y_{n-1}) \\ &= \sum_{j=1}^{4} p_{ij}\Pr(X_{n-1} = j|Y_{n-1}). \end{aligned} \tag{16.14}$$

[Filtering]

$$
\begin{aligned}
\Pr(X_n = i \mid Y_n) &= \Pr(X_n = i \mid Y_{n-1}, y_n) \\
&= \frac{p(y_n|X_n = i, Y_{n-1})\Pr(X_n = i \mid Y_{n-1})}{p(y_n|Y_{n-1})}
\end{aligned} \tag{16.15}
$$

where $p(y_n|X_n = i, Y_{n-1})$ and $p(y_n|Y_{n-1})$ are respectively given by

$$
\begin{aligned}
p(y_n|X_n = i, Y_{n-1}) &= \frac{1}{\sqrt{2\pi\sigma^2}} \exp\Big\{-\frac{1}{2\sigma^2}(y_n - a_1^{(i)} y_{n-1} - a_2^{(i)} y_{n-2})^2\Big\}, \\
p(y_n|Y_{n-1}) &= \sum_{j=1}^{4} p(y_n|X_n = j, Y_{n-1})\Pr(X_n = j \mid Y_{n-1}).
\end{aligned} \tag{16.16}
$$

The unknown parameter vector for this model is, $\theta = \{\sigma^2, \alpha, \beta, \gamma, a_1^{(i)}, a_2^{(i)}, i = 1, \ldots, 4\}$. The likelihood of θ given the observation vector Y_n is

$$
L(\theta|y) = p(y_1)p(y_2|y_1) \prod_{n=2}^{N-1} p(y_{n+1}|Y_n). \tag{16.17}
$$

The terms in (16.14)–(16.17) were evaluated by the non-Gaussian filter.

The values of the estimated parameters are $\hat{\sigma}^2 = 0.09888$, $\hat{\alpha} = 0.005614$, $\hat{\beta} = 0.02267$, $\hat{\gamma} = 0.132 \times 10^{-5}$. The values of the estimated AR coefficients of the four models are shown in Table 16.4.

Table 16.4. Estimated parameters of the models

	AR_1	AR_2	AR_3	AR_4
a_{i1}	1.8095	1.2848	1.2848	0.8219
a_{i2}	−0.9272	−0.8866	−0.5970	−0.6387

These estimates were obtained by using the theoretical model values as initial estimates. Subsequently, we perturbed those initial estimates quite significantly and discovered that the final parameter estimates were reasonably consistent with respect to variations of the initial estimates.

The satisfactory results that were obtained are most easily communicated graphically. Figures 16.2B, C, ..., F respectively show the generated time series and, the smoothed values of the posterior probability of being in state i at time n. The agreement of the posterior probability of being in state $i, i = 1, \ldots, 4$, Figures 16.2C–F and the event, being in state $i, i = 1, \ldots, 4$, Figure 16.2A, is clearly very close.

We regard these results as very promising. Several additional studies to possibly extend the practical usefulness of the non-Gaussian Markov chain modeling method come to mind. These include, allowing the transition probabilities to become time dependent, allowing the hidden Markov chain states to be governed by multivariate time series models and increasing the dimensionality of the modeling by allowing the model coefficients to depend upon previous states as well as the current state.

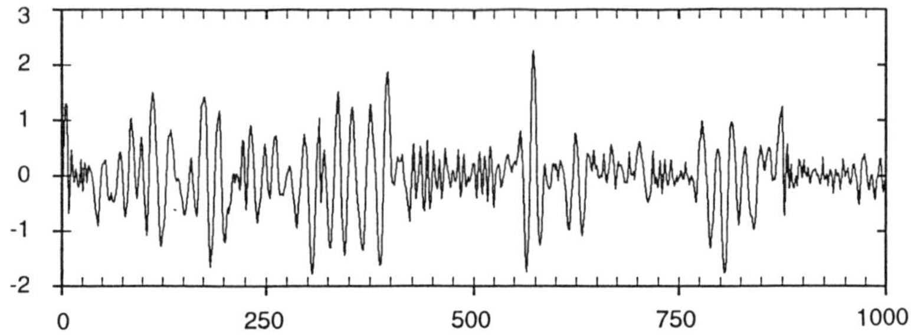

Figure 16.2 B: The observed data.

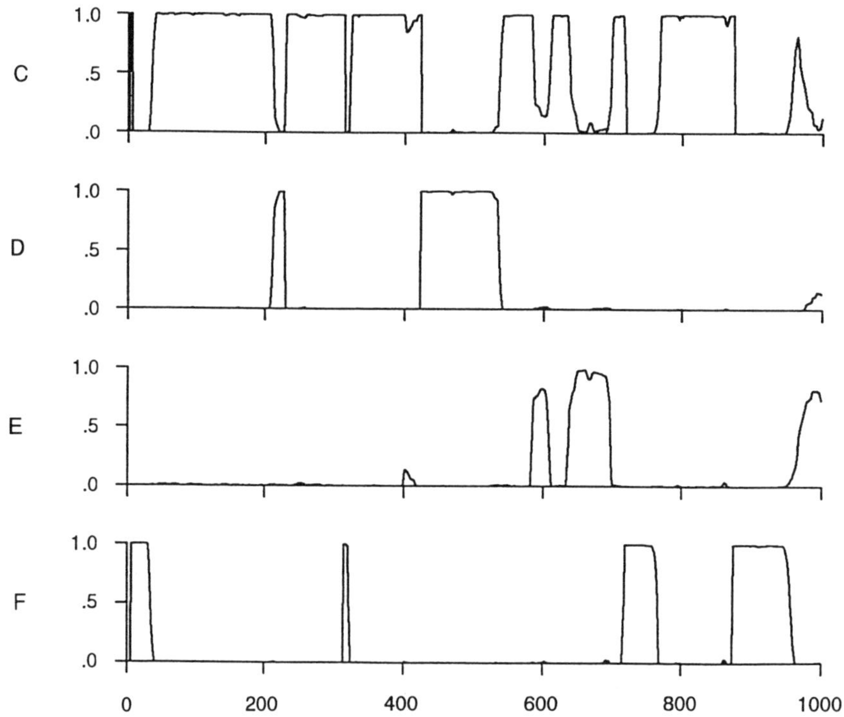

Figure 16.2 C, ..., F: The filtered values of the probabilities of the states.

16.3 SPVAR Modeling for Spectrum Estimation

In this section, the one channel at-a-time paradigm, (Chapter 12), is exploited in conjunction with the smoothness priors scalar long AR model for the purpose of achieving multivariate autoregressive spectrum estimation for stationary covariance time series. We refer to this smoothness priors vector autoregressive modeling by "SPVAR". Our objective was to examine the conjecture that the one channel at a-time smoothness priors modeling approach might yield superior power spectrum estimation than the more conventional approach in which the autoregressive modeling is achieved via the solution of the multivariate version of the Yule-Walker equations. This topic is treated more fully in Stone (1993).

16.3.1 Background

The use of VAR models for spectrum estimation is mentioned in Jenkins and Watts (1968). Early uses of VAR for spectrum estimation appear in Gersch (1970), Gersch and Goddard (1970), (applied to the analysis of electroencephalograms), Akaike (1971, 1988), (in control theory applications), and Jones (1974), (also in electroencephalogram analysis). The analysis of the statistical properties of multivariate spectrum estimation has primarily been concentrated on Fourier analysis methods, (Brillinger 1974, 1975, Koopmans 1974). Those methods lead more naturally to nice analytic inferential results. Gersch (1978) is an early empirical treatment of the statistical performance of the VAR model method for multidimensional spectral analysis. Recent treatment of VAR model spectral estimation in engineering texts, Kay (1988) and Marple (1987) for example, confine their discussion only to the simplest, bivariate autoregressive model of order 1 examples, so additional study of this topic is well warranted.

Given the N observations of a d-vector zero-mean stationary time series x_n, $n = 1, \ldots, N$, the power spectral density matrix may be estimated by first fitting the VAR model,

$$x_n = \sum_{i=1}^{p} A_i x_{n-i} + w_n, \quad E(w_n) = 0, \quad \mathrm{cov}(w_n, w_{n'}) = W\delta_{n,n'}. \tag{16.18}$$

In (16.18), the AR model coefficient matrices A_i and the innovations covariance matrix W are $d \times d$. The model order p along with the unknown matrix parameters $W, A_1, \ldots, A_p$ and W are estimated from the data. The AR matrix coefficients may be obtained by solving the multivariate Yule-Walker equations (Luktepohl 1993) or by the computationally more efficient, Whittle (1963) recursive algorithm. Akaike's AIC may be used to determine the model order p. As in Chapter 12, the power spectral density matrix P_f of the stationary time series X_n, $n = 1, \ldots, N$ is given by

$$P_f = A_f^{-1} W A_f^{-1*} \tag{16.19}$$

with A_f, the frequency response matrix of the whitening filter

$$A_f = I - \sum_{k=1}^{p} A_k \exp(-2i\pi fk) \tag{16.20}$$

the normalized frequency domain over which the spectrum is computed is $-\frac{1}{2} \leq f \leq \frac{1}{2}$.

A justification for attempting to achieve multivariate spectrum estimation by means other than a direct application of (16.18) and (16.20) with parameters estimated by least squares or maximum likelihood methods can be appreciated from the same argument that we suggested for scalar time series. Namely, from a Bayesian point of view, the spectral density computed from the (non-Bayesian) fitted parametric model cannot possibly be optimum. A Bayesian estimate of power spectral density would require that the spectral density of parametric models of different model orders be weighted in accordance with the likelihood and the prior of the model order of different models. Also, for relatively short data spans, the sample variability of the statistic used for determining the best order of the parametric model fitted to the data makes ordinary parametric model fitting statistically unreliable. Thus we are motivated in the problem of multivariate spectral estimation to attempt to exploit the quasi-Bayesian smoothness priors method that worked very well in scalar spectrum estimation as described in Chapter 4.

16.3.2 The Approach and an Example

Our solution to the estimation of the power spectrum density using VAR model is realized by extending the smoothness priors long AR modeling method for scalar time series to the multivariate case. Such an extension is made possible by exploiting the one channel at-a-time paradigm that was introduced in Chapter 12. As in Chapter 12 the data from the individual time series is interlaced to permit an instantaneous response-orthogonal innovations representation of the VAR model. In that way the data can be modeled by a set of d scalar AR models to each of which we fit a smoothness priors long AR order model, (Chapter 4). The individual scalar AR models are then combined to yield the SPVAR model which in turn yields the estimation of the power spectral density matrix, (16.19).

Simulation study results of SPVAR power spectrum estimation performance are shown here. The study was designed to explore whether or not the performance of the SPVAR modeling method was superior to the more conventional VAR-AIC modeling method for multivariate stationary time series power spectrum estimation. Experiments were conducted using two theoretical models, an AR(2) and a ARMA(2,1), with simulation data lengths $N = 200$. In this way, the SPVAR model performance was examined under two different modeling conditions. In one, the fitted VAR models are in the same class as the generated data, (the AR(2) model). In the other, ARMA(2,1) model case, the theoretically equivalent VAR model is of infinite order so the fitted VAR models are approximations to the data generating model and are not in the same modeling class as the generated data.

Only results for the ARMA model simulations are shown here. Additional details and examples are in Stone (1993).

ARMA Simulation Examples

First, some details of the computational results from the Whittle-AIC and SPVAR modeling of a single realization of data simulated from an ARMA(2,1) model are

Table 16.5 VAR model order selection

order	1	2	3	4	5
AIC	1386.635	1042.949	971.205	951.686	947.401
order	6	7	8	9	10
AIC	941.209	940.880	944.824	939.630	944.929

shown. That is followed by the results of a more extensive simulation study.

The theoretical ARMA(2,1) considered was the bivariate time series model

$$x_n - A_1 x_{n-1} - A_2 x_{n-2} = \varepsilon_n - B_1 \varepsilon_{n-1}, \tag{16.21}$$

with $E(\varepsilon_n) = 0$, $\operatorname{cov}(\varepsilon_n, \varepsilon_n') = V\delta_{n,n'}$ and

$$\begin{aligned} A_1 &= \begin{bmatrix} 0.130 & 0.130 \\ -0.130 & -1.130 \end{bmatrix}, \quad A_2 = \begin{bmatrix} 0.850 & 0.085 \\ -0.085 & 0.850 \end{bmatrix}, \\ B_1 &= \begin{bmatrix} 0.600 & -0.300 \\ -0.300 & 0.600 \end{bmatrix}, \quad V = \begin{bmatrix} 1.000 & 0.500 \\ 0.500 & 1.250 \end{bmatrix}. \end{aligned} \tag{16.22}$$

The simulated data from the ARMA model was fitted by both the Whittle-AIC MCAR method and the SPVAR modeling method. The resulting estimated VAR models were used in the computation of the matrix power spectral density. A sum of the relative squares of the difference between the estimated model function and the true model function was computed as a goodness of fit criterion. Results were computed for the four functions, power spectral density in channels one and two, coherence squared and transfer function as functions of frequency.

A Detailed Worked Example

For this example, $N = 200$ data points were generated using the bivariate ARMA(2,1) theoretical model. The conventional Yule-Walker Whittle-AIC VAR algorithm was applied to the simulated data considering a maximum order 10 VAR model. The values of the AIC computed for each VAR order model fitted are shown in Table 16.5. The AIC best model selected for this data was AR(9).

The SPVAR modeling method was also used to model this data. The resulting two scalar time series are modeled individually by the smoothness priors long AR method. For an order p VAR model the resulting scalar AR model orders are respectively, $p_1 = p$ and $p_2 = p+1$. Two different SPVAR models, SPVAR(10) and SPVAR(13) were fit by the method described in Chapter 4.

Because it is informative, the resulting AR coefficient matrices and the innovations matrix computed for the SPVAR(10) model are shown below.

$$\hat{V}_1 = \begin{bmatrix} 1.199 & 0.505 \\ 0.505 & 1.320 \end{bmatrix}, \quad \hat{A}_1 = \begin{bmatrix} -0.248 & -0.132 \\ -0.416 & -0.647 \end{bmatrix}, \quad \hat{A}_2 = \begin{bmatrix} 0.777 & 0.111 \\ -0.564 & 0.496 \end{bmatrix},$$

Table 16.6 ARMA(2,1) VAR model, Whitle-AIC VAR and SPVAR spectral function modeling

	Simulation results: Relative MSE		
	Whittle	$SPVAR_{10}$	$SPVAR_{13}$
Channel 1, mean	37.768	18.369	21.060
std. dev	5340.213	8.581	10.410
Channel 2, mean	28.935	7.732	8.102
std. dev	2543.022	8.591	14.830
Coherence, mean	14.396	14.168	14.396
std. dev	74.773	39.074	43.410
Trans.fcn, mean	10.735	7.556	7.814
std. dev	85.775	9.613	10.310

$$\hat{A}_3 = \begin{bmatrix} -0.402 & 0.084 \\ 0.117 & 0.305 \end{bmatrix}, \quad \hat{A}_4 = \begin{bmatrix} 0.079 & 0.012 \\ -0.094 & 0.124 \end{bmatrix}, \quad \hat{A}_5 = \begin{bmatrix} -0.055 & 0.092 \\ 0.025 & 0.078 \end{bmatrix},$$

$$\hat{A}_6 = \begin{bmatrix} -0.002 & 0.041 \\ 0.005 & 0.083 \end{bmatrix}, \quad \hat{A}_7 = \begin{bmatrix} 0.011 & 0.004 \\ 0.007 & 0.048 \end{bmatrix}, \quad \hat{A}_8 = \begin{bmatrix} 0.001 & -0.002 \\ 0.008 & 0.008 \end{bmatrix},$$

$$\hat{A}_9 = \begin{bmatrix} -0.004 & -0.001 \\ 0.005 & 0.002 \end{bmatrix}, \quad \hat{A}_{10} = \begin{bmatrix} 0.000 & -0.000 \\ -0.001 & 0.001 \end{bmatrix}.$$

As expected, examination of the values in the AR coefficient matrices reveals that the effect of the priors is to produce successively higher order coefficient matrices whose parameter values diminish.

Monte Carlo Results

The bivariate ARMA(2,1) model was used again as the theoretical model from which simulation data are generated. Data from 30 simulations was computed with $N = 200$. The goodness-of fit of the estimates to the theoretical functions was measured as the sum of relative squared errors between the true and the estimated value, $\sum_{i=1}^{N}\{(y_i - \hat{y}_i)/y_i\}^2$ where y_i is the theoretical value and $\hat{y}_i$ is an estimate of y_i. The sum of relative squared errors are tabulated in Table 16.6.

The smaller mean values and very much smaller standard deviations of the relative MSE results of the SPVAR(10) method clearly indicate that the SPVAR(10) model is superior to the Whittle-AIC method of multivariate spectrum estimation by VAR modeling methods. We also note that the degradation in performance in going from the SPVAR(10) model to the SPVAR(13) model is quite small. That evidence helps confirm our conjecture that the performance of the SPVAR spectral estimation method is relatively insensitive to model order.

REFERENCES

Akaike, H. (1968), "On the use of a linear model for the identification of feedback systems", *Ann. Inst. Statist. Math.*, 20, 425–439.

Akaike, H. (1969), "Fitting autoregressive models for prediction", *Ann. Inst. Statist. Math.*, 21, 243–247.

Akaike, H. (1971), "Autoregressive model fitting for control", *Ann. Inst. of Statist. Math.*, 23, 163–180.

Akaike, H. (1973), "Information theory and an extension of the maximum likelihood principle", in *Second International Symposium in Information Theory*, eds. B.N. Petroc and F. Caski, Budapest, Akademiai Kiado, 267–281.

Akaike, H. (1974a), "A new look at the statistical model identification", *IEEE Trans. Auto. Control*, AC-19, 716–723.

Akaike, H. (1974b), "Markovian representation of stochastic processes and its application to the analysis of autoregressive moving average processes", *Ann. Inst. Statist. Math.*, 26, 363–387.

Akaike, H. (1977a), "On entropy maximization principle", in *Applications of Statistics*, P.R. Krishnaiah, Ed. North Holland, The Netherlands, 27–41.

Akaike, H. (1977b), in *Discussion of the papers by Mr. Campbell and Professor Walker*, Dr. Morris and Dr. Tong, J. R. Statist. Soc. A140, 456.

Akaike, H. (1978), "A new look at the Bayes procedure", *Biometrika*, 65, 53–59.

Akaike, H. (1979), "A Bayesian extension of the minimum AIC procedure of autoregressive model fitting", *Biometrika*, 66, 237–242.

Akaike, H. (1980a), "Likelihood and the Bayes procedure", in *Bayesian Statistics*, J.M. Bernardo, M.H. De Groot, D.V. Lindley and A.F.M. Smith eds., University Press, Valencia , Spain, 143–166.

Akaike, H. (1980b), "Seasonal adjustment by a Bayesian modeling", *J. Time Series Anal.*, 1,1–13.

Akaike, H. (1981a), "Likelihood of a model and information criteria", *J. Econometrics*, 16, 3–14.

Akaike, H. (1981b), "Modern developments of statistical methods", in *Trends and Progress in System Identification*, ed. P. Eykoff, Pergamon Press, New York, 169–184.

Akaike, H. (1985), "Prediction and Entropy", in *A Celebration of Statistics*, eds. A.C. Atkinson and S.E. Feinberg, Springer-Verlag, New York, 1–24.

Akaike, H. (1986), "The selection of smoothness priors for distributed lag estimation", in *Bayesian Inference and Decision Techniques*, eds. P.K. Goel and A. Zellner, North-Holland, Amsterdam, 109–118.

Akaike, H. (1987a), "Some reflections on the modeling of time series", *Time Series and Econometric Modeling*, Vol. 3, eds. I.B. MacNeil and G.J. Umphrey, Reidel, Dordrecht, 13–28.

Akaike, H. (1987b), "Factor analysis and the AIC", *Psychometrika*, 52, 317–332.

Akaike, H., G. Kitagawa, E. Arahata and F. Tada (1979), *TIMSAC-78, Computer Science Monograph*, No. 11, Inst. Statist. Math., Tokyo.

Akaike, H., T. Ozaki, M. Ishiguro, Y. Ogata, G. Kitagawa, Y-H. Tamura, E. Arahata, K. Katsura and Y. Tamura, (1985), *TIMSAC-84, Computer Science Monograph*, No. 22, Inst. Statist. Math., Tokyo.

Alspach, D.L. (1974), "Gaussian sum approximation in nonlinear filtering and control", *Information Science*, 7, 271–290.

Alspach, D.L. and H.W. Sorenson (1972), "Nonlinear Bayesian estimation using Gaussian sum approximations", *IEEE Trans, Auto. Control*, 17, 439–447.

Andel, J. (1993), "A time series with suddenly changing parameters", *J. Time Series Anal.*, 6, 35–52.

Anderson, B.D.O. and J.B. Moore (1979), *Optimal Filtering*, Prentice Hall, New Jersey.

Anderson, J.A. and A. Senthiselvan (1980), "Smooth estimates for the hazard function", *J. R. Statist. Soc.*, B 42, 332–327.

Anderson, T.W.A. (1971), *The Statistical Analysis of Time Series*, Wiley, New York.

Andrade Netto, M.L., L. Gimeno, and M.J. Mendes, (1978), "On the optimal and suboptimal nonlinear filtering problem for discrete-time systems", *IEEE Trans. Auto. Control*, AC-23, 1062–1067.

Ansley, C.F. (1979), "An algorithm for the exact likelihood of a mixed autoregressive-moving average process", *Biometrika*, 66, 59–65.

Ansley, C.F. and W.E. Wecker W.E. (1983), "Extensions and examples of signal extraction approach to regression", in *Proceedings ASA-CENSUS-NBER Conference on Applied Time Series Analysis of Economic Data*, ed. A. Zellner, U.S. Bureau of the Census, Washington D.C. 181–192.

Aoki, M. (1990), *State Space Modeling of Time Series (2nd Edition)*, Springer-Verlag, Berlin.

Åstrom, K.J. and T. Bohlin (1965), "Numerical Identification of a Linear Dynamical System from Normal Operating Records", IFAC Symposium on the Theory of Self-Adaptive Control, Teddington, 96–111.

Åstrom, K.J. and B. Wittenmark (1973), "On self-tuning regulators", *Automatica*, 9, 185–199.

Atkinson, A.C. (1980), "A note on the generalized information criterion for choice of a model", *Biometrika*, 67, 413–418.

Bartlett, M.S. (1950), "Periodogram analysis and continuous spectra", *Biometrika*, 37, 1–16.

Bartlett, M.S. and D.G. Kendall (1946), "The statistical analysis of variance-heterogeneity and the logarithmic transformation", *Supplement, J. Roy. Statist. Soc*, Vol VIII, 128–133.

Beltrão, K.I. and P. Bloomfield (1987), "Determining the bandwidth of a kernel spectrum estimate", *J. Time Ser. Anal*, 8, 21–38.

Berger, J.O. (1980), *Statistical Decision Theory and Bayesian Analysis, 2nd Edition*, Springer-Verlag, New York.

Beutler, F.J. (1970), "Alias free randomly timed sampling of stochastic processes", *IEEE Trans. Info. Theory*, IT-16, 147–152.

Bhansali, R.J. (1979), "A mixed spectrum analysis of the lynx data", *J. R. Statist. Soc.*, A 142, 199–209.

Bhansali, R.J. (1986), "A derivation of the information criterion for selecting autoregressive models", *Adv. Appl. Prob.*, 18, 360–387.

Bhansali, R.J. and D.Y. Downham (1977), "Some properties of the order of an autoregressive model selected by a generalization of Akaike's FPE criterion", *Biometrika*, 64, 547–551.

Bierman G.J. (1977), *Factorization Methods for Discrete Sequential Estimation*, Academic Press, New York.

Blackman, R.B. and J.W. Tukey (1959), *The measurement of Power Spectra from the Viewpoint of Communications Engineering*, Dover, New York.

Bloomfield, P. (1976), *Fourier Analysis of Time Series: An Introduction*, Wiley, New York.

Boahash, B. (1992) "Estimating and interpreting the instantaneous frequency of a signal, Part I: Fundamentals, Part II: Algorithms and applications", *Proc. IEEE*, 80, 520–538, 540–568.

Bodenstein, G. and H.M. Praetorious (1977), "Feature extraction from the encephalogram by adaptive segmentation", *Proc. IEEE*, 65, 642–652.

Bohlin, T. (1976), "Four cases of identification of changing systems", in *System Identification Advances and Case Studies*, eds. R. Mehra R. and D.G. Lainotis, Academic Press, New York.

Bolt, B.A. (1987), *Earthquakes*, W.H. Freeman, San Francisco.

Box, G.E.P, S.C. Hillmer and G.C. Tiao (1978) "Analysis and moding of seasonal time series", in *Seasonal Analysis of Time Seres*, ed. A. Zellner, US Bureau of the Census, Economic Research Report ER-1, 309–334.

Box, G.E.P. and G.M. Jenkins (1970), *Time Series Analysis, Forecasting and Control*, Holden-day, San Francisco.

Bozdogan, H. (1987), "Model selection and Akaike's Information criterion (AIC): The general theory and its analytical extensions", *Psychometrika*, 52, 345–370.

Bozdogan, H. (ed.), (1994) *Proceedings of the First US/Japan Conference on the Frontiers of Statistical Modeling: An Information Approach*, Kluewer, The Netherlands.

Bozdogan, H. and Sclove S. L. (1987), "Multi-sample cluster analysis using Akaike's information criterion", *Psychometrika*, 52, 345–370.

Brazier, M.A.B. (1972), "Spread of seizure discharges in epilepsy: Anatomical and electrophysiological considerations", *Exp. Neurol.*, 36, 263–272.

Brillinger, D.R. (1972), " The spectral analysis of stationary interval functions", in *Proceedings Sixth Berkeley Symposium Probability and Statistics*, 483–513.

Brillinger, D.R. (1974), "Fourier analysis of stationary processes", *Proc. IEEE*, 62, 1628–1643.

Brillinger, D.R. (1975), *Time Series: Data Analysis and Theory*, Holt, Rhinehart and Winston, New York.

Brillinger, D.R. (1988), The 1983 Wald Memorial Lectures: "Some statistical methods for random process data from seismology and neurophysiology", *Annals of Statistics*, 16, 1–54.

Brillinger, D.R. (1993a), "The digital rainbow: Some history and applications of numerical spectrum analysis", *Canad. J. Statist.*, 21, 1–19.

Brillinger, D.R. (1993b), "An application of statistics to seismology: dispersion and modes", in *Developments in Time Series Analysis*, ed. T. Subba Rao, Chapman and Hall, London, 331–340.

Brillinger, D.R. (1994), "Some asymptotics of wavelets in the stationary error case", *Technical Report*, Department of Statistics, University of California, Berkeley.

Brillinger, D.R., J. Guckenheimer, P. Guttorp and G. Oster (1980), "Empirical modeling of population time series data: The case of age and density dependent vital rates", *Lectures on Mathematics in the Life Sciences*, 13, 65–90.

Brillinger, D.R. and M. Hatanaka (1969), "An harmonic analysis of nonstationary multivariate economic processes", *Econometrika*, 131–141.

Brillinger, D.R. and M. Hatanaka (1970), "A permanent income hypothesis relating to the aggregate demand for money (An application of spectral and moving spectral analysis)", *Economic Studies Quarterly*, XXI, 44–71.

Brockwell, P.J. and R.A. Davis, (1991), *Time Series: Theory and Methods, Second Edition*, Springer-Verlag, New York.

Brotherton, T. and W. Gersch (1981), "A data analytic approach to the smoothing problem and some of its variations", *Proc. 20th IEEE Conf. on Decision and Control*, San Diego, California, 1061–1069.

Bulmar, M.G. (1974), "A statistical analysis of the 10-year cycle in Canada", *J. Anim. Ecol.*, 43, 701–715.

Bucy, R.S. (1969), "Bayes' theorem and digital realization for nonlinear filters", *J. Astronaut. Sci.*, 17, 80–94.

Bucy, R.S. and K.D. Senne (1971), "Digital synthesis of nonlinear filters", *Automatica*, 7, 287–298.

Buja, A., Hastie T. and Tibshirani R. (1989), "Linear smoothers and additive models, (with discussion)", *Ann. Statist.*, 17, 453–555.

Caines, P.E. (1988), *Linear Stochastic Systems*, Wiley, New York.

Caines, P.E. and J. Rissanen (1974), "Maximum likelihood estimation of parameters in multivariate Gaussian stochastic processes", *IEEE Trans. Info. Theory*, IT-20, 102–104.

Campbell, M. and A. Walker (1977), "A survey of statistical work on the Mackenzie river series of annual lynx trappings for years 1821–1934 and a new analysis", *J. R. Statist. Soc.*, A 140, 411–432.

Carlin, B.P., N.G. Polson and D.S. Stoffer (1992), "A Monte Carlo approach to nonnormal and nonlinear state space modeling", *J. Amer. Statist. Assoc.*, 75, 493–500.

Carter, C.K. and R. Kohn (1993), "A comparison of Markov chain Monte Carlo sampling schemes for linear state space models", in *Proceedings American Statistical Association Business and Economic Statistics Section*, 131–136.

Carter C.K. and R. Kohn (1994), "On Gibbs sampling for state space models", *Biometrika*, 81, 541-553.

Chan, S.W., G.C. Goodwin and K.S. Sin (1984), "Convergence properties of the Ricatti difference equation in optimal filtering of nonstabilizable systems", *IEEE Trans. Auto. Control*, AC-29, 110–118.

Choi, B. S. (1992) *ARMA Model Identification*, Springer-Verlag.

Chow, G.C. (1981), "A comparison of the information and posterior probability criteria for model selection", *J. Econometrics*, 16, 21–33.

Churchill, G.A. (1989), "Stochastic models for heterogeneous DNA sequences", *Bull. Math. Biology*, 51, 4451–4460.

Churchill, G.A. (1989), "Hidden Markov chains and the analysis of genome structure", *Computers Chem.*, 16, 107–115.

Cleveland, R.B., W.S. Cleveland, J.E. McRae and I.J. Terpenning (1990), "STL: A seasonal trend decomposition procedure based on Loess", *J. Official Statist.*, 6, 3–73.

Cleveland, W.S. and S.J. Devlin (1980), "Calendar effects in monthly time series; Detection by spectrum analysis and graphical methods", *J. Amer. Statist. Assoc.*, 75, 487–496.

Cleveland, W.S., S.J. Devlin and I. Terpenning (1982), "The SABL seasonal adjustment and calendar adjustment procedures", *Time Series Analysis: Theory and Practice*, 1, 539–564.

Cox, D.R. (1981), "Statistical analysis of time series", *Scand. J. Statist.*, 8, 93–118.

Cox, D.R. and D.V. Hinkley (1974), *Theoretical Statistics*, Chapman and Hall, London.

Craddock, J.M. (1967), "An experiment in the analysis and prediction of time series", *The Statistician*, 17, 257–268.

Cramér, H. (1946), *Mathematical Methods of Statistics*, Princeton University Press, Princeton, New Jersey

Dahlhaus, R. (1993), "Fitting time series models to nonstationary processes", *Beiträge zur Statistik*, 4, Universität Heidelberg.

Dahlhaus, R., M.H. Neumann and R. von Sachs (1995), "Nonlinear wavelet estimation of time-varying autoregressive processes", *Technical Report*, Department of Statistics, Stanford University.

Daniell, P.J. (1946), "Discussion on the Symposium on Autocorrelation in Time Series", *J. R. Statist. Soc. (Suppl.)*, 8, 88–90.

Davis, H.T. and R.H. Jones (1968), "Estimation of the innovations variance of a stationary time series", *J. Amer. Statist. Assoc.*, 63, 141–149.

De Boor, C. and R.E. Lynch (1966), "On splines and their minimum properties", *J. Math. and Mechanics*, 15, 953–969.

de Figueirido, R.J.P. and Y.G. Jan, (1971), "Spline filters", *Proceedings of the 2nd Symposium on Nonlinear Estimation Theory, and Its Applications*, San Diego, 127–141.

De Sousa and P.J. Thomson, (1982), "LPC distance measures and statistical tests with particular reference to the likelihood ratio", *IEEE Trans. Acoust. Speech, Signal Processing*, ASSP-30, 304–315.

Dennis, J.E. Jr. and R.B. Schnabel (1983), *Numerical Methods for Unconstrained Optimization and Nonlinear Equations*, Prentice Hall, New Jersey.

Doob, J.L. (1953), *Stochastic Processes*, Wiley, New York.

Dorfman, J.H. and A.M. Havenner (1992), "A Bayesian approach to state space multivariate time series modeling", *J. Econometrics*, 52, 315–346.

Donoho, D.L. and I.M. Johnstone (1992), "Minimax estimation via wavelet shrinkage", *Technical Report*, No. 402, Department of Statistics, Stanford University.

Donoho, D.L. and I.M. Johnstone (1993), "Adapting to unknown smoothness via wavelet shrinkage", *Technical Report*, Department of Statistics, Stanford University.

Duncan, D.B. and S.D. Horn (1972), "Linear dynamic recursive estimation from viewpoint of regression analysis", *J. Amer. Statist. Assoc.*, 67, 815–821.

Dyer, P. and S. McReynolds (1969), "Extension of square-root filtering to include process noise", *J. Opt. Theory and Appl.*, 3, 444–459.

Dzhapridze, K. (1986), *Parameter Estimation and Hypothesis Testing in Spectral Analysis of Stationary Time Series*, Springer-Verlag, New York.

Elton, C. and M. Nicholson (1942), "The ten year cycle in numbers of the lynx in Canada", *J. Anim. Ecol.*, 11, 215–244.

Ettore, M. (1987), "Statistical inference in doubly stochastic point processes", Ph.D. Thesis, University of California, Berkeley, California.

Findley, D.F. (1985a), "On the use of multiple models for multi-step forecasting", *Proc. Bus. and Eco. Statist., Amer. Statist. Assoc.*, 3, 528–531.

Findley, D.F. (1985b), "On the unbiasedness property of the AIC for linear stochastic time series models", *J. Time Series Anal.*, 6, 229–252.

Findley, D.F. (1988), "Comparing not necessarily nested models with the minimum AIC and maximum Kullback-Leibler entropy criteria: New perspectives and connections", *Proc. Bus. Econ Sec., Amer. Statist. Assoc.*, 110–118.

Fornell, C. and R.T. Rust (1989), "Incorporating prior theory in covariance structure analysis: A Bayesian approach", *Psychometrika*, 54, 249–259.

Franke, J. and W. Hardle (1992), "On bootstrapping kernel spectral estimates", *Ann. Statist.*, 20, 121–145.

Franklin, J.N. (1970), "Well-posed stochastic extensions of ill-posed problems", *J. Math. Anal. and Applications*, 31, 682–716.

Friedlander, B. (1982a), "Lattice filters for adaptive processing", *Proc. IEEE*, 70, 829–867.

Friedlander, B. (1982b), "Lattice methods for spectral estimation", *Proc. IEEE*, 70, 990–1017.

Frühwirth-Schnatter, S. (1994), "Data augmentation and dynamic linear models", *J. Time Series Anal.*, 183–202.

Gardner, G.A., A.C. Harvey and G.D.A. Phillips, (1980), "An algorithm for exact estimation of ARMA models by means of Kalman filtering", *J. Appl. Statist.*, 29, Algorithm AS154, 311–322.

Gelfand, A.E. and A.F.M. Smith (1990), "Sampling-based approaches to calculating marginal densitiers", *J. Amer. Statist. Assoc.*, 85, 398–409.

Gelfand, A.E., S.E. Hills, A. Racine-Poon and A.F.M. Smith, (1990), "Illustration of Bayesian inference in normal data models using Gibbs sampling", *J. Amer. Statist. Assoc.*, 85, 972–985.

Gersch, W. (1981), "Nearest neighbor rule classification of stationary and nonstationary time series", in *Applied Time Series Analysis II*, D.F. Findley, Ed., Academic Press , New York, 221–270.

Gersch, W. (1985), "Modeling nonstationary time series and inferring instantaneous dependency, feedback and causality: An application to human epileptic seizure event data", *Identification and System Parameter Estimation 1985, 7 th IFAC/IFORS Symposium*, York, UK, 737–742.

Gersch, W. (1987a), "Non-Stationary Multichannel Time Series Analysis", Chapter 10, *EEG Handbook, Revised Series*, Vol 1, ed. A. Gevins, Academic Press, 261–296.

Gersch, W. (1987b), "The linear dependency structure of covariance nonstationary time series", *Technical Report*, NPS55-87-006, Naval Postgraduate School, Monterey California.

Gersch, W. (1990), "Smoothness priors modeling of seafloor bathymetric data", *Geophys. Research Letters*, 17, 97–100.

Gersch, W. (1991), "Nonstationary covariance time series: Modeling by smoothness priors methods", in *Systems and Control*, eds. T. Ono and F. Kozin, Mita Press, Tokyo, 73–84.

Gersch, W. (1992), "Smoothness Priors", in *New Directions in Time Series Part II*, eds.D. Brillinger, P. Caines, J. Geweke, E. Parzen, M. Rosenblatt and M. S. Taqqu, The IMA Volumes in Mathematics and Its Applications Vol. 46, Springer-Verlag 113–146.

Gersch, W. and G.V. Goddard (1970), "Epileptic focus location, spectral analysis method", *Science*, 150, 701–702.

Gersch, W. and G. Kitagawa (1983a), "The prediction of time series with trends and seasonalities", *J. Bus. Econ. Statist.*, 1, 253–264.

Gersch, W. and G. Kitagawa (1983b), "A time varying multivariate autoregressive modeling of econometric time series", *Proc. Amer. Statist. Assoc., Bus. and Econ. Stats.*, 403–408.

Gersch, W. and G. Kitagawa (1985), "A time varying AR coefficient model for modeling and simulating earthquake ground motion", *Earthquake Eng. Structural Dynamics*, 13, 243–254.

Gersch, W. and Kitagawa, G. (1987), "Smoothness Priors Transfer Function Estimation", *Technical Report* NPS55-87-005, Naval Postgraduate School, Monterey CA.

Gersch, W. and Kitagawa G. (1988), "Smoothness priors in time series", in *Bayesian Analysis of Time Series and Dynamic Systems*, ed. J.C. Spall, Marcel Dekker, New York, 431–476.

Gersch, W. and Kitagawa G. (1989), "Smoothness priors transfer function estimation", *Automatica*, 25, 603–608.

Gersch, W., F. Martinelli, J. Yonemoto, M.D. Low, and J.A. McEwen, (1980), "Kullback Leibler Nearest neighbor rule classification of EEGs: The EEG population screening problem; an anesthesia level EEG classification problem", *Computers Biomed. Research*, 13, 283–296.

Gersch, W. and D.R. Sharpe (1973), "Estimation of power spectra with finite-order autoregressive models", *IEEE Trans. Auto. Control*, AC-18, 367–369.

Gersch, W. and M. Sklarz (1986), "Scalar and multivariate increasing horizon AR modeling and prediction", *Amer. Statist. Assoc. 1986 Proc. Bus. Econ. Stats.*

Section, 154–159.

Gersch, W. and D. Stone, (1994), "One channel at-a-time multichannel autoregressive modeling: Applications to stationary and nonstationary covariance time series", in *The Frontiers of Statistical Modeling*, H. Bozdogan, Ed., Kluever Press.

Gersch, W. and D. Stone, (1995), "Multivariate autoregressive time series modeling: One scalar autoregressive model at-a-time", *Commun. in Statist. Theory and Methods*, 24, 2715–2733.

Gersch, W. and B.R. Tharp, (1976), "Spectral regression-amount of information analysis of seizures in humans", in *Qualities and Quantitative Advances in Epilepsy* , (ed. P. Kellaway), Raven Press, 509–532.

Geweke, J. (1982), "The measurement of linear dependence and feedback between multiple time series", (with discussion), *J. Amer. Statist. Assoc.*, 77, 304–324.

Ghysels, E., (1993a), "On seasonality and switching-regime models", *Amer. Statist. Assoc. 1993 Proc. Bus. Econ. Stats. Section*, 308–312.

Ghysels, E., (ed.), (1993b), "Seasonality and econometric models", *J. Econometrics*, 55, 1–357.

Gilbert, L.E. and A. Malinverno (1988), "A characterization of the spectral density of residual ocean floor topography", *Geophys. Res. Letters*, 15, 1401–1404.

Goldfield, S.M. and R.E .Quandt (1973), "A Markov model for switching regressions", *J. Econometrics*, 1, 3–16.

Golub, G. (1965), "Numerical methods for solving linear least-square problems", *Num. Math.*, 7, 206–216.

Good, I.J. (1965), *The Estimation of Probabilities*, M.I.T. Press, Cambridge, Mass.

Good, I.J. and J.R. Gaskins (1980), "Density estimation and bump hunting by the penalized likelihood method exemplified by scattering and meteorite data", *J. Amer. Statist. Assoc.*, 75, 42–73.

Goodwin, G.C. and R.E. Payne (1977), *Dynamic System Identification: Experimental Design and Data Analysis*, Academic Press, New York.

Gordon, K. and A.F.M Smith (1988), "Modeling and monitoring discontinuous changes in time series", in *Bayesian Analysis of Time Series and Dynamic Linear Models*, ed. J.C. Spall, Marcel Dekker, New York, 359–392.

Gotman, J. (1983) "The measurement of small time differences between EEG channels: Methods and applications to epileptic seizure propagation", *Electroenceph. and Clinical Neurophysiology*, 50, 501–514.

Graham, A. (1981), *Kronecker Products and Matrix Calculus: with Applications*, Halstead press, New York.

Granger, C.W.J and M. Hatanaka (1964), *Spectral Analysis of Economic Time Series*, Princeton University Press, New Jersey.

Gray, R.M., A. Buzo, A.H. Gray and Y. Matsuyama, (1980), "Distortion measures for speech processing", *IEEE Trans. Acoust. Speech & Signal Process.*, ASSP–28, 367–376.

Green, P.J. and B.W. Silverman (1994), *Nonparametric Regression and Generalized Linear Models: A Roughness Penalty Approach*, Chapman and Hall, London.

Grenander, U. and M. Rosenblatt (1953a), "Statistical spectral analysis of time series arising from stationary stochastic processes", *Ann. Math. Stat.*, 24, 537–558.

Grenander, U. and M. Rosenblatt (1953b), *Statistical Analysis of Stationary Time Series*, Wiley, New York.

Grenier, Y. (1982), "Time varying lattices and autoregressive models: Parameter estimation", in *Proc. of ICASSP '82*, Paris France.

Greville, T.N.T. (1957), "On smoothing a finite table: A matrix approach", *SIAM J. Appl.*, 5, 137–154.

Gu, C. (1990), "Adaptive spline smoothing in non-Gaussian regression models", *J. Amer. Statist. Assoc.*, 85, 801–807.

Gu, C.(1992), "Penalized likelihood regression: A Bayesian analysis", *Statistica Sinica*, 2, 255–264.

Gu, C.(1993a), "Smoothing spline density estimation: A dimensionless automatic algorithm", *J. Amer. Statist. Assoc.*, 88, 495–504.

Gu, C.(1993b), "Penalized likelihood hazard estimation: Algorithms and examples", in *Statistical Decision Theory and Related Topics V*, eds. S.S. Gupta and J.O. Bergers, Springer-Verlag.

Gu, C. and C. Qiu (1993), "Smoothing spline density estimation: Theory", *Ann. Statist.*, 21, 217–234.

Guo, L. (1990), "Estimating time-varying parameters by the Kalman filter based algorithm: Stability and convergence", *IEEE Transactions on Automatic Control*, AC-35, 141–147.

Guo, L., H.F. Chen and J.F. Zhang, (1992), "Identification of stochastic time-varying parameters", in *New Directions in Time Series Analysis Part II*, eds. D. Brillinger, P. Caines, J. Geweke, E. Parzen, M. Rosenblatt and M. S. Taqqu, The IMA Volumes in Mathematics and its Applications, Vol. 46, Springer-Verlag, 211–223.

Hamilton, J.D. (1989), "A new approach to the economic analysis of nonstationary time series and the business cycle", *Econometrica*, 57, 357–384.

Hannan, E.J. (1960), *Time Series Analysis*, Methuen, London

Hannan, E.J. (1970), *Multiple Time Series*, Wiley, New York

Hannan, E.J. (1986), "Remembrance of things past", *The Craft of Probabilistic Modeling*, ed. J. Gani, Springfer-Verlag, New York.

Hannan, E.J. (1987), "Rational transfer function approximation", (with discussion), *Statistical Science*, 2, 135–161.

Hannan, E.J. and B.G. Quinn (1979), "The determination of the order of an autoregression", *J. R. Statist. Soc.*, B 41, 190–195.

Harrison, P.J. and C.F. Stevens (1976), "Bayesian forecasting", (with discussion), *J. R. Statist. Soc.*, B 38,205–247.

Harvey, A.C. (1989), *Forecasting, structural time series models and the Kalman filter*, Cambridge University Press, Victoria, Australia.

Harvey, A.C and Fernandes (1989), "Time series models for count data or qualitative observations", *J. Bus. Econ. Statist.*,

Harvey, A.C and G.D.A. Phillips (1979), "The estimation of regression models with autoregressive-moving average disturbances", *Biometrika*, 66, 49–58.

Hastie, T. and R. Tibshirani (1993), "Varying-coefficient models", *J. R. Statist. Soc.*, B 55, 757–796.

Haykin, S (1991), *Adaptive Filter Theory, 2nd edition*, Prentice Hall, New Jersey.

Hecht, (1973), "System identification using Bayesian estimation", in *4th Symposium on Nonlinear Estimation*, 151–160.

Higuchi, T., K. Kita and T. Ogawa (1988), "Bayesian statistical inference to remove periodic noise in the optical observations aboard a spacecraft", *Appl. Optics*, 27, 4514–4519.

Higuchi, T. (1991), "Method to subtract an effect of the geocorona EUV radiation from the low energy particle (LEP) data by the Akebono (EXOS-D) satellite", *J. Geomag. Geoelectr.*, 43, 957–978.

Higuchi, T, (1995), "On the resampling scheme in the filtering procedure of Kitagawa's Monte Carlo filter", *Res. Memo #556*, Inst. Statist. Math., Tokyo.

Hillmer, S.C. (1982), "Forecasting time series with trading day variation", *J. Forecasting*, 1, 385–395.

Ho, Y.C. and R.C.K. Lee (1964), "A Bayesian approach to problems in stochastic estimation and control", *IEEE Trans. Auto. Control*, AC-9, 333–339.

Hodges, P.E. and D.F. Hale, (1993), "A computational method for estimating densities of non-Gaussian nonstationary univariate time series", *J. Time Series Anal.*, 14, 163–178.

Houacine, A. and G. Demoment, (1987), "Fast adaptive spectrum estimation: Bayesian approach and long AR models", *Intl. Conf. Acoust. Speech, Signal Proc. (ICASSP)*, 4, 2085–2088.

Huber, P.J. (1965), "The behavior of maximum likelihood estimators under nonstandard conditions", in *Proceedings Fifth Berkeley Symposium Probability and Statistics, Vol. 1*, 221–233.

Hunt, B.R. (1970), "The inverse problem of radiography", *Math. Biosciences*, 8, 161–179.

Hunt, B.R. (1973), "The application of constrained least-squares estimation to image restoration by digital computer", *IEEE Trans. Computers*, C-22, 805–812.

Hurukawa, N. and M. Imoto (1989), "A Bayesian approach to the detection of temporal changes in P wave velocity", *J. Geophys. Res.*, 94, 1649–1654.

Hurvich, C. M. (1992), "Selection of time series models and spectrum estimates using a bias-corrected generalization of the AIC", in *New Directions in Time Series Analysis Part I*, eds. D. Brillinger, P. Caines, J. Geweke, E. Parzen, M. Rosenblatt and M. S. Taqqu, The IMA Volumes in Mathematics and its Applications, Vol. 46, Springer-Verlag, 155–168.

Hurvich, C.M. and K.I. Beltrão (1990), "Cross-validatory choice of a spectrum estimate and its connections with the AIC", *J. of Time Series Anal.*, 11, 121–137.

Hurvich, C.M. and C-L Tsai, (1989), "Regression and time series model selection in small samples", *Biometrika*, 76, 297–307.

Hurvich, C.M. and C-L Tsai (1993), "A corrected Akaike Information criterion for vector autoregressive model selection", *J. Time Series Anal.*, 14, 271–279

Igarashi, and Wakita (1991), "Tidal responses and earthquake-related changes in the water level of deep wells", *J. Geophysical Res.*, 96, 4269–4278.

Ishiguro, M. and Y. Sakamoto (1983), "A Bayesian approach To binary response curve estimation", *Ann. Inst. Statist. Math.*, 35B, 115–137.

Ishiguro, M., H. Akaike H., M. Ooe and S. Nakai (1981), "A Bayesian approach to the analysis of earth tides", *Proc. 9th Inst. Conference on Earth Tides.*

Ishii, N, A. Iwata and N. Suzumura (1979), "Segmentation of non-stationary time series", *Int J. Syst. Sci.*, 10, 883–894.

Ishiguro, M. and Sakamoto Y. (1983), "A Bayesian approach to binary response curve estimation", *Ann. Inst. Statist. Math.*, 35,B, 115–137.

Itakura, F. (1975), "Minimum prediction residual principle applied to speech recognition", *IEEE Trans. Acoust., Speech, Signal Proc.*, ASSP-23, 67–72.

Izenman, A. (1985), "J.R. Wolf and the Zurich sunspot relative numbers", *Mathematical Intelligence*, 7, 27–33.

Jaynes, E.T. (1957), "Information theory and statistical mechanics I", *Physical Review*, 106, 620–630.

Jazwinski, A.H. (1970), *Stochastic Processes and Filtering Theory* , Academic Press, New York.

Jenkins, G.M. and D.G. Watts (1968), *Spectral Analysis and Its Applications*, Holden-Day, San Francisco.

Jiang X-Q, and G. Kitagawa (1993), "A time varying vector autoregressive modeling of nonstationary time series", *Signal Processing*, 33, 315–331.

Johnson, R.W. (1979), "Axiomatic characterizations of the directed divergences and their linear combinations", *IEEE Trans. Info. Theory*, IT-25, 709–716.

Johnson, N.L. and S. Kotz (1970), *Distributions In Statistics, Continuous Univariate Distributions, 1*, Wiley, New York.

Johnstone, I.M. and B.W. Silverman (1994), "Wavelet threshold estimators for data with correlated noise", *Technical Report*, University of Bristol.

Jones, R.H. (1974), "Identification and autoregressive spectrum estimation", *IEEE Trans. Auto. Control*, AC-19, 894–897.

Jones, R.H. (1976), "Autoregressive order selection", *Geophy.*, 41, 771–773.

Jones, R.H. (1977), "Spectrum estimation from unequally spaced data", *Preprints: Fifth Conference on Probability and Statistics in Atmospheric Sciences, American Meteorological Society*, 277–282.

Jones, R.H. (1980), "Maximum likelihood fitting of ARMA models to time series with missing observations", *Technometrics*, 22, 389–395.

Jones R.H. (1981), "Fitting a continuous time autoregression to discrete data", in *Applied Time Series Analysis II* , ed. D. Findley, 651-682.

Jones, R.H. (1993), *Longitudional Data With Serial Correlation*, Chapman and Hall, London.

Jöreskog, K. G. (1992), "Applications of some new fit measures for structural equation models", Paper presented at *the Intnl. Conf. on Social Science Methodology in Trento. Italy*, June 22–26, 1992.

Kay, S.M. (1988), *Modern Spectral Estimation: Theory and Applications*, Prentice Hall.

Kailath, T. (1980), *Linear Systems*, Prentice-Hall, Englewood Cliffs, New Jersey.

Kalman, R.E. (1960) "A new approach to linear filtering and prediction problems", *Trans. Amer. Soc. Mech. Eng., J. Basic Engineering*, 82, 35–45.

Kalman, R.E. and R.S. Bucy (1961), "New results in linear filtering and prediction theory", *Trans of the ASME, J. Basic Engng.*, 83D, 95–108.

Kaminski, P.G., A.E. Bryson, and S.F. Schmidt (1971), "Discrete square root filtering: A survey of current technique", *IEE Trans. Auto. Control*, AC-16, 727–735.

Kashiwagi, N. (1993), "On the use of the Kalman filter for spatial smoothing", *Ann. Inst. Statist. Math.*, 45, 21–34.

Kashiwagi, N. and T. Yanigimoto (1992), "Smoothing serial count data through a state-space model", *Biometrics*, 48, 1187–1194

Katzoff, M. and R.H. Shumway (1993), "Nonlinear structural models for mortality series", *Proceeding of the 2nd U.S.-Japan Joint Seminar on Statistical Time Series Analysis*, Tokyo, Japan, 319–348.

Kavalieris L. (1993), "Transfer function estimation", *J. Time Series Anal.*, 14, 485–496.

Kimeldorf, G.S. and Wahba G. (1970a), "A correspondence between Bayesian estimation on stochastic processes and smoothing splines", *Ann. Math. Statist.*, 41, 495–502.

Kimeldorf, G.S. and Wahba G. (1970b), "Spline functions and stochastic processes", *Sankhya*, Ser. A, 32, 173–180.

Kimeldorf, G.S. and Wahba G. (1971), "Some results on Tchebycheffian splines", *J. Math. Anal. Appl.*, 33, 82–95.

Kitagawa, G. (1981), "A nonstationary time series model and its fitting by a recursive filter", *J. Time Series Anal.*, 2, 103–116.

Kitagawa, G. (1983), "Changing spectrum estimation", *J. Sound and Vibration*, 89, No. 4, 443–445.

Kitagawa, G. (1984), "State space modeling of nonstationary time series and smoothing of unequally spaced data", in *Time Series Analysis of Irregularly Observed Data*, ed. E.Parzen, Lecture Notes in Statistics, 25, 189–210. Springer-Verlag, New York,

Kitagawa, G. (1986), "Time-varying autoregressive model, – An introduction to the program TVCAR –", (in Japanese with English abstract), *Proc. of the Inst. Statist. Math.*, 34, 273–283.

Kitagawa, G. (1987), "Non-Gaussian state space modeling of nonstationary time series, (with discussion)", *J. Amer. Statist. Assoc.*, 82, 1032–1063.

Kitagawa, G. (1988), "Numerical approach to non-Gaussian smoothing and its applications", *20th Interface Symposium Computer Science and Statistics*, 379–388.

Kitagawa, G. (1989), "Non-Gaussian seasonal adjustment", *Computers & Mathematics with Applications*, 18, 503–514.

Kitagawa, G. (1991), "A nonlinear smoothing method for time series analysis", *Statistica Sinica*, 1, 371–388.

Kitagawa, G. (1993), "A Monte-Carlo filtering and smoothing method for non-Gaussian nonlinear state space models", *Proceedings of the 2nd U.S.-Japan Joint Seminor on Statistical Time Series Analysis*, 110–131.

Kitagawa, G. (1994), "The two-filter formula for smoothing and an implementation of the Gaussian-sum smoother", *Ann. Inst. Statist. Math.*, 46, 605–623.

Kitagawa, G. (1996), "Monte Carlo filter and smoother for non-Gaussian nonlinear state space models", *Journal of Computational and Graphical Statistics*, 5, 1–25.

Kitagawa, G. and H. Akaike (1978), "A procedure for the modeling of nonstationary time series", *Ann. Inst. Stat. Math.*, 30-B, 215–363.

Kitagawa, G. and H. Akaike (1981), On TIMSAC-78, in *Applied Time Series Analysis II*, ed. D.F. Findley, Academic Press, New York.

Kitagawa, G.and W. Gersch (1984), "A smoothness priors-state space modeling of time series with trend and seasonality", *J. Amer. Statist. Assoc.*, 79, 378–389.

Kitagawa, G.and W. Gersch (1985a), "A smoothness priors long AR model method for spectral estimation", *IEEE Trans. on Automatic Control*, AC-30, 57–65.

Kitagawa, G.and W. Gersch (1985b), "A smoothness priors time varying AR coefficient modeling of nonstationary time series", *IEEE Trans. on Automatic Control*, AC-30, 48–56.

Kitagawa, G. and N. Matsumoto (1996), "Detection of coseismic changes of underground water level", *Journal of the American Statistical Association*, 91, (to appear).

Kitagawa, G. and T. Takanami (1985), "Extraction of a signal by a time series model and screening out micro earthquakes", *Signal Processing*, 8, 303–314.

Kohn, R. and C.F. Ansley (1986), "Estimation, prediction and interpolation for ARIMA models with missing data", *J. Amer. Statist. Assoc.*, 81, 751–761.

Kohn, R. and C. Ansley (1987), "A new algorithm for spline smoothing based on smoothing a stochastic process", *SIAM J. Sci. Statist. Comput.*, 8, 33–48.

Kohn, R. and C.F. Ansley (1988), "Smoothness priors and optimal interpolation and smoothing", in *Bayesian Analysis of Time Series and Dynamic Systems*, ed. J.C. Spall, Marcel Dekker, New York.

Kohn, R. and C.F. Ansley (1989), "A fast algorithm for signal extraction influence and cross-validating in state space models", *Biometrika*, 76, 65–79.

Komaki, F. (1993), "State-space modeling of time series sampled from continuous processes with pulses", *Biometrika*, 80 417–429

Kooperberg, C., C.J. Stone and Y.K Truong (1995), "Logspline estimation of a possibly mixed spectral distribution", *J. Time Series Anal.*, 16, 359-388.

Koopmans, L.H. (1974), *The Spectral Analysis of Time Series*, Academic Press, New York.

Kotz, S. and N.L. Johnson (1992), *Breakthroughs in Statistics, Vol. 1*, Wiley, New York.

Kozin, F. (1977), "Estimation and modeling of nonstationary time series", in *Proceedings: Symposium Applied Computational Methods in Engineering*, University of Southern California, 603–612.

Kozin, F. and F. Nakajima (1980), "The order determination problem for linear time-varying AR models", *IEEE Transactions on Automatic Control*, AC-25, 250–257.

Kramer, S.C. and H.W. Sorenson (1988), "Recursive Bayesian estimation using point-wise constant approximation", *Automatica*, 24, 789–801.

Kullback, S. and R.A. Leibler (1951), "On information and sufficiency", *Ann. Math. Stat.*, 22, 79–86.

Kullback, S. (1958), *Information Theory and Statistics*, Wiley, New York.

Lawson, C.L and R.J. Hanson (1974), *Solving Least Squares Problems*, Prentice-Hall, New Jersey.

LeCam, L. (1990), "Maximum likelihood: An introduction", *Intl. Stat. Review*, 58, 153–171.

Lee, D.T.L., Morf M. and Friedlander B. (1981), "Recursive least squares ladder estimation algorithms", *IEEE Transactions Acoust., Speech, Signal Processing*, ASSP-29, 627–641.

Leonard, T. (1978), *Density estimation, stochastic processes and prior information (with discussion)*, *J. Roy. Statist. Soc.*, B 40, 113–146.

Lev-Ari, H. (1987), "Modular architectures for adaptive multichannel lattice algorithms", *IEEE Trans. Acoust. Speech & Signal Process.*, ASSP-35, 543–552.

Lev-Ari, H. and T. Kailath (1984), "Lattice filter parameterization and modeling of nonstationary processes", *IEEE Trans. Inform. Theory*, IT-30, 2–16.

Lindley, D.V. and Smith A.F.M. (1972), "Bayes estimate for the linear model", *J. R. Statist. Soc.*, B, 34, 1–41.

Ling, F. and Proakis, (1984), "A generalized multichannel least squares lattice algorithm based on sequential processing stages", *IEEE Trans. Acoust. Speech & Signal Process.*, ASSP–32, 381–390.

Linhart, H. and W. Zucchini (1986), *Model Selection*, Wiley, New York.

Loynes, R.M. (1968), "On the concept of the spectrum for nonstationary processes", *J. R. Statist. Soc.*, B 30, 1–30.

Luenberger, B. G. (1969), *Optimization by Vector Space Methods*, Wiley, New York.

Lutkepohl, H. (1985), 'Computation of criteria for estimating the order of a vector autoregressive process", *J. Time Series Anal.*, 6, 35–52

Lutkepohl, H. (1993), *Introduction To Multiple Time Series Analysis, 2nd Edition*, Springer-Verlag, Berlin.

Makhoul, J. (1975), "Linear prediction-Tutorial review", *Proc. IEEE*, 63, 561–580.

Mallows, C.L. (1973) "Some comments on the C_p", *Technometrics*, 15, 661–675.

Marple, M.L. (1987), *Digital Spectral Analysis*, Prentice Hall.

Martin, W. and P. Flandrin (1985), "Wigner-Ville spectral analysis of nonstationary processes", *IEEE Trans. Acoust., Speech, Signal Processing*, ASSP-33, 1461–1476.

Matsumoto, N. (1992), "Regression analysis for anomalous changes of ground water level due to earthquakes", *Geophysical Research Letters*, 19, 1193–1196.

McCulloch, R.E. and R.S. Tsay (1994), "Statistical analysis of macroeconomic time series via Markov switching models", *J. Time Series Anal.*, 15, 523–539.

McGowan, R. and R. Kuc (1982), "A direct relation between a signal time series and its unwrapped phase", *IEEE Trans. on Acoustics, Speech and Signal Processing*, 30, 719–725.

McReynolds, S.R. (1975), "Multidimensional Hermite-Gaussian quadrature formulae and their application to nonlinear estimation", *6th Symp. on Nonlinear Estimation*, 188–191.

Mehra, R. (1971), "Identification of stochastic linear dynamic system using Kalman filter representation", *AIAA J.*, 9, 28–31.

Meinhold R.J. and N.D. Singpurwalla (1983), "Understanding the Kalman filter", *Amer. Statistician*, 37, 123-127.

Meinhold, R.J. and N.D. Singpurwalla (1989), "Robustification of Kalman filter models", *J. Amer. Statist. Assoc.*, 84, 479–486.

Mittnik, S. (1989), "Multivariate time series analysis and state space models", *Computers and Math. with Appltns.*, 17, 1189–1201.

Mohammad-Djafari, A. (1993), "Bayesian approach with maximum entropy priors to imaging inverse problems, I Foundations, II Applications", *Technical Report LSS/GPI-94-01*, Laboratoire des Signaux et Systèmes, École Supeérieure d'Électricité, Centre Nationale de la Recherche Scientifique, Gif-sur-Yvette Cdex, France.

Moran, P.A.P. (1953a, b) "The statistical analysis of the Canadian lynx cycle I, II", *Austr. J. Zool.*, 1, 163–173 , 219–298.

Moran, P.A.P. (1954), "Some experiments on the prediction of sunspot numbers", *J. R. Statist. Soc.*, B 16, 112–117.

Morf. M., A. Viera, and T. Kailath (1978a), "Covariance characterization by partial autocorrelation matrices", *Ann. Statist.*, 6, 643–648.

Morf, M., A. Viera, D.T.L. Lee and T. Kailath (1978b), "Recursive multichannel maximum entropy spectral estimation", *IEEE Trans. Geosci. Electron.*, GE-16, 85–94.

Morris, J.A. (1977), "Forecasting the sunspot cycle", *J. R. Statist. Soc.*, A-140, 437–468.

Moura, J.M.F. and A.B. Baggeroer, (1988), "Phase unwrapping of signals propagated under the arctic ice crust: A statistical approach", *IEEE Trans. on Acoust., Speech and Signal Proc.*, 36, 617–630.

Murata, Y. (1993), "Estimation of optimum average surficial density from gravity data: An objective Bayesian approach", *J. Geophys. Res.*, 98, B7, 12097–12,109.

Nakamura, T.(1986), "Bayesian cohort models for general cohort table analysis", *Ann. Inst. Statist. Math.*, 38 B, 353–370.

Naniwa, S. (1987), "Nonstationary covariance structure of detrended economic time series: A time varying model approach", *Monetary and Economic Studies*, Inst. Monetary and Economic Studies, Bank of Japan, 5, 217–269.

Naniwa, S. (1986), "Trend estimation via smoothness priors-state space modeling", *Monetary and Economic Studies*, Inst. Monetary and Economic Studies, Bank of Japan, 4, 79–112.

Nashed, M. and G. Wahba (1974), "Generalized inverses in reproducing kernel spaces: An approach to regularization of linear operator equations", *SIAM J. Math. Anal.*, 5, 974–987.

Neumann, J. von (1941), "Distribution of the ratio of the mean square successive difference to the variance", *Ann. Math. Statist.*, 12, 367–395.

Newton, J.J. (1982), "Using periodic autoregressions for multiple spectral estimation", *Technometrics*, 24, 109–116.

Nicholls, D.F. and B.G Quinn (1981), "The estimation of random coefficient autoregessive model I", *J. Time Series Anal.*, 1, 37–46.

Nishimura, T. (1976), "New approach to square root smoothers", *J. Soc. Instrument and Control Engineers*, 13, 209–214.

Nychka, D. (1981), "Bayesian "confidence" intervals for smoothing splines", *J. Amer. Statist. Assoc.*, 83, 1134–1143.

O'Sullivan, F., B.S. Yandell and W.J. Raynor Jr. (1986), "Automatic Smoothing of Regression Functions in Generalized Linear Models", *J. Amer. Statist. Assoc.*, 81, 96–103.

Ozaki, T. and H. Tong (1975), "On the fitting of nonstationary autoregressive models in time series analysis", in *Proceedings 8th Hawaii Intnl. Conf. on System Sciences*, 224–246.

Pagano, M. (1978), "Periodic and multiple autoregressions, *Ann. Statist.*, 6, 1310–1317.

Page, C.H. (1952), "Instantaneous power spectra", *J. Appl. Phys.*, 23, 103–106.

Parsons, B. and G.G. Sclater (1977), "An analysis of the variations of ocean floor bathymetry and heat flow with age", *J. Geophys. Res.*, 82, 803–827.

Parzen, E. (1957a), "On consistent estimates of the spectrum of a stationary time series", *Ann. Math. Stat.*, 28, 329–348.

Parzen, E. (1957b), "On chosing an estimate of the spectral density function of a stationary time series", *Ann. Math. Stat.*, 28, 921–932.

Parzen, E. (1959), "Statistical inference on time series by Hilbert spaces methods", *Appl. Math. Stat. Lab.*, Stanford Univ., Tech. Rep. 23 . Reprinted in Parzen E. (1967), *Time Series Analysis Papers*, Holden-Day San Francisco.

Parzen, E. (1961), "An approach to time series analysis", *Ann. Math. Stat.*, 32, 951–989.

Parzen, E. (1963), "A new approach to the synthesis of optimal smoothing and prediction systems", in *Mathematical Optimization Techniques*, ed. R. Bellman, 75–108.

Parzen, E. (1967), *Time Series Analysis Papers*, Holden Day, San Francisco.

Parzen, E. (1970), "Statistical inference on time series by RKHS methods", *Proc. 12-th Biennial Seminar Canadian Mathematical Congress on Time Series and Stochastic Processes; Convexity and Combinatorics* , ed. R. Pyke, 1–39.

Parzen, E. (1971), "On the equivalence among time series parameter estimation, approximation theory and control theory", *Proceedings of the Fifth Princeton Conference on Information Science*, 1–5.

Parzen, E. (1978), "Some recent advances in time series modeling", in *Modern Spectrum Analysis*, ed. D.G.Childers, IEEE Press, New York, 221–247.

Parzen, E. (1983), "Autoregressive spectral estimation", in *Handbook of Statistics*, 3, eds. D.R. Brillinger and R. Krishnaiah, North-Holland, 221–247.

Pawitan, and Gangopadhyay (1991), "Efficient bias corrected nonparametric spectral estimation", *Biometrika*, 78, 825–832.

Phillips, D.L. (1962), "A technique for numerical solution of certain integral equations of the first kind", *J. Assoc. Comput. Mach.*, 9, 97–101.

Porat, B., B. Friedlander and M. Morf (1982), "Square-root covariance ladder estimation algorithms", *IEEE Transactions Automat. Control*, AC-17, 813–829.

Politis, D.N. and J.P. Romano (1992), "A general resampling scheme for triangular arrays of α-mixing random variables with application to the problem of spectral density estimation", *Ann. Statist.*, 20, 1985–2007.

Press, W. H., S.A. Teukolsky, W.T. Vetterling and B.P. Flannery (1992), *Numerical Recipes in FORTRAN, The art of Scientific Computing, Second Edition*, Cambridge.

Priestley M.B. (1962a) "Basic considerations in the estimation of spectra", *Technometrics*, 4, 551-564.

Priestley, M.B. (1962b), "Analysis of stationary processes with mixed spectrum I and II", *J. R. Statist. Soc.*, B 26, 215–233, 511–529.

Priestley, M.B. (1965), "Evolutionary spectra for non-stationary processes", *J. R. Statist. Soc.*, B 27, 204–229.

Priestley, M.B. (1981) *Spectral Analysis and Time Series* (2 Vols.), Academic Press, London.

Quinn, B.G. (1980), "Order determination for a multivariate autoregression", *J. R. Statist. Soc.*, B-42, 182–185.

Rabiner, L.R. and B.H. Juang (1986), "An introduction to hidden Markov models", *IEEE Acoustics, Speech and Signal Proc. Magazine*, 3, 4–16.

Rabiner, L.R. (1989), "A tutorial on hidden Markov models and selected applications in speech recognition", *Proc. IEEE*, 77, 257–285.

Raftery, A.E. (1995), "Change point and change curve modeling in stochastic processes and spatial statistics", in *Applied Change Point Problems in Statistics*, eds. B. Sinha, A. Rukhin and M. Ahsanullah, 81–102.

Rauch, H.H., F. Tung and C.T. Streibel (1965), "Maximum likelihood estimates of linear dynamic systems", *AIAA Journal*, 3, 1445–1450.

Rissanen, J. (1978) "Modeling by shortest data description", *Automatica*, 14, 465–471.

Rissanen, J. (1989) *Statistical Complexity in Statistical Inquiry*, World Scientific, Singapore.

Robinson, P.M. (1977), "Estimation of a time series model from unequally spaced data", *Stochastic Processes and Their Applications*, 6, 9–24.

Sage, A.P. and J.L. Melsa (1971), *Estimation Theory with Applications to Communication and Control*, McGraw-Hill, New York.

Sakai, H. (1982), "Circular lattice filtering using Pagano's method", *IEEE Trans. Acoustics, Speech and Signal Processing*, ASSP-30, 279–287.

Sakamoto, T., M. Ishiguro M. and G. Kitagawa (1986), *Akaike Information Criterion Statistics*, D. Reidel, Holland.

Sambamboorthi, N. (1989), "Information theoretic criterion approach to dimensionality reduction in multinomial regression models", *Commun. in Stat. Theory and Methodology*, 18, 2087–2106.

Sanderson, A.C., J. Segen and E. Richey, (1983), "Hierarchical modeling of EEG signals", *IEEE Trans. Pattern and Machine Intell.*, PAMI-2, 405–414.

Schnatter, S. (1992), "Integration-based Kalman-Filtering for a dynamic generalized linear trend model", *Computational Statistics and Data Analysis*, 13, 447–459.

Schoenberg, I.J. (1964), "Spline functions and the problems of graduation", *Proc. Natl. Acad. Sci. U.S.A.*, 52, 947–950.

Sclove, S.L. (1987), "Application of model-selection criteria to some problems in multivariate analysis", *Psychometrika*, 52, 333–343.

Schwabe, H. (1838), "Uber die flecken der sonne", *Astronomishe Nachricten*, 15, 223–248.

Schwartz, G. (1978), "Estimating the dimension of a model", *Ann. Statist.*, 6, 461–464.

Schweppe, C.F. (1965), "Evaluation of likelihood functions for Gaussian signals", *IEEE Transactions on Information Theory*, IT -11, 61–70.

Segen, J. and A.C. Sanderson (1980), "Detecting change in a time series", *IEEE Trans. Inform. Theory*, IT-26, 249–255.

Shapiro and Silverman (1960), "Alias-free sampling of random noise", *J. Soc. Indust. Appl. Math.*, 8, 225–248.

Shibata, R. (1976) "Selection of the order of an autoregressive model by Akaike's information criterion", *Biometrika*, 63, 117–126.

Shibata, R. (1980) "Asymptotically efficient selection of the order of the model for estimating parameters of a linear process", *Ann. Statist.*, 8, 147–164.

Shibata, R. (1981), "An optimal selection of regression variables", *Biometrika*, 68, 45–54.

Shibata, R. (1983), "A theoretical view of of the use of the AIC", in *Time Series Analysis: Theory and Practice, IV*, ed. O.D. Anderson, North-Holland, Amsterdam, 237–244.

Shibata, R. (1986), "Consistency of model selection and parameter estimation", in Essays in Time Series and Allied Processes, eds. J.M. Gani and M.B. Priestley, Applied Probability Trust, Sheffield, 127–141.

Shiller, R. (1973), "A distributed lag estimator derived from smoothness priors", *Econometrica*, 41,775–778.

Shishkin, J., A.H. Young and J.C. Musgrave, (1976), "The X-11 variant of the Census Method II seasonal adjustment program", *Technical Paper No. 15, Bureau of the Census, U.S. Dept of Commerce.*

Shore J.E. and R.W. Johnson, (1980), "Axiomatic derivation of the principle of maximum entropy and the principle of minimum cross-entropy", *IEEE Trans. Info. Theory*, IT-26, 26-37.

Shumway, R.H. and D.S. Stoffer (1991), "Dynamic linear models with switching", *J. Amer. Statist. Assoc.*, 86, 763–769.

Shuster, A. (1898), "On the investigation of hidden periodicities with application to a supposed 26-day period of meteorological phenomena", *Terr. Mag. Atmos. Elect.*, 3, 13–41.

Shuster, A. (1906), "On the periodicities of sunspots", *Philos. Trans. R. Society A*, 206, 69–100.

Silverman, B.W. (1982), "On the estimation of a probability density function by the maximum penalized likelihood method", *Ann. Statist.*, 10, 795–810.

Silverman, B.W. (1985), "Some aspects of the spline smoothing approach to non-parametric regression curve fitting", *J. R. Statist. Soc. B*, 36, 1–52.

Solo, V. (1986), *Topics in Advanced Time Series Analysis*, Lecture Notes in Mathematics, Springer-Verlag, New York.

Sorenson, H.W. (1974) "On the development of practical nonlinear filters", *Information Science*, 7, 253–270.

Sorenson, H.W. (1985), Ed. *Kalman Filtering: Theory and Application*, IEEE Press, Piscatawney, New Jersey.

Sorenson, H.W. and D.L. Alspach (1971), "Recursive Bayesian estimation using Gaussian sums", *Automatica*, 7, 465–479.
Sorenson, H.W. and A.R. Stubberud (1968), "Nonlinear filtering by approximation of the a posteriori density", *Intern. J. Contro*, 8, 33–51.
Srinivasan, K. (1970), "State estimation by orthogonal expansion of probability distributions", *IEEE Trans. on Automatic Control*, AC–15, 3–10.
Stoica, P., R.L. Moses, T. Soderstrom and J. Li, (1991), "Optimal high-order Yule-Walker estimation of sinusoidal frequencies", *IEEE Trans. on Acoustics*, 39, 1360–1368.
Stone, D. (1993), "One channel at-a-time multichannel autoregressive modeling: Applications to stationary and nonstationary covariance time series", Ph.D. Thesis in Communication and Information Sciences, University of Hawaii.
Sugiura, N. (1978) "Further analysis of the data by Akaike's information criterion and the finite corrections", *Commun. in Statist. Theory*, A7, 13–26.
Suzuki, S., T. Takanami, Y. Motoya, M. Kazahara and I. Nakanashi (1986), "Automatic processing system for microearthquake network of Hokkaido University", *Proceedings of the Annual Meeting of Seismological Society of Japan*, (in Japanese).
Swanepol, J.W. and J.W.J. van Wyk (1986), "The bootstrap applied to spectral density function estimation", *Biometrika*, 73, 135–142.
Tanabe, K., M. Sagae and S. Ueda (1988), "BNDE, Fortran subroutines for computing Bayesian nonparametric univariate and bivariate density estimator", *Computer Science Monograph #24*, Institute of Statistical Mathematics, Tokyo, Japan.
Takanami, T. (1991), "ISM data 43-3-01: Seismograms of foreshocks of 1982 Urakawa-Oki earthquake", *Ann. Inst. Statist. Math.*, 43, 605.
Takanami, T. and G. Kitagawa (1991), "Estimation of the arrival times of seismic waves by multivariate time series model", *Ann. Inst. Statist. Math.*, 43, 407–433.
Takane, Y., Bozdogan H. and Shibayama T. (1987), "Ideal point discriminant analysis", *Psychometrika*, 52, 371–392.
Tanizaki, H. (1993), *Nonlinear filters: Estimation and Applications*, Springer-Verlag, New York.
Terasvirta, T. (1987), "Smoothness in regression: Asymptotic considerations", in *Time Series and Econometric Modeling*, eds. I. B. MacNeil, G.J. Umphrey, Reidel, Holland, 47–64.
Tiao, G.C. and Tsay R.S. (1985), "A canonical correlation approach to modeling multivariate time series", *Proc. Econ. Bus. Stat. Sec., Amer. Stat. Assoc.*, 112–120.
Tiao, G.C. and Tsay R.S. (1989), "Model specification in multivariate time series (with discussion),", *J. R. Statist. Soc.*, B 51, 157–213
Tikhonov, A. N. (1963), "Solution of incorrectly formulated problems and the regularization method", *Soviet Math. Dokl.*, 4, 1035–1038.
Titterington, D.M. (1985), "Common structure of smoothing techniques in statistics", *Int. Statist. Rev.*, 53 141–170.
Tong, H. (1977), "Some comments on the Canadian lynx data", *J. R. Statist. Soc.*, A-140, 432–436.

Tong, H. (1982), "A note in using threshold autoregressive models for multi-step-ahead prediction of cyclical data", *J. Time Series Anal.*, 2, 279–284.

Tong, H. (1983), *Threshold Models in Non-linear Time Series*, Springer-Verlag, New York.

Tong, H. and Z.M. Wu (1982), "Multi-step ahead forecasting of cyclical data by threshold autoregression", in *Time Series Analysis*, ed. O.D Anderson, North-Holland, Amsterdam, 733-753.

Tjostheim, D. and J. Paulsen (1983), "Bias of some commonly used time series estimates", *Biometrika*, 50, 129–134.

Tsay, R.S. (1988), "Outliers, level shifts and variance changes in time series", *J. Forecasting*, 7, 1–20.

Ville, W. (1948), "Theorie et aplication de la notion de signal analytique", *Cables et Transmissions*, 2A, 61–74.

Wahba, G. (1978), "Improper priors, spline smoothing and the problem of guarding against model errors in regression", *J. R. Statist. Soc.*, B-40, 364–372.

Wahba, G. (1980), "Automatic smoothing of the log periodogram", *J. Amer. Statist. Assoc.*, 75, 122–132.

Wahba, G. (1983), "Bayesian confidence intervals for the cross-validated smoothing spline", *J. R. Statist. Soc.*, B-45, 133–150.

Wahba, G. (1990), *Spline Models for Observational Data*, Soc. Indust. Appl. Math., Philadelphia.

Wahba, G. and Wold S. (1975), "A completely automatic French curve: Fitting spline functions by cross validation", *Commun. Statist.*, 4(1), 1–17.

Waldmeier, M (1961), "The sunspot activity in the years 1610–1960", Schultess & Co. Zurich.

Walker, A.M. (1973), "On the estimation of a harmonic component in a time series with stationary dependent residuals", *Adv. Appl. Prob.*, 5, 217–241.

Wecker, W.E. and C.R. Ansley (1983), "The signal extraction approach to nonlinear regression and spline smoothing", *J. Amer. Statist. Assoc.*, 78, 81–89.

Weinert, H.L. (1972), "A reproducing kernel Hilbert space approach to spline problems, with applications in estimation and control", Ph. D. thesis, Dept. of Electrical Engineering, Stanford University.

Weinert, H.L., R.H. Byrd, and G.S. Sidhu, (1980), "A stochastic framework for recursive computation of spline functions: Part II. Smoothing splines", *J. Optim. Theory Appl.*, 30, 255–268.

West, M. (1981), "Robust sequential approximate Bayesian estimation", *J. R. Stat. Soc.*, B, 43, 157–166.

West, M., P.J. Harrison and H.S. Mignon (1985), "Dynamic generalized linear models and Bayesian forecasting, (with discussion)", *J. Amer. Statist. Assoc.*, 80, 73–97.

West, M. and P.J. Harrison (1986), "Monitoring and adaption in Bayesian forecasting models", *J. Amer. Statist. Assoc.*, 81, 714–750.

West, M. and P.J. Harrison (1989), *Bayesian Forecasting and Dynamic Models*, Springer Series in Statistics, Springer-Verlag, Berlin.

Whittaker, E.T. (1923), "On a new method of graduation", *Proc. Edinborough Math. Assoc.*, 78, 81–89.

Whittaker, E.T. and G. Robinson (1924), *Calculus of Observations, A Treasure on Numerical Calculations*, Blackie and Son, Lmtd., London, 303–306.

Whittle, P. (1952), "On the estimation of a harmonic component in a time series' harmonic components and covariance structure", *Trabajos Estadist.*, 43–57.

Whittle, P. (1954), "The statistical analysis of a seiche record", *Sears Foundn. J. of Marine Research*, 13, 76–100.

Whittle, P. (1963), "On the fitting of multivariable autoregressions and the approximate canonical factorization of a spectral density matrix", *Biometrika*, 50, 129–134.

Whittle, P. (1965), "Recursive relations for predictors of non-stationary processes", *J. R. Statist. Soc.*, B 27, 523–532.

Wiener, N. (1949), *The Extrapolation, Interpolation and Smoothing of Stationary Time Series*, Wiley, New York.

Wiggins, R.A. and Robinson E.A. (1965), "Recursive solution to the multichannel filtering problem", *J. Geophys. Res.*, 70, 1885–1891.

Willman, W.W. (1981), "Edgeworth expansion in state perturbation estimation", *IEEE Trans. Auto. Control*, AC-20, 493–498.

Woodroofe, M. (1982), "On model selection and the arc sine laws", *Ann. Statist.*, 10, 1182–1194.

Yanagimoto, T. and N. Kashiwagi (1990), "Empirical Bayes methods for smoothing data and for simultaneous estimation of many parameters", *Environmental Health Perspectives*, 87, 109–114.

Yabuki, T. and M. Matsu'ura (1992), "Geodetic data inversion using a Bayesian information criterion for spatial distribution of a fault slip", *Geophys. J. Int.*, 363–375.

Yoshimura, H. (1979), "The solar-cycle period-amplitude relation as evidence of hysteresis of the solar cycle non-linear magnetic oscillation and the long term (55-year) cycle modulation", *Astrophys. J.*, 227, 1047–1058.

Young, M.R. (1993), "Robust Bayesian seasonal decomposition", *Proc. Intl. Society of Bayesian Analysis.*

Yule, G. U. (1927), "On the method of investigating periodicities in disturbed series, with special reference to Wolfer's sunspot numbers", *Philos. Trans. Roy. Soc. London*, A, 226, 267–298.

Zadeh, L. and C. Desoer (1963), *Linear System Theory*, McGraw-Hill, New York.

Zellner, A. (1971), *An Introduction to Bayesian Inference in Economics*, Wiley, New York.

Zellner, A. (ed.), (1978), *Seasonal Analysis of Economic Time Series*, U.S. Department of Commerce, Bureau of the Census, Washington D.C.

Zellner, A. (ed.), (1981), *Applied Time Series Analysis of Economic Data*, U.S. Department of Commerce, Bureau of the Census, Washington D.C.

Zellner, A. (ed.), (1983), *Proceedings ASA-CENSUS-NBER Conf. on Applied Time Series Analysis of Economic Data*, U.S. Bureau of the Census, Washington D.C..

Author Index

Subject Index

Lecture Notes in Statistics

For information about Volumes 1 to 33 please contact Springer-Verlag

Vol. 34: D.E. Critchlow, Metric Methods for Analyzing Partially Ranked Data. x, 216 pages, 1985.

Vol. 35: T. Calinski and W. Klonecki (Editors), Linear Statistical Inference. Proceedings, 1984. vi, 318 pages, 1985.

Vol. 36: B. Matern, Spatial Variation. Second Edition. 151 pages, 1986.

Vol. 37: R. Dykstra, T. Robertson and F.T. Wright (Editors), Advances in Order Restricted Statistical Inference. Proceedings, 1985. viii, 295 pages, 1986.

Vol. 38: R.W. Pearson and R.F. Boruch (Editors), Survey Research Designs: Towards a Better Understanding of Their Costs and Benefits. v, 129 pages, 1986.

Vol. 39: J.D. Malley, Optimal Unbiased Estimation of Variance Components. ix, 146 pages, 1986.

Vol. 40: H.R. Lerche, Boundary Crossing of Brownian Motion. v, 142 pages, 1986.

Vol. 41: F. Baccelli, P. Brémaud, Palm Probabilities and Stationary Queues. vii, 106 pages, 1987.

Vol. 42: S. Kullback, J.C. Keegel, J.H. Kullback, Topics in Statistical Information Theory. ix, 158 pages, 1987.

Vol. 43: B.C. Arnold, Majorization and the Lorenz Order: A Brief Introduction. vi, 122 pages, 1987.

Vol. 44: D.L. McLeish, Christopher G. Small, The Theory and Applications of Statistical Inference Functions. vi, 124 pages, 1987.

Vol. 45: J.K. Ghosh (Editor), Statistical Information and Likelihood. 384 pages, 1988.

Vol. 46: H.-G. Müller, Nonparametric Regression Analysis of Longitudinal Data. vi, 199 pages, 1988.

Vol. 47: A.J. Getson, F.C. Hsuan, {2}-Inverses and Their Statistical Application. viii, 110 pages, 1988.

Vol. 48: G.L. Bretthorst, Bayesian Spectrum Analysis and Parameter Estimation. xii, 209 pages, 1988.

Vol. 49: S.L. Lauritzen, Extremal Families and Systems of Sufficient Statistics. xv, 268 pages, 1988.

Vol. 50: O.E. Barndorff-Nielsen, Parametric Statistical Models and Likelihood. vii, 276 pages, 1988.

Vol. 51: J. Hüsler, R.-D. Reiss (Editors). Extreme Value Theory, Proceedings, 1987. x, 279 pages, 1989.

Vol. 52: P.K. Goel, T. Ramalingam, The Matching Methodology: Some Statistical Properties. viii, 152 pages, 1989.

Vol. 53: B.C. Arnold, N. Balakrishnan, Relations, Bounds and Approximations for Order Statistics. ix, 173 pages, 1989.

Vol. 54: K.R. Shah, B.K. Sinha, Theory of Optimal Designs. viii, 171 pages, 1989.

Vol. 55: L. McDonald, B. Manly, J. Lockwood, J. Logan (Editors), Estimation and Analysis of Insect Populations. Proceedings, 1988. xiv, 492 pages, 1989.

Vol. 56: J.K. Lindsey, The Analysis of Categorical Data Using GLIM. v, 168 pages, 1989.

Vol. 57: A. Decarli, B.J. Francis, R. Gilchrist, G.U.H. Seeber (Editors), Statistical Modelling. Proceedings, 1989. ix, 343 pages, 1989.

Vol. 58: O.E. Barndorff-Nielsen, P. Bl¾sild, P.S. Eriksen, Decomposition and Invariance of Measures, and Statistical Transformation Models. v, 147 pages, 1989.

Vol. 59: S. Gupta, R. Mukerjee, A Calculus for Factorial Arrangements. vi, 126 pages, 1989.

Vol. 60: L. Gyorfi, W. Härdle, P. Sarda, Ph. Vieu, Nonparametric Curve Estimation from Time Series. viii, 153 pages, 1989.

Vol. 61: J. Breckling, The Analysis of Directional Time Series: Applications to Wind Speed and Direction. viii, 238 pages, 1989.

Vol. 62: J.C. Akkerboom, Testing Problems with Linear or Angular Inequality Constraints. xii, 291 pages, 1990.

Vol. 63: J. Pfanzagl, Estimation in Semiparametric Models: Some Recent Developments. iii, 112 pages, 1990.

Vol. 64: S. Gabler, Minimax Solutions in Sampling from Finite Populations. v, 132 pages, 1990.

Vol. 65: A. Janssen, D.M. Mason, Non-Standard Rank Tests. vi, 252 pages, 1990.

Vol 66: T. Wright, Exact Confidence Bounds when Sampling from Small Finite Universes. xvi, 431 pages, 1991.

Vol. 67: M.A. Tanner, Tools for Statistical Inference: Observed Data and Data Augmentation Methods. vi, 110 pages, 1991.

Vol. 68: M. Taniguchi, Higher Order Asymptotic Theory for Time Series Analysis. viii, 160 pages, 1991.

Vol. 69: N.J.D. Nagelkerke, Maximum Likelihood Estimation of Functional Relationships. V, 110 pages, 1992.

Vol. 70: K. Iida, Studies on the Optimal Search Plan. viii, 130 pages, 1992.

Vol. 71: E.M.R.A. Engel, A Road to Randomness in Physical Systems. ix, 155 pages, 1992.

Vol. 72: J.K. Lindsey, The Analysis of Stochastic Processes using GLIM. vi, 294 pages, 1992.

Vol. 73: B.C. Arnold, E. Castillo, J.-M. Sarabia, Conditionally Specified Distributions. xiii, 151 pages, 1992.

Vol. 74: P. Barone, A. Frigessi, M. Piccioni, Stochastic Models, Statistical Methods, and Algorithms in Image Analysis. vi, 258 pages, 1992.

Vol. 75: P.K. Goel, N.S. Iyengar (Eds.), Bayesian Analysis in Statistics and Econometrics. xi, 410 pages, 1992.

Vol. 76: L. Bondesson, Generalized Gamma Convolutions and Related Classes of Distributions and Densities. viii, 173 pages, 1992.

Vol. 77: E. Mammen, When Does Bootstrap Work? Asymptotic Results and Simulations. vi, 196 pages, 1992.

Vol. 78: L. Fahrmeir, B. Francis, R. Gilchrist, G. Tutz (Eds.), Advances in GLIM and Statistical Modelling: Proceedings of the GLIM92 Conference and the 7th International Workshop on Statistical Modelling, Munich, 13-17 July 1992. ix, 225 pages, 1992.

Vol. 79: N. Schmitz, Optimal Sequentially Planned Decision Procedures. xii, 209 pages, 1992.

Vol. 80: M. Fligner, J. Verducci (Eds.), Probability Models and Statistical Analyses for Ranking Data. xxii, 306 pages, 1992.

Vol. 81: P. Spirtes, C. Glymour, R. Scheines, Causation, Prediction, and Search. xxiii, 526 pages, 1993.

Vol. 82: A. Korostelev and A. Tsybakov, Minimax Theory of Image Reconstruction. xii, 268 pages, 1993.

Vol. 83: C. Gatsonis, J. Hodges, R. Kass, N. Singpurwalla (Editors), Case Studies in Bayesian Statistics. xii, 437 pages, 1993.

Vol. 84: S. Yamada, Pivotal Measures in Statistical Experiments and Sufficiency. vii, 129 pages, 1994.

Vol. 85: P. Doukhan, Mixing: Properties and Examples. xi, 142 pages, 1994.

Vol. 86: W. Vach, Logistic Regression with Missing Values in the Covariates. xi, 139 pages, 1994.

Vol. 87: J. Müller, Lectures on Random Voronoi Tessellations.vii, 134 pages, 1994.

Vol. 88: J. E. Kolassa, Series Approximation Methods in Statistics.viii, 150 pages, 1994.

Vol. 89: P. Cheeseman, R.W. Oldford (Editors), Selecting Models From Data: AI and Statistics IV. xii, 487 pages, 1994.

Vol. 90: A. Csenki, Dependability for Systems with a Partitioned State Space: Markov and Semi-Markov Theory and Computational Implementation. x, 241 pages, 1994.

Vol. 91: J.D. Malley, Statistical Applications of Jordan Algebras. viii, 101 pages, 1994.

Vol. 92: M. Eerola, Probabilistic Causality in Longitudinal Studies. vii, 133 pages, 1994.

Vol. 93: Bernard Van Cutsem (Editor), Classification and Dissimilarity Analysis. xiv, 238 pages, 1994.

Vol. 94: Jane F. Gentleman and G.A. Whitmore (Editors), Case Studies in Data Analysis. viii, 262 pages, 1994.

Vol. 95: Shelemyahu Zacks, Stochastic Visibility in Random Fields. x, 175 pages, 1994.

Vol. 96: Ibrahim Rahimov, Random Sums and Branching Stochastic Processes. viii, 195 pages, 1995.

Vol. 97: R. Szekli, Stochastic Ordering and Dependence in Applied Probability. viii, 194 pages, 1995.

Vol. 98: Philippe Barbe and Patrice Bertail, The Weighted Bootstrap. viii, 230 pages, 1995.

Vol. 99: C.C. Heyde (Editor), Branching Processes: Proceedings of the First World Congress. viii, 185 pages, 1995.

Vol. 100: Wlodzimierz Bryc, The Normal Distribution: Characterizations with Applications. viii, 139 pages, 1995.

Vol. 101: H.H. Andersen, M.Højbjerre, D. Sørensen, P.S.Eriksen, Linear and Graphical Models: for the Multivariate Complex Normal Distribution. x, 184 pages, 1995.

Vol. 102: A.M. Mathai, Serge B. Provost, Takesi Hayakawa, Bilinear Forms and Zonal Polynomials. x, 378 pages, 1995.

Vol. 103: Anestis Antoniadis and Georges Oppenheim (Editors), Wavelets and Statistics. vi, 411 pages, 1995.

Vol. 104: Gilg U.H. Seeber, Brian J. Francis, Reinhold Hatzinger, Gabriele Steckel-Berger (Editors), Statistical Modelling: 10th International Workshop, Innsbruck, July 10-14th 1995. x, 327 pages, 1995.

Vol. 105: Constantine Gatsonis, James S. Hodges, Robert E. Kass, Nozer D. Singpurwalla(Editors), Case Studies in Bayesian Statistics, Volume II. x, 354 pages, 1995.

Vol. 106: Harald Niederreiter, Peter Jau-Shyong Shiue (Editors), Monte Carlo and Quasi-Monte Carlo Methods in Scientific Computing. xiv, 372 pages, 1995.

Vol. 107: Masafumi Akahira, Kei Takeuchi, Non-Regular Statistical Estimation. vii, 183 pages, 1995.

Vol. 108: Wesley L. Schaible (Editor), Indirect Estimators in U.S. Federal Programs. iix, 195 pages, 1995.

Vol. 109: Helmut Rieder (Editor), Robust Statistics, Data Analysis, and Computer Intensive Methods. xiv, 427 pages, 1996.

Vol. 110: D. Bosq, Nonparametric Statistics for Stochastic Processes. xii, 169 pages, 1996.

Vol. 111: Leon Willenborg, Ton de Waal, Statistical Disclosure Control in Practice. xiv, 152 pages, 1996.

Vol. 112: Doug Fischer, Hans-J. Lenz (Editors), Learning from Data. xii, 450 pages, 1996.

Vol. 113: Rainer Schwabe, Optimum Designs for Multi-Factor Models. viii, 124 pages, 1996.

Vol. 114: C.C. Heyde, Yu. V. Prohorov, R. Pyke, and S. T. Rachev (Editors), Proceedings of the Athens Conference on Appplied Probability and Time Series Volume I: Applied Probability. viii, 424 pages, 1996.

Vol. 115: P.M. Robinson, M. Rosenblatt (Editors), Proceedings of the Athens Conference on Applied Probability and Time Series Volume II: Time Series. viii, 448 pages, 1996.

Vol. 116: Genshiro Kitagawa and Will Gersch, Smoothness Priors Analysis of Time Series. x, 261 pages, 1996.